Logic, Methodology and Philosophy of Science

Proceedings of the 15th International Congress (Helsinki)

Models and Modelling

Logic, Methodology and Philosophy of Science
Proceedings of the 15th International Congress (Helsinki)

Models and Modelling

Edited by
Hannes Leitgeb,
Ilkka Niiniluoto,
Päivi Seppälä,
and
Elliott Sober

© Individual author and College Publications 2017
All rights reserved.

ISBN 978-1-84890-229-9

College Publications
Scientific Director: Dov Gabbay
Managing Director: Jane Spurr

http://www.collegepublications.co.uk

Printed by Lightning Source, Milton Keynes, UK

All rights reserved. No part of this publication may be reproduced, stored in a retrieval system or transmitted in any form, or by any means, electronic, mechanical, photocopying, recording or otherwise without prior permission, in writing, from the publisher.

Contents

Preface	iv
Editors' preface	iv
Preface of the Division of Logic, Methodology, and Philosophy of Science (DLMPS)	v
Preface of the Local Organizing Committee	vii
Congress Programme	xiv

Articles

A Logic — 1

1. **Logic in play** — 3
   JOHAN VAN BENTHEM

2. **Models in geometry and logic: 1870-1920** — 41
   PATRICIA BLANCHETTE

3. **Squeezing arguments and strong logics** — 63
   JULIETTE KENNEDY AND JOUKO VÄÄNÄNEN

4. **Global reflection principles** — 85
   PHILIP WELCH

5. **Logic revision** — 105
   EDWIN MARES

6. **An introduction to logical nihilism** — 125
   GILLIAN RUSSELL

	7	Intensional logic before Leibniz	137
		PAUL THOM	

B General Philosophy of Science — 153

	8	On some French probabilists of the twentieth century	155
		MARIA CARLA GALAVOTTI	
	9	Patrick Suppes: from logic to probabilistic metaphysics	175
		ANNE FAGOT-LARGEAULT	
	10	Logical empiricist reconstructions of theoretical knowledge	189
		WILLIAM DEMOPOULOS	
	11	How values can influence science without threatening its integrity	207
		SVEN OVE HANSSON	

C Philosophical Issues of Particular Disciplines — 223

	12	Prospects for an integrated history and philosophy of composition	225
		HASOK CHANG	
	13	A new look at the history of science	243
		ANA BARAHONA	
	14	Comparing causes	261
		ARNAUD POCHEVILLE, PAUL E. GRIFFITHS AND KAROLA STOTZ	
	15	What is action-oriented perception?	287
		ZOE DRAYSON	
	16	Philosophy of science in practice	301
		MIEKE BOON	
	17	Patchwork narratives for tumour heterogeneity	323
		GIOVANNI BONIOLO	

D CLMPS 2015 conference theme: Models and Modelling 347

18 Models and modeling in formal epistemology 349
ELEONORA CRESTO

19 Unrealistic models, mechanisms, and the social sciences 367
HAROLD KINCAID

20 Modelling failure 381
USKALI MÄKI

E International Council for Science (ICSU) special session: Future Earth 401

21 Transformative research for a sustainable future Earth 403
GORDON MCBEAN AND HEIDE HACKMANN

22 Biodiversity and triage 421
MARK COLYVAN

Index 436

Editors' preface

This volume contains papers based on invited lectures from the 15th International Congress of Logic, Methodology and Philosophy of Science, in Helsinki, Finland, on August 3-8, 2015. The CLMPS 2015 was held under the auspices of the International Union of History and Philosophy of Science (IUHPS), Division of Logic, Methodology, and Philosophy of Science (DLMPS), by the invitation of the Finnish National Committee for the Philosophy of Science, and hosted by the University of Helsinki. We have included in this book the opening statements by the representatives of the DLMPS, and the Local Organizing Committee.

In the traditional way, the Congress included six plenary lectures, 39 invited papers and 476 contributed papers distributed in 17 sections, and 172 papers in special symposia (including affiliated meetings and commission sessions). A complete list of the plenary lectures and invited papers appears on pages ix - xiii. This book includes 23 of the invited papers. Abstracts of all papers of the CLMPS, together with abstracts of papers of the Logic Colloquium 2015, were published in the Book of Abstracts in 2015 (clmps.helsinki.fi/materials/CLMPS_LC_bookofabstracts29.7.2015.pdf).

The process of editing the manuscripts into a camera ready version was led in Helsinki by Päivi Seppälä, the Congress Secretary of the CLMPS. Her effective team consisted M.Soc.Sc. Juho Pääkkönen, B.Phil. Leena Tulkki, and Ms. Hanna Pankka. We are grateful to all of them for their careful work. We also express our gratitude to King's College Publications and Managing Director Jane Spurr who wished to continue the tradition of publishing the Proceedings of the CLMPS in their series.

Munich, Helsinki, and Madison, WI, January 2017
Hannes Leitgeb, Ilkka Niiniluoto, Päivi Seppälä, Elliott Sober

Preface of the Division of Logic, Methodology, and Philosophy of Science (DLMPS)

Welcome to the 15th Congress on Logic, Methodology, and Philosophy of Science (CLMPS). The CLMPS is one of the most important activities of the Division of Logic, Methodology and Philosophy of Science (DLMPS) of the International Union for History and Philosophy of Science (IUHPS).

The DLMPS was founded in 1955, and currently has 31 ordinary members (representing countries) and seven international societies as international members. The first Congress took place in Stanford, California in 1960. One of its organizers was Patrick Suppes (1922-2014) who later became President of the DLMPS from 1975-1979; he passed away recently. Given his importance for the Division and our field, it is fitting that there will be a symposium on his work at the Helsinki Congress (Saturday, 8 August, 10am-12pm).

Since that first congress in Stanford in 1960, the CLMPS has been one of the most important global meetings of logicians and philosophers of science, taking place every four years and complementing national and regional meetings that occur more frequently. The Helsinki CLMPS continues this fine tradition. The programme for this Congress, and the abstracts in this volume, show that this Congress will provide a vibrant, probing, and multi-dimensional picture of our research field.

The DLMPS functions as the umbrella organization for logic and philosophy of science at the global level. The DLMPS facilitates discussion of developments in our fields in a number of commissions, some of which are shared with our sister division in the IUHPS, the Division of History of Science and Technology (DHST). In recent years, the two divisions have intensified their collaboration, and this will be visible at the Helsinki Congress in the form of a meeting of the Joint Commission (Friday, 7 August, 2:30pm-7pm; Saturday, 8 August, 10am-11am and 1:30pm-3:30pm), as well as symposia of commissions that are or are planning to be shared between the two divisions: the Teaching Commission, the International Association for Science and Cultural Diversity (IASCUD), and the Commission on History and Philosophy of Computing (HaPoC).

The IUHPS is part of the International Council for Science (ICSU), for which the IUHPS plays a special role by providing a meta-perspective on the scientific enterprise, reflecting on scientific practices and institutions. DLMPS has taken this task seriously at the Helsinki Congress by organizing two special ICSU sessions relating to current global ICSU research projects. One is on "Future Earth and Models of Climate Change"; the other is on "Health and Welfare". In these sessions, scientists working on these projects will interact with philosophers of science. In 2014 and 2015, IUHPS has coordinated a project for ICSU entitled "Cultures of Mathematical

Research Training"; there will be a report on that project as part of the already mentioned meeting of the International Association for Science and Cultural Diversity (IASCUD).

In conclusion we, as representatives of the Executive Committee of DLMPS, want to thank Hannes Leitgeb, who chaired the Program Committee, and Ilkka Niiniluoto, who chaired the Local Organizing Committee, for their excellent work in planning this Congress. They and the members of their committees have worked hard and deserve our gratitude. We also want to thank the scholars who will be giving talks at the Congress. The success of our Congress is now in their hands!

Elliott Sober
President, DLMPS

Benedikt Löwe
Assistant Secretary General, DLMPS

Preface of the Local Organizing Committee

On behalf of the Local Organizing Committee (LOC) of the 15th international Congress of Logic, Methodology, and Philosophy of Science (CLMPS), held in Helsinki, August 3-8, 2015, I wish to thank all of those who came to the beautiful capital of Finland on August 3-8, 2015.

The great tradition of international congresses of LMPS, under the auspices of the Division of Logic, Methodology and Philosophy of Science (DLMPS), was started in 1960 at Stanford University. Every four years these meetings bring together logicians and philosophers of science from all over the world to present and discuss their current work. In addition to 45 plenary and invited speakers, the Call for Papers attracted 762 philosophers to submit their contributed papers and symposium proposals to the congress. The number of registered participants from 55 countries was 713.

The programme covered all systematic and historical aspects of formal logic, general philosophy of science, and philosophical issues of special sciences. The theme of the 15th Congress was "Models and Modelling". A special feature of the CLMPS in 2015 was the co-location of the Logic Colloquium (LC), the European Summer Meeting of the Association for Symbolic Logic (ASL), in Helsinki, which allowed the participants also to follow a rich supply of lectures in mathematical logic.

The hosting University of Helsinki was established in 1640 as the Royal Academy of Turku. Its staff included a Professor of Theoretical Philosophy whose task was to teach logic or the art of thinking. When the Academy was moved to Helsinki 1828, it adopted the Humboldtian model of research-based education where philosophy played a leading academic role. Modern philosophy of science reached the University a century later, when Eino Kaila as the Professor of Theoretical Philosophy introduced the principles of logical empiricism. Kaila's students – among them Georg Henrik von Wright, Oiva Ketonen, and Erik Stenius - and von Wright's student Jaakko Hintikka transformed Helsinki to an important international centre of logic and philosophy of science. This philosophical approach was complemented in 1973 by the establishment of a new chair of Mathematical Logic at the Department of Mathematics.

Both von Wright (in 1963-65) and Hintikka (in 1965) are past presidents of the International Union of History and Philosophy of Science. Hintikka was an active participant as a speaker in both CLMPS and LC, but sadly passed away only a few days after the Congress.

The local organizers of the 15th Congress of LMPS included the Philosophical Society of Finland (founded in 1873 by Professor Thiodolf Rein), the section of Theoretical Philosophy of the Department of Philosophy, History, Culture and Art Studies, and the Finnish Centre of Excellence in the Philosophy of the Social Sciences (TINT). I wish to thank the members of our team, especially the congress secretary Päivi Seppälä, for their excellent and skilful work.

We are most grateful to the sponsors of the Congress: DLMPS, the University of Helsinki, the Finnish Federation of Scientific Societies, the Finnish Academy of Sciences and Letters, the Council of Academies of Science, the Finnish Cultural Foundation, the Wihuri Foundation, TINT, and the City of Helsinki.

On behalf of the local organizers, I wish to thank Elliott Sober and Benedikt Löwe (DLMPS), Hannes Leitgeb (chair of the Programme Committee), and Jouko Väänänen (chair of the Local Organizing Committee of LC) for their effective co-operation.

The venue of the Congress was the neoclassical main building of the University of Helsinki, located in the middle of the compact downtown area. Summertime in Finland provided pleasant sunshine and white nights, and the participants were able to enjoy the blend of logic and philosophy with a friendly intellectual and cultural atmosphere.

Ilkka Niiniluoto
Chair of the Local Organising Committee

Plenary Lectures

Logic in Play
Johan van Benthem, Amsterdam, Stanford, and Tsinghua University

Repertoires: How to Transform a Project into a Research Community
Rachel A. Ankeny, University of Adelaide, AUSTRALIA
Sabina Leonelli, University of Exeter, UNITED KINDOM

Scientific Realism, Models and the Social and Behavioral Sciences
Harold Kincaid, University of Cape Town, SOUTH AFRICA

Cubical homotopy type theory and univalence
Steve Awodey, Carnegie Mellon University, USA

Spacetime Functionalism
Eleanor Knox, King's College, London, UNITED KINGDOM

Models in Geometry and Logic: 1880-1920
Patricia Blanchette, University of Notre Dame, USA

Invited Papers

A1. Mathematical Logic

Constructing normal numbers
Verónica Becher, University of Buenos Aires, ARGENTINA

Entanglement and Formalism Freeness: Templates from Logic and Set Theory
Juliette Kennedy, University of Helsinki, FINLAND

Global Reflection Principles
P.D. Welch, School of Mathematics, University of Bristol, UNITED KINGDOM

A2. Philosophical Logic

Could there be no logic?
Gillian Russell, Department of Philosophy, Washington University St Louis, UNITED STATES

Logic Revision: Some Formal and Semi-Formal Techniques for Logic Choice
Edwin Mares, Victoria University of Wellington, NEW ZELAND

A3. Computational Logic and Applications of Logic

Syntactic Epistemic Logic
Sergei Artemov, City University of New York, USA

A Logical Revolution
Moshe Y. Vardi, Rice University, USA

A4. Invited Session: Historical Aspects of Logic

Intensional logic before Leibniz
Paul Thom, University of Sydney, AUSTRALIA

B1. Methodology

From nowhere, from here now, or from there then. A tale of success-to-truth inferences along perspectivalist lines.
Michela Massimi, University of Edinburgh, UNITED KINGDOM

Fact, Fiction, and Finance: Methodological Aspects of Econophysics
Dean Rickles, University of Sydney, AUSTRALIA

B2. Formal Philosophy of Science and Formal Epistemology

Bayesian Philosophy of Science
Stephan Hartmann, LMU Munich, GERMANY

The credit economy and the economic rationality of science
Kevin Zollman, Carnegie Mellon University, USA

B3. Metaphysical Issues in the Philosophy of Science

What gives Direction to Time?
Barry Loewer, Rutgers University, USA

On the Prospects of an Effective Metaphysics
Kerry McKenzie, UC San Diego, USA

B4. Ethical and Political Issues in the Philosophy of Science

Rationality at the science-policy interface
Sven Ove Hansson, KTH Royal Institute of Technology, Stockholm, SWEDEN

Biodiversity and Bio-patenting: Constructive Challenges of Scientific Research
Sang-Wook Yi, Hanyang University, Seoul, SOUTH KOREA

B5. Historical Aspects in the Philosophy of Science

On theories
William Demopoulos, The University of Western Ontario London, CANADA (presented by Thomas Uebel)

At the Roots of Probabilistic Epistemology
Maria Carla Galavotti, University of Bologna, ITALY

C1. Philosophy of the Formal Sciences

Explanation in Mathematics
Mark Colyvan, University of Sydney, AUSTRALIA

Three degrees of Imprecise Probability [IP] Theory
Teddy Seidenfeld, Carnegie Mellon University, USA

C2. Philosophy of the Physical Sciences

The quantum origin of statistical-mechanical probability
David Wallace, University of Oxford, UNITED KINGDOM

C3. Philosophy of the Life Sciences

The Transnational Turn in the History of Science
Ana Barahona, National Autonomous University of Mexico UNAM, MEXICO

Information and Causation in Biology
Paul E. Griffiths, University of Sydney, AUSTRALIA

C4. Philosophy of the Cognitive and Behavioural Sciences

The Rewards of Associative Modeling
Cameron Buckner, University of Houston, USA

What is action-oriented perception?
Zoe Drayson, University of Stirling, UNITED KINGDOM

C5. Philosophy of the Humanities and the Social Sciences

Scaffolding and Bootstrapping: How Archaeological Evidence Bites Back
Alison Wylie, University of Washington, UNITED STATES & Durham University, UNITED KINGDOM

C6. Philosophy of the Applied Sciences and Technology

How to bring philosophy back into science – Epistemological constructivism as a viable picture of science?
Mieke Boon, University of Twente, The Netherlands

C7. Philosophy of Medicine

Molecular medicine: the clinical method enters the lab. What primary tumor culture teaches us?
Giovanni Boniolo, University of Milan & IEO, ITALY

C8. Metaphilosophy

Intuition and Replication
Jennifer Nagel, University of Toronto, CANADA

Joint Commission - International Union of History and Philosophy of Science (IUHPS)

Prospects for an Integrated History and Philosophy of Composition
Chang Hasok, University of Cambridge, UNITED KINGDOM

CLMPS 2015 conference theme: Models and Modelling

Confirmation Theory for Idealized Models
Michael Weisberg, University of Pennsylvania, USA

Models and Modelling in Formal Epistemology: Some Thoughts on Probability Aggregation
Eleonora Cresto, University of Buenos Aires & The National Scientific and Technical Research Council, ARGENTINA

Modelling failure
Uskali Mäki, University of Helsinki, FINLAND

International Council for Science (ICSU): Health and Wellbeing in the Changing Urban Environment

Health and Well-Being in the Changing Urban Environment – An interdisciplinary program of the International Council for Science (ICSU)
Dov Jaron, Drexel University, USA & ICSU, FRANCE

Well-being and Health: A Perspective from Philosophy of Science
Anna Alexandrova, University of Cambridge, UNITED KINGDOM

Preface   xiii

International Council for Science (ICSU): Future Earth

Transformative Research for a Sustainable Future Earth
Gordon McBean, International Council for Science & Western University, CANADA

Biodiversity and Triage
Mark Colyvan, University of Sydney, AUSTRALIA

Climate Models and Calibration and Confirmation: The Need for a More Nuanced Picture of Use-Novelty and Double-Counting
Charlotte Werndl, London School of Economics, UNITED KINGDOM

Models and Empirical Philosophy: A Session in Honor of Patrick Suppes

Pat Suppes : from logic to probabilistic metaphysics
Anne Fagot-Largeault, The Collège de France, FRANCE

Congress Programme

MONDAY, 3 August	TUESDAY, 4 August	WEDNESDAY, 5 August
9.00-12.00/14.30 Registration Main Building lobby	9.00-10.30 Room 1 Plenary lecture: Rachel Ankeny	9.00-10.30 Great Hall Plenary lecture: Harold Kincaid
	10.30-11.00 Coffee	10.30-11.00 Coffee
	11.00-13.00 Contributed Papers & Symposia +A3. Artemov & Vardi +B4. Hansson & Yi	11.00-13.00 Contributed Papers & Symposia +B2. Hartmann & Zollman + B5. Demopoulos & Galavotti
12.00-13.00 Room 1 Models: Weisberg		
13.00-14.30 Lunch Main Building	13.00-14.30 Lunch	13.00-14.30 Lunch
14.30-16.30 Opening Ceremony Great Hall	14.30-16.30 Contributed Papers & Symposia +ICSU: Health & Wellbeing: Jaron & Alexandrova	14.30-16.30 Contributed Papers & Symposia +A1. Becher & Kennedy +B3. Loewer & McKenzie
16.30-17.00 Coffee	16.30-17.00 Coffee	16.30-17.00 Coffee
17.00-18.30 Great Hall Plenary lecture: Johan van Benthem	17.00-19.00 Contributed Papers & Symposia + C3. Barahona & Griffiths	17.00-18.30 Contributed Papers +A1. Welch (17.00-18.00)
18.30-20.00 University Reception Great Hall Foyers		18.30-20.00 City Hall Reception
	19.30-21.30 City Visit	

Congress Programme

THURSDAY, 6 August	FRIDAY, 7 August	SATURDAY, 8 August
9.00-10.30, Pl Plenary lecture Steve Awodey	9.00-10.30 Great Hall Plenary lecture: Eleanor Knox	
10.30-11.00 Coffee	10.30-11.00 Coffee	10.00-12.00 Contributed Papers & Symposia + Joint Commision: Chang + Session in honor of Suppes
11.00-13.00 Contributed Papers & Symposia + A2. Mares & Russell + C6. Boon & C7. Boniolo	11.00-13.00 Contributed Papers & Symposia +C4. Buckner & Drayson + B1. Massimi & Rickles	
13.00-14.30 Lunch	13.00-14.30 Lunch (+ Executive lunch)	12.00-13.30 Lunch
14.30-16.30 Contributed Papers & Symposia + ICSU: Future Earth: McBean & Colyvan	14.30-16.30 Contributed Papers & Symposia + Models & Modelling Cresto & Mäki	13.30-15.30 Contributed Papers & Symposia +C1: Colyvan & Seidenfeld +C5. Wylie & C8. Nagel
16.30-17.00 Coffee	16.30-17.00 Coffee	15.30-16.00 Coffee
17.00-.19.00 Contributed Papers & Symposia + ICSU: Future Earth: Werndl (17.00-18.00)	17.00-19.00 Contributed Papers & Symposia +C2. Wallace (17.00-18.00) + A4. Thom (17.00-18.00)	16.00-18.00 Great Hall Plenary lecture: Patricia Blanchette + Closing Ceremony
18.00-20.00 DLMPS General Assembly Consistory's Hall		*Scientific Program* *Scientific w. LC 2015*
	19.30-24.00 Dinner, Restaurant Bank	*Social Program* *Social w. LC 2015*

Part A

Logic

1 Logic in play

JOHAN VAN BENTHEM *

Abstract. Going beyond the traditional focus on consequence and inference, logic can be broadened to an exact theory of general information-driven agency drawing on many sources, without giving up on its well-established mathematical modus operandi. We show how this broader agenda involves the design of new kinds of dynamic logics for action, information, knowledge update, and belief change, and eventually, entangled with these, agents' preferences and goals. In particular, we explore several active interfaces of logic and games, where all these themes come together in natural concrete scenarios, ending up with advocating a move from game theory to a theory of play. Our presentation throughout takes the form of discussing typical examples, identifying major themes and suggesting new open problems concerning logic and agency. Finally, we explore what taking this agency-based perspective means for a variety of fields, including philosophy, linguistics, and game theory—and we conclude with some thoughts on what the field of logic is, or can become, in the perspective presented in this article.

Keywords: Dynamic-epistemic logic, games, agency, information.

1 Introduction: Two perspectives on logic

What logic is about and what it is up to can be viewed in two different ways. One can see logic as describing the core structures of reality, with logical constants mirroring the structure of compound facts. In this perspective, logic is close to metaphysics, though some people soften this stance by letting logical consequence refer to infor-

---

*Amsterdam, Stanford & Tsinghua, http://staff.fnwi.uva.nl/j.vanbenthem

mational dependencies in the world, and logical constants to how we classify that information in ways that are useful to us. On either variant, logic is about structures out there in the world, structures that would be there even if no living being ever existed in any planet of this universe. This view of logic has a long and distinguished history going back to metaphysical or scientific inquiry. But there is also a second view where agents are of the essence, and it has an equally long pedigree in the history of the field. Logic is also typically embodied in human activities such as conversation or argumentation, and some historians believe that this is even how the scientific subject arose in the first place, out of reflection on this practice in philosophical, legal or political settings. Logical constants are then about structured moves that can take place in such social scenarios, while logical consequence has to do with forcing one's interlocutor to accept certain propositions. In other words, logic can be about the world out there, but just as well about the agents perceiving that world and acting in it.

The two perspectives are not at odds. Clearly, agents will only behave successfully in the long run if their modes of representation and reasoning fit the facts of the world. Also, the two views occur entangled in natural language, the culture medium for all we say and do. One instance is the pervasive 'product–process ambiguity' (van Benthem, 1996) between expressions for activities and their products in the world. Consider Carnap's famous book *Der Logische Aufbau der Welt* (Carnap, 1928). In German, "Aufbau" is ambiguous between 'structure' of the world and our 'construction' of it— and something similar is true for other natural languages. And these two viewpoints are clearly involved in what might be seen as an intriguing conceptual dance.

2 Logic and games: A natural combination

Despite endorsing this lofty balance, my purpose in this paper is to explore the activity or process perspective on logic, as it still has not received the full attention than it deserves (van Benthem, 2011). And going that way, actors come to the fore, that is the agents employing logic. So, my main topic might be called logic and agency.

In pursuing this line, I am going to first restrict attention to an area where many issues become concrete because we have vivid intuitions about them, namely, *games*. Games are a natural prism for the themes of this paper (van Benthem, 2014a). First of all, they are a natural practice where we hone our logical skills, and in particular, major logical activities such as argumentation have clear game-like features, such as choices of what to say, long-term strategies for dealing with opponents, and preferences as to the outcome of a debate. But also, games are a concrete model of intelligent social interaction, and they exhibit many structures that invite logical analysis. The interface of logic and games (and game theory) is rich and growing, with computer science as a 'Dritter im Bund', and we will show how. After that, I will explore more general agency-related themes and their connections with logic, and having done that, I conclude with some consequences of taking this agent-oriented perspective for a range of

Logic in play   5

issues from philosophy to the sciences, and for logic itself.[1]

3 Evaluation games in logic

Let us make a simple connection first between a basic notion in logic and one in games, that has been proposed by a range of authors since the 1960s. Truth and falsity for formulas in models can be analyzed in terms of a two-player game, where we pull apart logical notions into different roles—a standard pattern in creating 'logical games'. Roles allow the mind to play against itself, testing things to the utmost.

Games, roles, and moves Let φ be a first-order formula, and M a model. Verifier claims that φ is true in M, Falsifier claims that it is false. To be fully precise qua first-order semantics, we would also need an assignment sending individual variables to objects in the model, but in our exposition here, we will mostly downplay this finesse. Now the logical structure of the formula φ induces a scheduling for the game:

> At disjunctions, Verifier has to choose a disjunct for further play, while at conjunctions, it is Falsifier who has to make this choice. The game for a negation $\neg\varphi$ is the dual of the game for φ, all marks for turns, winning and losing are reversed between the two players. Verifier has to choose an object in M for an existential formula $\exists x\varphi$, while Falsifier chooses an object in M for a universal formula $\forall x\varphi$.

Each round drops a logical operator. When we reach an atom, a check takes place against the model M, and the game ends: Verifier wins if the atom is true, and Falsifier wins if it is false.

Example A formula in a network.

Consider the network depicted in Figure 1, with five nodes, and a connected relation: that is, there is an arrow between every two points (though we will not draw the resulting reflexive arrows, for convenience). The first-order formula $\forall x \forall y \, (Rxy \lor \exists z \, (Rxz \land Rzy))$ says that one can get from any point in the graph to any other point by at most two directed arrows. This assertion is false in the given network, and the game will show how. For instance, Falsifier can pick the objects $x = 5$ and $y = 4$, and can then win against any counter-play by Verifier.[2]

There is something general going on here that connects logic and game theory.

Truth, falsity and winning strategies The following equivalence can be proved easily by induction on the structure of first-order formulas, and the definition of the game in a

---

[1] This paper is largely a programmatic survey, with an emphasis on ideas and suggestive examples. Hence, I will not include extensive references, which are given, for instance, in the books cited here.

[2] Note how games end in finitely many steps, since the formula in play gets smaller in each round.

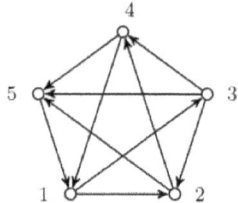

Figure 1.

model (the details of this proof again involve the use of variable assignments).

Fact A first-order formula φ is true in a model M iff Verifier has a winning strategy in game(φ, M).

This ties truth, a basic logical notion, to the game-theoretic notion of a strategy (for Verifier). Likewise falsity matches existence of a winning strategy for Falsifier.

Discussion Simple as it is, the preceding result suggests radical thoughts. The notion of a strategy is more fine-grained than truth, as there can be more than one winning strategy for a player. In the preceding example, another winning strategy for Falsifier is $x = 2$, $y = 3$. Thus, a game semantics is more fine-grained than mere truth values, a feature not exploited much so far, which fits intuitions about there being a natural hierarchy of less or more fine-grained denotations for sentences of a language.

Also, in the definition of the game, the clause for the quantifiers looks different from that for the connectives, challenging a standard analogy. A quantifier episode consists of two things: the choice of an object, and then, using that, playing the rest of the game. The real game operation here seems to be the sequential composition of a separate quantifier sub-game and the one for the remaining formula. This rearranges the standard geography of logical constants: the basic games are now atomic tests and object selection, while the general game constructions over these are choice, role switch, and composition. We refer to van Benthem, 2014a for the effects of this reappraisal. For instance, strikingly, the basic proof system underneath first-order logic from a game-theoretic point of view is then decidable.

Evaluation games exist for many logical languages. However, not all of these are as simple as the one we just showed. For instance, evaluation games for so-called fixed-point logics that can represent recursive definitions, say, of transitive closure or of wellfounded relations, allow for infinite histories of the game, since unfoldings of fixed-point variables in a formula under consideration may return to a larger formula of a shape encountered before. For such infinite games, the above lemma still holds, but the proof becomes much more delicate. This setting leads to deep connections with Automata Theory that we cannot go into here.

*Excursion: **model comparison*** There are many games for other logical purposes, such as testing satisfiability, or comparing models. We add an example of the latter to show

Logic in play

another aspect of the intimate connection between games and logical syntax.

Consider two models *M*, *N*. Player *D* (*Duplicator*) claims that *M*, *N* are similar, while (*Spoiler*) maintains that they are different. Players set some finite number *k* of rounds for the game, the severity of the probe. In each round, *S* chooses a model, and picks an object *d* in its domain. *D* then chooses an object *e* in the other model, and the pair (d, e) is added to the current list of matched objects. After *k* rounds, the object matching is inspected. If it is a *partial isomorphism*, *D* wins; otherwise, *S* does.

A telling example compares the ordering of the integers and rationals: the latter a dense structure, the former discrete. Here is how this comes to light in the game:

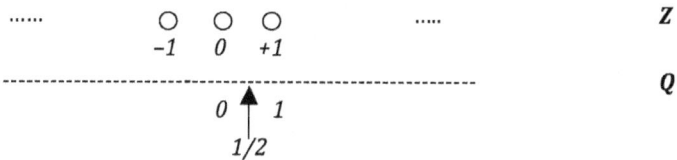

Figure 2.

By choosing objects well, *D* has a winning strategy here for the game over two rounds. But *S* can always win the game in three rounds, as suggested by our picture.

Here is a typical play:

Round 1	*S* chooses 0 in *Z*	*D* chooses 0 in *Q*
Round 2	*S* chooses 1 in *Z*	*D* chooses 1 in *Q*
Round 3	*S* chooses 1/2 in *Q*	any response for *D* is loosing

In playing the games, winning strategies for *S* are tightly correlated with first-order formulas φ that bring out a difference between the models. It is easy to see how *S*'s winning strategy in the preceding example matches step by step with the logical formula defining density:

$$\forall x \forall y \, (x < y \rightarrow \exists z \, (x < z \wedge z < y))$$

The analogy goes right down to how quantifier alternations mark model switches.[3]

Game operations and logical constants The structure of logical games may well involve further operations beyond the simple choices, switches and sequential composition that we have seen so far. In particular, there are also *parallel* constructions where different games are played at the same time. In fact, the preceding model comparison games are already a sort of parallel composition of games played in different models, with switches between these models initiated by one of the players. In general, on

---

[3] Model comparison games can also be continued infinitely, but we do not pursue this here.

the current view, the traditional set of logical constants can be naturally extended to mirror a broad spectrum of game operations.

4 Logic and game theory, basic encounters

One immediate effect of the above junction is that logical laws acquire game-theoretic import, and start connecting up with basic issues in game theory.

Excluded middle Consider the classical law of Excluded Middle $\varphi \vee \neg\varphi$. Its validity says that in every game of the form $\varphi \vee \neg\varphi$ over any model M, Verifier has a winning strategy. But by our game rules, that strategy consists of a choice which game to play, and in which role. Unpacking this information, always, either Verifier or Falsifier has a wining strategy in the φ–game. Games having this very special property that one of the two players has a winning strategy are called *determined*. And we can see why first-order evaluation games have this property by referring to what may well be the earliest result in game theory.

Zermelo's Theorem The following result by Zermelo dates back to the 1910s.

> *Fact* Two-player zero-sum games with finite depth are determined.
>
> *Proof.* For each specific depth, and two players i, j, this is Excluded Middle unpacked into its game meaning. To see this, here is the case with two rounds: $\exists x \forall y WIN_i xy \vee \neg \exists x \forall y WIN_i xy$ is equivalent by pure logic to $\exists x \forall y WIN_i xy \vee \forall x \exists y \neg WIN_i xy$, which in its turn is equivalent, at least in zero-sum games, to a determinacy statement $\exists x \forall y WIN_i xy \vee \forall x \exists y WIN_j xy$.

However, there is also a generic proof across models solving the game by an algorithm that computes colors White for nodes where player i has a winning strategy and Black for nodes where the opponent j has a winning strategy. The algorithm colors end nodes according to who wins the game there. Next, working upwards in the game tree, it colors a node that is a turn for player i white if there is at least one white daughter node, and black if all daughter nodes were colored black. The coloring rule for turns of player j is the obvious dual.

Equilibrium and fixed-point logic First-order evaluation games satisfy the conditions of Zermelo's Theorem, and so, their determinacy is explained. For later reference though, note one can also look differently at the structure of the game solution process going on here. The coloring algorithm itself has a logical form that can be read as an inductive definition of the eventual winning-strategy predicates WIN_i, WIN_j. The stages of the algorithm correspond to steps of unfolding the recursive definition, until the first fixed-point is reached where the predicates no longer change. The shape of the driving formula here is as follows:

$$WIN_i \leftrightarrow (end \wedge win_i) \vee (turn_i \wedge \langle move_i \rangle WIN_i) \vee (turn_j \wedge [move_j] WIN_i)$$

Logic in play

The right-hand formula has only positive syntactic occurrences of the variable for the strategic winning predicate WIN_i being defined here. Therefore, in standard fixed-point logics, we can prefix a greatest fixed-point operator to the right-hand formula to describe the eventual solution predicate.

This is not just a technical observation. Games are considered solved in game theory when we have an equilibrium between what players can achieve. Algorithms solving games often approach these equilibrium states in a stepwise manner. Thus, game-theoretic equilibrium is naturally connected with not just logic but also computer science, and the latter two fields meet in fixed-point logics of recursion and computation such as the modal μ–calculus, or $LFP(FO)$: the extension of first-order logic with operators for smallest and greatest fixed-points.

Logic and game equivalence But there is much more to the interplay of logical laws and games. Our second example again starts from a simple law of propositional logic, this time, Distribution of conjunction over disjunction. We display the two formulas involved in this law, and draw the shape of their game trees in a picture:

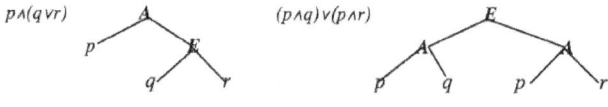

Figure 3.

Now we ask a new question, triggered by the logical equivalence.

Are these two games the same?

Distribution suggests that the two games are the same, and one can see that game-theoretically by focusing on the *powers* of players, i.e., those sets of outcomes that they can force the game to end in when they play one of their strategies against any possible counter-play by the opponent. It is easy to calculate that players have the same powers in both games.[4]

But there are natural alternative views when we look at players' actions and choices. The two games depicted are clearly different qua structure of moves and turns. For instance, there is no intermediate node in the game on the right that matches the situation in E's choice point on the left. To match this richer view of the 'how' rather than just the 'what' of control over outcomes, a more discriminating structural equivalence for games is needed. Good candidates for such a finer view are versions of the modal process equivalence of *bisimulation* that correlate available moves at each pair of matched nodes. Using a bisimulation, players can match their moves in two games

---

[4]In both games, A has the powers $\{p\}$, $\{q,r\}$, while E has the powers $\{p,q\}$, $\{p,r\}$. Here we use a condition of Monotonicity: powers of players are closed under taking supersets. Incidentally, powers in infinite games are sets of histories, but the idea remains the same.

step by step, thereby simulating the strategies themselves, not just the powers over final outcomes that they give rise to.

Further equivalences These are just two natural levels. Dropping monotonicity, van Benthem, Bezhanishvili, and Enqvist, 2016 identifies strategies with 'exact powers', where an outcome set shows what a player can force the game to end in, while the elements of that set show which choices are left to the *other player*. This notion of game equivalence differs from the earlier two: e.g., it does not validate Distribution. Thus, it induces an interesting weaker propositional logic yet to be explored. It also supports new game languages of the sort explored in the next section.

Thus, in this simple setting, we encounter the intriguing possibility that logics with different validities may match up with different structure levels for looking at games. In the following section, we will connect logic and games in one more manner.

5 Invariances, languages, and zoom levels

What we see here is an instance of a general phenomenon in mathematics and other fields. A subject can be studied at various levels of detail, and these levels are specified in terms of *invariance* under suitably chosen transformations, as proposed by Helmholtz and Klein in the 19th century, and in 20th century logic, by Tarski. There need not be a unique best choice here: Euclidean geometry is not 'better' than topology, it all depends on one's purpose. The same is true for games, and indeed, game theory has various natural structure levels, from extensive game trees to strategic matrix forms closer to the above power level.

Of special interest to logicians is that setting a level of detail corresponds to introducing a *language* that defines just the invariant properties characteristic for the chosen level.[5] The finer the invariance relation, the richer the matching language.

A typical illustration are the above two views of games. The richer action–choice level suggests a standard *modal language* over game trees viewed as relational models, where we have many results showing that (with some technical caveats) bisimilar models share the same modal theory. By contrast, focusing on powers suggests a coarser (but interesting) modal 'neighborhood language' for invariant properties, with 'forcing modalities' describing properties of players' powers. So, logical analysis of games is not embodied in one unique formal system: there is a hierarchy of logical languages and their logical laws matching a hierarchy of natural levels for representing games, or social interaction generally.[6]

---

[5]Helmholtz' own original motivation was finding an underpinning for the language of geometry.

[6]Our earlier example of Distribution suggested that logical equivalence is based on the power view. This

One can also restate these points in terms of 'zoom'. Many people believe that logic is organized pedantry. We take a given reasoning practice, and then supply more and more details until all arguments are fully spelled out, say, the way mathematical proofs might be spelled out to machine-readable first-order formulas. This is the *zooming-in* direction of supplying ever more detail, allowing us to see new phenomena at a microscopic level, such as small proof steps that can be automated. On this view, logicians are like moles digging in the soil below the observable cognitive behavior. But there is also a *zooming-out* direction where logical analysis does exactly the opposite: one looks at a reasoning practice, but only considers some global features in a rough formalism—the way, say, modal logic can yield a bare theory of the topological interior operation. This time, we see new things precisely because we ignore details, and soar, free as birds, far above the given reasoning practice. Both directions of zoom occur in logic, and there are interesting conceptual and technical questions concerning a systematic back-and-forth among various levels of analysis.[7]

In the light of Sections 4 and 5, developing a rich multi-level view of games and corresponding logics seems a well-worth enterprise.[8]

6 Entanglement in two directions

Our discussion so far has shown that logic and games form a natural combination. But, stepping back, we have really presented a mixture of two different directions.

Logic of games In one direction, often called logic *of* games, we use techniques from logic to analyze structures in games. In this contact, logic is applied as it stands, and the resulting systems are sometimes called 'game logics'. This is the direction that can be found in logical theories of multi-agent systems (Shoham & Leyton-Brown, 2009), or in current work on the logical foundations of game theory, as in 'epistemic game theory' that analyzes players' reasoning and strategic equilibria by logical means (Branderburger, 2014; Perea, 2012).

Logic as games But there is also a converse strand through many of the examples presented so far, where games themselves are used to analyze basic notions of logic, or in a current phrase, we pursue logic *as* games. This second direction of thought can run from weak claims, where 'logic games' are just used as convenient didactical tools, to strong methodological programs where games are viewed as embodying the essential meaning of logical systems.

---

may pose an initial perplexity, but we will return to this issue in Section 18 below.

[7] See van Benthem, 2016 for definitions of the phenomenon of tracking between levels, and some general results on when it works and when it does not, including connections with Category Theory.

[8] We will revisit this issue later in our discussion of 'Theory of Play', since our intuitions about game equivalence may have a hidden parameter: the type of agents that are playing the game.

Cycles The two directions are not at odds, although one should beware of confusion. For instance, in our game-theoretic analysis of first-order formulas, we can view these formulas as statements about some given model, but also as algebraic terms defining a kind of game playable on any model. These games have properties which can themselves be stated in some further logical (meta-)language, which could itself have evaluation games, and so on.

Thus, there is a productive cycle here. For instance, one can start from a class G of games, introduce a logical description language $L(G)$, and then consider games for evaluating the formulas of that logic, or for comparing its models. These activities in $G(L(G))$-mode are not always disjoint: a model comparison game for a logic of games may be close to a notion of structural equivalence for games in G. But $L(G(L))$-mode makes sense just as well. For instance, we defined evaluation games for first-order logic, but in studying those games from a solution perspective, we found patterns that suggests a natural fixed-point logic for those games, whose expressive power is known to exceed that of first-order logic itself.

The two directions are not as disjoint as they may seem. The monograph van Benthem, 2014a devotes two whole parts to hybrid systems merging motivations from logic games and game logics.[9]

Computation Our main point here is that, while logic and games form a natural combination, they do so in different entangled ways. Nevertheless, this two-component picture may also be misleading as it underplays what may well be an essential third ingredient: the role of *computation*.

Much of the basic work on game logics and logic games today is happening in the foundations of computation (Grädel, Thomas, and Wilke, 2002, Abramsky, 2008), where infinite games, Automata Theory, and Co-Algebra enter the fray, and where the emphasis in studying computation is shifting from extensional input-output views to the production of behavior by interactive systems. This is congenial to the later turn to be made in this paper toward general agency, since today's interactive computation is really a form of social agency (van Benthem, 2015), blurring the border line between games and computation just as much as that with logic.

Perhaps the best eventual picture is a triangle of *Logic, Games, and Computation*.

7 From logic and games to intelligent agency

In the rest of this paper, logic of games played by agents will be the central theme.

However, let us emphasize that logic as games is a rich area, too. In particular, our few examples should not be taken to convey the whole flavor. Evaluation games are

---

[9]There are even challenges to the whole scheme: for instance, evolutionary games seem hard to fit in.

Logic in play   13

just one way of casting logical notions as games. Argumentation, too, is a game, and winning strategies in suitably defined formal argumentation or dialogue games can be identified with proofs, in a tradition going back to Lorenzen. Argumentation or dialogue is also a powerful model for interactive computation, and the resulting work in the foundations of computation is producing deep results about games with a general thrust. For instance, the category-theoretic treatment of 'game semantics' in Abramsky, 2008 brings to light new logical constants, such as sequential versus parallel readings of disjunction that obey different laws. All this links up with resource logics such as linear logic, Proof Theory, Type Theory, and many other fields. Again we see how computation is a natural partner here.[10]

Having found natural interactions between logic and the study of games, we will now broaden our scope. Games are played by *agents*, and these agents are involved in a wide variety of activities. These range from deliberation before the game starts to processing information, or even dealing with unexpected surprises, during the game, all the way to post-game analysis, and perhaps rationalization of one's behavior.

In the coming sections we discuss a wide range of basic abilities of agents, relevant to playing games but also much more broadly, that can all be studied in logic.[11]

8 Dealing with many information sources

Inference and questions Inference is an important source of information, but there are others on a par with it. Suppose you are in a restaurant.

> Three people have ordered drinks, one each: water, beer, and wine.
> A new waiter comes carrying three glasses. What happens?

There are six ways the glasses can be distributed over the three customers, and here is a scene that plays out every day in many places. The new waiter needs to reduce the 6 options to 1, and solves his information problem as follows. He asks who has the water, puts that glass, then asks who has the beer, puts that, and then, without asking, puts the wine. What we see here is questions and answers reducing uncertainty from 6 options to 2 and then to 1, with the last step just being an inference (either explicit, or implicit in the act of placing the glass of wine).

There is a unity in this scenario: questions and inferences go together.

---

[10]For a broad panorama of the two directions at the interface between logic and games, with a landscape of game logics at different zoom levels, a presentation of different ways in which laws of logic can embody game-theoretic principles, and a survey of game-like systems that show features of both directions distinguished here, we refer to van Benthem, 2014a.

[11]As always, our exposition is a survey, for further details and results we refer to van Benthem, 2011.

Three sources of information There are further natural informational acts. This is already clear in the natural sciences: experiments and observations count for as much as mathematical deductions. An elegant compact statement from the world of practical common sense can be found in ancient Chinese Moist texts (400 BC):

知 闻 说 亲   zhi wen shuo qin

This elegant sparse phrase says that knowledge (zhi) comes from: hearing from others (wen), proof (shuo), or experience (qin). The Moist illustration is of someone seeing an object inside a dark room, wondering about its color. He sees a white object outside of the room, and someone tells him that the object inside the room has the same color as the one inside. He then infers that the object inside is white. What we see is a cooperation between observation, communication, and inference.[12]

Our basic informational abilities have at least this threefold range, and logical theory should deal with modeling all of this. As we will see soon, it can. But for now, let us continue with one more striking aspect of human reasoning abilities.

9 Social interactive reasoning

To some people, the high point of rationality is embodied in Rodin's 'Thinker': eyes closed, all on one's own. But in reality, intelligence seldom comes alone. In nature, many-body problems are the key to how the universe works, rather than the 'natural place' of individual objects. Likewise, our intelligent abilities usually unfold interactively in contacts with others. This so-called 'theory of mind' is seen as a crucial human ability in cognitive science, and it reaches far beyond the world of practical common sense: even the purest sciences themselves are a social enterprise.

Multi-agent knowledge 'Many-mind problems' are already key to asking the questions of our earlier base example. In addition to ground-level facts, we also communicate what we know or do not know about others. When I ask you a question, I normally convey to you that I do not know the answer, and also, that I think that you may know the answer: the latter is iterated two-agent knowledge.[13] Answering a question conveys a fact but, if done in public, it also makes sure that both participants know that they now know the answer, and this knowledge even goes to higher depths of entanglement, all the way to 'common knowledge' in the group.[14] It is this sort of iterated and entangled knowledge and ignorance of various sorts that keeps communication flowing, but also holds it in place and makes it successful.

---

[12]Frege famously gave up on natural language in favor of his "Begriffsschrift" because of the 'prolixity' of the former. But by 'natural language' he meant German: what if he had known Chinese?

[13]Exceptions are rhetorical questions, say, by teachers. But we are also attuned to when these occur.

[14]Of course, there is more to questions than just conveying information: tyically, questions also raise *issues*, and they set or modify a current agenda of conversation or inquiry.

Logic in play

There is logical structure underneath all this. We can reason about knowledge of agents about ground facts and each other in systems of *epistemic logic*, even group notions like common knowledge then turn out to have a precise logical behavior.

Just for illustration, we state an important valid equivalence relating knowledge of individuals and that of groups. Here we write $E_G\varphi$ for 'everyone in the group G knows that φ' and $C_G\varphi$ for 'φ is common knowledge in the group G':

$$C_G\varphi \leftrightarrow (\varphi \wedge E_G C_G \varphi)$$

This is not exactly a game-theoretic formula of our earlier fixed-point type, but it has a similar logical form, and it expressess an informational equilibrium in a group.

This example also illustrates a much more general point that is often under-appreciated. Action and information can exhibit very similar patterns, at least in the light of logic.

Dynamics with cards But here is another fundamental aspect to be recognized in the above. The whole point of communication as a social process is that what agents know keeps changing all the time as informational events occur. Again, what happens in the course of a game provides many concrete appealing examples of this flux. Consider the following baby card game:

> Three players, John, Mary, Paul get one card each: John gets **red**, Mary **white**, Paul **blue**. Now Mary asks John: "Do you have the blue card?" Who knows what now? John answers: "No". Who knows what now?

Audiences differ on answers to this little puzzle, but many people manage to figure out that after the question, John (but not Paul) has learnt who has which card, while after the answer, both John and Mary (but not Paul) know the cards. However, the fact that John and Mary know is common knowledge, so even Paul has learnt a lot from the information exchange in this scenario.

10 Information change, update, and dynamic logic

As we shall see, logic has the resources for describing scenarios like this: its scope extends beyond inference to questions—and to observation, and other basic acts.

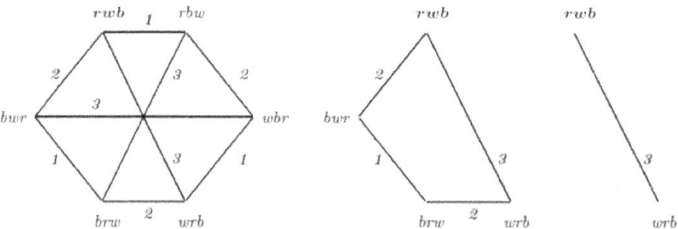

Figure 4.

Art of modeling We can represent the initial situation as a simple epistemic model with 6 states for the possible deals of the cards, and uncertainty lines indicating what agents cannot tell apart. For instance, in the actual state ***RWB***, Mary knows she has the white card, but cannot tell whether she is in ***RWB*** or in ***BWR***—and so on.

Now, crucially, further informational acts or events will change this initial model, following a widespread intuition in the literature that reflects common sense: new information decreases the current range of possibilities. For instance, learning that Mary does not have the blue card removes 2 options (***RBW***, ***WBR***) to yield the second picture shown in the sequence, where we can see graphically that in ***RWB***, 1 knows the cards: no uncertainty line departs there for him. The picture also shows many more facts in a direct visual manner: e.g., Paul knows that John knows the cards. The answer removes two more worlds (***BWR***, ***BRW***), resulting in the final stage depicted where the only remaining uncertainty line is for Paul. [15]

More generally, the informational action here is a public event $!\varphi$ telling everyone that φ is the case. (This is often called a 'public announcement' or 'public observation' of φ.) Its effect is the following change in the current group information state:

> it takes a current epistemic model (M, s)—with an actual situation s—to a new model $M|\varphi, s$, where only those points remain that satisfy φ in M.

Public announcement is about the simplest informational event imaginable. Understanding its logical behavior is the key to understanding a wide range of more sophisticated informational model updates.

Dynamic logics of information Informational acts $!\varphi$ satisfy precise logical laws, in suitably chosen languages allowing us to state what agents learn. The key principles of such systems describe the basic 'recursion equations' for the information flow triggered by acts or events, stating what happens to one's existing knowledge through an informational event. We display one such law:

$$[!\varphi]K_i\psi \leftrightarrow (\varphi \to K_i(\varphi \to [!\varphi]\psi))$$

This says that agent i will (would) know that ψ after receiving the truthful information that φ is the case if and only if, assuming φ holds, the agent had conditional knowledge of the implication that ψ will be the case after the φ–update.[16]

The reader may be used to laws of logic saying what agents know automatically when they know certain other things. That is an extreme static case, with information com-

---

[15] The semantic model drawn here matches sketches that many people make of the scenario, being a graphical representation of crucial information. (Incidentally, making good sketches from scenarios stated in natural language is an art going far beyond the routine translations into 'logical form' that we drill our students in.) But while we have used update only as the engine of information flow, actual behavior in problem solving also includes inference steps. Human reasoning seems a hybrid of many informational facilities, crossing boundaries that we would normally keep separate as theorists.

[16] The latter proviso with truth after the update, not simpliciter, is truly needed in a dynamic setting. We cannot just say $K_i(\varphi \to \psi)$, because updates may change truth values of epistemic statements ψ.

Logic in play 17

ing 'for free'. Dynamic laws like the preceding generalize this to the wider informational setting highlighted here, telling us what agents know after they have taken the trouble to explore or communicate.

Multi-culturality Also noteworthy is the cooperation of several disciplines embodied in this one formula, the same way common artifacts of modern life, say, your reading glasses, are often little crystallized pieces of a long history of cross-cultural collaboration. The idea of having logics for knowledge ($K_i\psi$) comes from philosophy, representing speech acts $!\varphi$ explicitly can be seen as the essence of linguistics, and the methodology of describing change after events with modalities comes from dynamic logic of programs in computer science. This reflects the fact that for the program of this paper, traditional border lines between philosophical logic, computational logic, or mathematical logic make no sense. We need all the insight we can get if we are to understand the full realm of the logical.

11 Proven methods, broader scope

The preceding observation suggests a continuity that may be worth stressing. Extending the scope of logic as advocated here is not a break with the past. The standard methods and standards of rigor of logic still apply, and in fact, they are needed to get a grip on what is going on. Information dynamics has laws extending our usual repertoire, but its theory is still logic. In Section 21 of the paper we will return to the issue of how to view all this.[17]

12 From information processing to agency

So far, the agents that we describe are mere information-recording devices. But human agency is more than information processing. Behavior is driven by goals and preferences. Now to some people this means entering the realm of taste and arbitrariness, but in fact, there is logical structure even here when we analyze styles of reasoning. Tastes may be arbitrary, but the world of fashion can have very rigid laws.

Here is our running example of the complexities arising in the interplay of information and preference, even in extremely simple game scenarios:

Reasoning with preference In the following game, player A can go left, and get 1 Euro, while E gets nothing, or A can go right, giving the turn to player E. Then E can

---

[17] As the reader will see, our main line does not involve changing the standard laws of classical logic: dynamics does not call for alternative logics, but for extended vocabularies. While this may provide fewer thrills than the vertiginous joy rides into the deviant, vague or inconsistent that some philosophers prefer, we believe that alternative logics should not be multiplied beyond necessity.

go left, getting 100 Euro while *A* gets nothing, or right, in which case both players get 99 Euro. What might happen in this scenario?

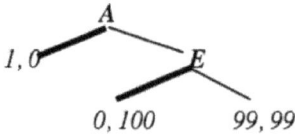

Figure 5.

One way of thinking is this. At her choice point, *E* faces a standard decision problem, and if she is rational, she will go left since she prefers that outcome over the one to the right. But *A* can see this coming, and realizes he will get 0 in that case, whereas going left will give him 1. So, *A* goes left, and it does not matter what *E* does.

This style of reasoning is called *Backward Induction*, and it describes a style of play for rational competing agents. Backward Induction has become a benchmark for logics of games, and we will return to its analysis below.

However, there are alternatives. Cooperative players might reach the endpoint (99, 99) by other plausible kinds of reasoning. There are many footholds for this. For instance, one weakness of the above argument is that it ignores the fact that *E* does not make her choice ex nihilo: she can also take into account the history of the game that led up to it, in terms of what she believes about the type of player that *A* is, or even if she has no such belief, she may just feel that 'she owes him'. In longer games than the one displayed, this could certainly matter.

In general, in most games we do not know which type of player we are up against—and the common distinction between 'competitive' versus 'cooperative' games does not help. For instance, academic life is a subtle (and sometimes not so subtle) mixture of both. So, we need an abstract stance that accommodates variety of behavior.

Philosophical plus computational logic Therefore, it makes sense to design logics that can account for any reasoning of the above kind about social scenarios, mixing information, action, belief, and preference. And when we analyze the ingredients for that reasoning, they read like a compact agenda for all of philosophical logic, involving knowledge, preference, belief (after all, *A* will never find out what *E* would do, so he cannot know it, but only have a belief about her decision), counterfactual conditionals (what would *E* have done, had *A* moved right), but also notions from computational logic in single actions or complex strategies, and fixed-points corresponding to various game-theoretic equilibria, such as the ones computed by Backward Induction, or by cooperative scenarios.

Again, we see our point of merging disciplinary agendas. Logic of games is an area where many different strands in earlier literature meet in concrete scenarios.

13 Benchmark for game logics: Backward induction

For a concrete illustration, we will look at the logical form of Backward Induction. Whether we endorse this style of reasoning or not, what is the logical form underlying its particular take on rationality? We will use this a showcase of our general approach in Sections 4 and 5: different zoom levels make sense for different purposes.

Backward Induction algorithm To define the algorithm, we first need an auxiliary notion. We say that a node x strictly dominates a sibling node y (siblings are immediate successors of some shared parent node z) for player i if all further outcomes of the game reachable after x are preferred by i to all outcomes of the game reachable after y. Now we can define a relational version of the BI algorithm as follows (the earlier numerical values were just added for drama): one keeps removing transitions from parent nodes to strictly dominated children. In general, this process must remove transitions, when we look at points whose only children are end nodes. Moreover, iteratively, the algorithm will move upward in the tree (it is really a refined version of the earlier Zermelo coloring algorithm) until we reach the root. Since the 'available move' relation BI can only get smaller in the process, it must stop by some stage, and this fixed-point is the solution produced by the algorithm.

Theorem
BI is the largest sub-relation of the move relation in finite game tree satisfying (a) the relation has a successor at each intermediate node, (b) CF:

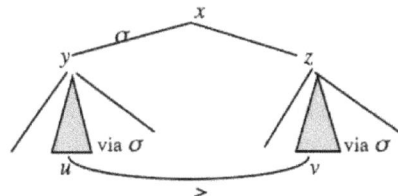

Figure 6.

Confluence (CF)
$\&_i \forall x \forall y ((Turn_i(x) \land x\sigma y) \to (x \text{ move } y \land \forall z (x \text{ move } z \to \exists u \exists v (end(u) \land end(v) \land y\sigma^* u \land z\sigma^* v \land v \leq_i u))))$

The logical form of rationality In a typical logician's modus operandi, we can now look at a (first-order) syntactic recursive description of the algorithm. It is easy to see then that the relation R that is produced only occurs *positively* in the above formula with ∀∀∃∃ syntax. Thus we have an observation which again reflects the close connection between game-theoretic equilibria and fixed-point logics.

Fact The Backward Induction strategy, viewed as a sub-relation of the total *move* relation, is definable in first-order fixed-point logic $LFP(FO)$.

One can view the fixed-point formula found here as bringing to light the 'logical form' of rationality as conceived of by Backward Induction. Other game solution methods may induce other logical forms.

Domination and belief But there are further interesting features worth pointing out, on the following intuitive interpretation. What is 'dominated' can change at each stage since the total set of available moves gets smaller. This set of available moves, and hence available histories of the game, may be said to represent the players' *current beliefs* about how the game might still go. Avoiding dominated moves at the current stage, as prescribed by the algorithm, is then a form of optimal decision making *given one's beliefs*, or in game-theoretic terms: players are assumed to be 'rational-in-beliefs', regardless of whether they are doing what is objectively best for them.

This belief interpretation can be made into a dynamic reanalysis of Backward Induction as an iterated belief revision procedure for players engaging in pre-game deliberation. We will look at this mechanism later on, since it has uses in many places.

As one more pointer to later themes, note that the Backward Induction solution is *uniformly definable* for both players, who are assumed to exhibit the same kind of reasoning and global beliefs about how other players operate – even though their base-level preferences and the available moves at their turns may differ completely. Once we drop this assumption of uniformity, however, we have to start thinking about games in which players can be of different types, raising the issue of how much diversity one can tolerate before social interaction, or at least, reasoning underlying social interaction, breaks down. This theme will return in our section on 'Theory of Play'.

Zoom levels Finally, recall our earlier theme of different zoom levels for logical analysis. Our definition of the Backward Induction strategy in fixed-point logic is extremely fine-grained. It is hard to imagine that reasoning in such detail informs all our behavior in daily interactions. Can we zoom out to higher description levels?

One way of zooming out uses a modal language for 'best actions' and preferences only, with Backward Induction kept under the hood. How will such surface level reasoning go? The earlier rationality principle is now expressed by this modal axiom, where σ denotes optimal moves (best actions) recommended by the algorithm:

$$(turn_i \wedge \langle \sigma^* \rangle (\textbf{\textit{end}} \wedge \varphi)) \rightarrow [move_i] \langle \sigma^* \rangle (\textbf{\textit{end}} \wedge \langle pref_i \rangle \varphi)$$

However, the complete modal logic of best action in this sense is not known, and it is not even clear that it is completely axiomatizable.[18] There are interesting analogies

---

[18] The technical reason is an insight from computational logic. The 'confluence patterns' in game trees induced by our BI analysis create a regular grid-like structure that may support embedding of so-called 'tiling problems' of high complexity in the logic. If this is so, then, though rational behavior itself may be simple and predictable, the logical theory of rationality might be highly complex.

between this modal level of analyzing best action and that of *deontic logic*, viewed as a high-zoom specification language for regulating and evaluating behavior.

The natural language of decision There may be other zoom levels here, and logic need not always have the best description language for interactive social behavior lying on the shelf. Indeed, natural language itself seems to do a good job in this respect. Our rich vocabulary concerning best or better actions, our hopes and fears, what we ought or are permitted to do, and other evaluative expressions form a rich and subtle medium for describing and influencing behavior. So far, logics of agency have only scratched the surface of this.

14 Players do much more

Three phases What we have looked at so far is only a tiny slice of the intelligent activities that players exhibit. Let us recall an earlier distinction. There are at least three different phases where logical reasoning plays a role: *before*, *during*, and *after* a game. The above Backward Induction procedure, and many 'game solution' procedures seem to belong to a pre-game *deliberation* phase. But also very important is the post-game phase of analysis and usually, *rationalization*, where we fix what is the 'lesson learnt' for future play. And perhaps most excitingly, many things happen during a game. Moves are played, obviously, but also, information is received and gets processed, as we saw in our cards scenario.

Adding the past One immediate instance of in-play reasoning occurs with Backward Induction itself. What if, in the course of a game, we see that our opponent deviates from the expectations generated by the BI algorithm? This time, in general, we have two inputs: what we expected beforehand, and what has actually happened.[19]

This opens up a wide space of options. We might consider the deviation an error. We might take it as a signal that the other player wants to cooperate with us, at least to some extent. We might also take a repeated simple pattern of deviations as information of a very different kind. Perhaps we are witnessing a drastic case of non-uniformity, and we are playing against an automaton that always plays one sort of move.

Belief revision It will be clear that the usual mathematical notion of a game tree, as a record of all possible runs of a game, does not suffice to model all these processes going on in play. Accordingly, various additional aspects have started appearing in studies of games, such as explicit modeling of 'player types' in epistemic game theory, or of various kinds of automata playing games in computational logic. We will return to this shift later, but for now, let us just note a more base-level common denominator behind the preceding scenarios. What surprising events in a game will do in general

---

[19] By contrast, our running example of the Backward Induction algorithm only looked at the future at any node, and then, we might just as well throw away the past play leading up to that node.

is force players to engage in on-line *belief revision*, leading to new expectations about the future course of the game.

This broader area is the realm of game-theoretic solution methods like 'Forward Induction'.[20] However, in this paper, we will look at things in a more general logical perspective, tying up with one more general aspect of agency In the coming two sections, we discuss two additional aspects of agency that we did not high-light before.

15 Dynamic logic of belief revision

From knowledge to belief Our actions are driven by belief as much as knowledge. And beliefs are not whimsical attitudes, or the soddy paper money that mimics the gold standard of knowledge.[21] Beliefs are crucial triggers for most of our actions, and their formation and maintenance may even have a more creative aspect than the knowledge in one's savings account.

Beliefs are generated by a much wider spectrum of informational events than the indubitable public announcements that we have considered so far. In particular, they can be triggered by signals carrying what might be called 'soft information' as much as by the hard information that we modeled earlier in our discussion of knowledge update in card games. More generally, changing beliefs is not a vice but an epistemic and cognitive virtue: our cognitive abilities often show at their best when we are confronted with a surprise, and need to re-adjust what we thought before.

Logic of belief revision Belief revision is crucial to the realm of agency that we are charting, and techniques similar to our earlier ones apply. First, we need a simple static base model for beliefs, and one common such structure are plausibility models $M = (W, \sim, \leq, V)$, where the agents' ranges in an epistemic model as defined earlier are now ordered by a binary relation of relative plausibility.[22] These models are a qualitative pilot setting on which to develop the main themes to follow—though they can be enriched with further structure (evidence, or probability) when needed.

Belief in the truth of a proposition φ (written $B\varphi$) can now be defined as truth of φ in all the most plausible worlds. But we need more expressive power. Given that information about the relevant set of worlds may change, we also need a notion of *conditional belief* $B^{\psi}\chi$ saying that χ is true in all most plausible worlds within the

---

[20] Such alternative solution methods may no longer work on simple annotated game trees. For a systematic hierarchy of logical models for games with increasing complexity, cf. van Benthem, 2014a. Richer models are needed as we relax what players know about the game and each other's strategies.

[21] Conceptually, paper money may have been the more innovative historical invention.

[22] We drop agent indices in what follows, but this is merely for notational convenience. There is no barrier toward dealing with multi-agent scenarios.

set of ψ–worlds.[23] Plausibility models also support other epistemic attitudes, such as 'safe belief' and 'strong belief', but we will not pursue this theme here.

Hard information The static logic of belief defined in this way is much like conditional logic in the semantic tradition. Of interest to us here is that our earlier theme of information flow and change forms a natural continuation of such systems from philosophical logic. First, consider the earlier events $!\varphi$ that produce the hard information that φ is the case. These sinple events can already drive quite interesting scenarios, witness the 'misleading with the truth' cases discussed in van Benthem, 2011, whose details we forego here.

> *Fact* The logic of changes in absolute and conditional beliefs under hard information is completely axiomatizable using suitable recursion axioms.

We merely display the two basic recursion laws for new beliefs:

> *Fact* The following equivalences are valid for hard belief revision:
> $[!\varphi]B\psi \leftrightarrow (\varphi \to B^\varphi[!\varphi]\psi)$
> $[!\varphi]B^\psi\chi \leftrightarrow (\varphi \to B^{\varphi \wedge [!\varphi]\psi}[!\varphi]\chi)$

Soft information Belief is not just one more attitude that changes under hard information. There can now also be events that do not eliminate worlds (every existing option remains available), but modify the plausibility pattern. There are many actions of this kind in the literature, but it will suffice to mention one characteristic example:

> A radical upgrade $\Uparrow \varphi$ puts all φ–worlds on top of all $\neg\varphi$–worlds in the ordering, and within these two zones, it keeps the old ordering.

Radical upgrade is a strong move in favor of φ, of a sort that has been studied in belief revision theory (Gärdenfors & Rott, 1995) and formal learning theory (Baltag, Smets, & Gierasimczuk, 2011).[24]

Again the complete dynamic logic of such events, viewed as denoting matching model changes, can be described by introducing appropriate modalities.

> *Fact* The logic of changes in absolute and conditional beliefs under soft information is completely axiomatizable using suitable recursion axioms.

This time, we display just one, formidable-looking, recursion law, where 'E' stands for the existential epistemic modality 'somewhere in the current epistemic range':

$[\Uparrow \varphi]B^\psi\chi \leftrightarrow (E(\varphi \wedge [\Uparrow \varphi]\psi) \wedge B^{\varphi \wedge [\Uparrow \varphi]\psi}[\Uparrow \varphi]\chi)$
$\vee (\neg E(\varphi \wedge [\Uparrow \varphi]\psi) \wedge B^{[\Uparrow \varphi]\psi}[\Uparrow \varphi]\chi)$

---

[23] More complex truth clauses are needed for infinite models. Note also that, unlike with the earlier conditional knowledge, conditional belief is not definable in terms of absolute belief.

[24] Alternatively, a radical upgrade can be seen as a strong deontic command to see to it that φ.

For an explanation of this law, and a method for its systematic derivation from the definitions of radical upgrade and conditional belief, see van Benthem, 2011.

Laws of learning The principle displayed just now shows that logical laws can describe the formation of new beliefs, and even of new conditional beliefs, as new information comes in. An alternative interpretation is as laws of *learning*, since much learning consists in modifying beliefs so as to improve their fit with the truth, or at least with reliable new information. However, there is nothing peculiar about radical upgrade that enables us to do this. Plausibility changes can come in a great variety of formats or 'learning policies'. And there exist several general methods for dealing with the induced dynamic logics of belief change. The survey van Benthem and Smets, 2015 references many classical contributions to the literature.

Finally, while the above law may look much more complex than standard axioms for epistemic or doxastic logic, this is to be expected. Belief is a subtle notion—and the greater complexity may also be seen as greater richness of content.

We have seen now, at least as a sketch, how logical techniques can deal with describing beliefs and belief changes just as for knowledge. Since beliefs are fallible, and can be wrong, this shows that logic is not tied exclusively to truth and knowledge, it also makes sense as a guide when we live in a world of error and confusion. We will return to this theme of *correction* rather than (just) *correctness* in a later section.

Coda Our themes here do not belong exclusively to logic. Learning and improving theories have long been major themes in the philosophy of science, and the same broad origins can be seen with counterfactuals, or belief revision theories. But then, at an Auld Alliance Congress like the DLMPS, who cares about exclusive labels like 'logician' or 'philosopher of science'?

16 Long term phenomena and limit behavior

A second prominent feature of games with logical import is their temporal horizon. *Strategies* are usually not single moves, but methods for achieving some effect only after many steps. Strategies in infinite games need not even work toward any finite apotheosis, but serve to produce particular kinds of never-ending *histories*. And in the temporal long run over histories, phenomena may emerge that are sui generis.

Limit phenomena A temporal perspective enriches our earlier dynamics. Single informative events, whether with hard or soft information, form longer histories. These may contain surprising emergent structure of their own, such as success or failure in converging to a fixed-point when running an update or deliberation procedure.

Consider the famous 'Muddy Children' puzzle (cf. van Benthem, 2011 and the references therein) where repeated announcement of the children's ignorance of their status (as long as this ignorance holds) results eventually in a flip-flop: some children do know their status after the last announcement of ignorance. Here, in addition to immediate effects of the individual statements, we see 'self-refuting' limit behavior of iterated announcements: at the final stage, the statement becomes false in the actual world. But this is just a start. Limit behavior of assertions can also be 'self-fulfilling'. Then, at the first fixed-point of iterated announcements, the statement announced is true everywhere in the remaining model, making it common knowledge among the agents involved. Here is a game scenario where this happens.

Example Backward Induction, hard scenario.

Consider our earlier analysis of the Backward Induction algorithm. Let **rat** be the statement — which can be true or false at any given node — that no player played a strictly dominated move when coming to this node. Announcing this information in our earlier public announcement style will leave certain nodes, but may remove others. Thus, afterwards, new nodes may satisfy, or fail to satisfy, **rat**—and hence iterating this assertion makes sense.

Consider our earlier running example. The following figure depicts what happens in a hard scenario of events !**rat** removing nodes from the tree that are strictly dominated by siblings as long as this can be done:

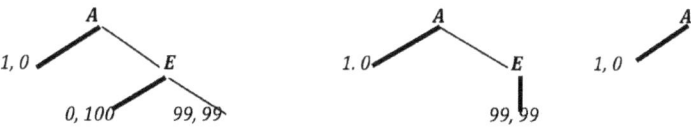

Figure 7.

At the final stage, all nodes satisfy the assertion **rat**, and we have a fixed-point.

> *Fact* The preceding limit procedure of announcing rationality always generates the Backward Induction path.

Example, continued Backward Induction, soft scenario.

By contrast, a scenario with soft information as input does not remove nodes but it modifies the plausibility relation. Here is how we can analyze Backward Induction as a deliberation procedure forming expectations. We start with all endpoints of the game tree incomparable qua plausibility. Next, at each stage, we compare sibling nodes, using the following notion.

A move x for player i *dominates* its sibling y *in beliefs* if the *most plausible* end nodes reachable after x along any path in the whole game tree are all better for the active player than all the most plausible end nodes reachable in the game after y. Rationality* (**rat***) is the assertion that no player plays a move that is dominated in beliefs. Now we perform a relation change that is like a radical upgrade ⇑**rat***: 'If x dominates y in beliefs, we make all end nodes from x more plausible than those reachable from y, keeping the old order inside these zones'.

This changes the plausibility order, and hence the dominance pattern, so that an iteration can start. Here are the stages for this procedure in the above example, where we use the letters x, y, z to stand for the end nodes or histories of the game:

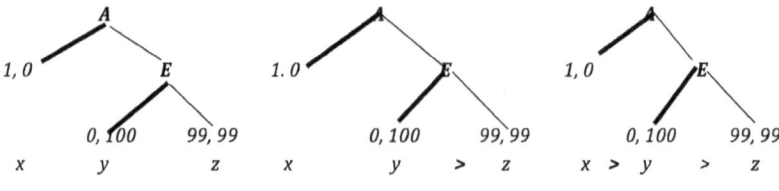

Figure 8.

In the first game tree, going right is not yet dominated in beliefs for A by going left. And so *rat** only has bite at E's turn, and the update makes $(0, 100)$ more plausible than $(99, 99)$. After this ordering change, however, going right has become dominated in beliefs, and a new update takes place, making A's going left most plausible.

> *Fact* On finite trees, the Backward Induction strategy is encoded in the plausibility order for end nodes created by iterated radical upgrade with rationality-in-belief.[25]

This can be proved by induction, using a natural equivalence between relational strategies as subsets of the total *move* relation and ('tree-compatible') plausibility orders on endpoints of the game tree, van Benthem and Gheerbrant, 2010. In particular, computation in our upgrade scenario for belief and plausibility and the earlier relational algorithm *BI* for Backward Induction are in harmony stage by stage:

> *Fact* For any game tree M and any k, $rel((\Uparrow rat^*)^k, M)) = BI^k$.

Thus, the algorithmics of Backward Induction and its analysis in terms of forming beliefs amount to the same thing. Still, the belief limit scenario also has some interesting features of its own. One clear benefit is that it yields fine-structure for the plausibility relations that are usually treated as primitives in doxastic logic.[26]

Limit learning Limit behavior need not always stabilize in this simple way. It can get more complex with plausibility updates for other statements. Baltag and Smets, 2011 show how iterated radical upgrades for complex formulas can oscillate forever on truth of statements with beliefs occurring under dynamic modalities, though absolute beliefs will converge. Baltag et al., 2011 show how this behavior extends to 'learning in the limit' as eventual alignment of belief to the true hypothesis. They show that public announcement and radical upgrade are universal learning methods from an input stream of hard information, given a suitable prior plausibility order en-

---

[25] Moreover, at the end of this procedure, players have acquired *common belief in rationality*.

[26] Many of these facts can be explained by the form of the relevant statements in fixed-point logics.

coding the chosen learning method. But iterated radical upgrade is the only universal learning method under input containing a finite number of errors.[27]

Digression: social networks Beyond games, this setting applies to many other social scenarios with irreducible collective group behavior. In particular, it fits a recent trend of describing social networks (Liu, Seligman, & Girard, 2014) where agents' opinions are influenced by those of their neighbors, and long-term dynamical system behavior emerges. Such systems need not reach equilibrium in the sense of the earlier fixed-points, but they can still show other forms of stability, where the group cycles recurrently through certain patterns of epistemic states.[28]

Many earlier themes make sense here, including different zoom levels and matching logical languages. Described at one level, agents' opinions may keep oscillating, but at a level of percentages for and against, a group may be stable, as happens in equilibria studied in evolutionary game theory (Osborne & Rubinstein, 1994). These levels are entangled with the issue of automatic group behavior versus individual decisions: many unique intelligent decision makers may still add up to one statistical mob.

Returning to our focus on games, a temporal long-term perspective is also natural for players in infinite games. We can think of their strategies as compounding available individual local moves, but also at a higher zoom level, in terms of players' powers for forcing the game to produce specific sets of histories satisfying certain properties. We present one instance of a logical principle that governs the latter setting.

Example Weak Determinacy.

We noted earlier that most logic games are 'determined' in the sense that one of the players has a winning strategy. But not all games are determined (the Zermelo conditions are quite strong), and in fact, most interesting games are not. However, here is a fact from descriptive set theory that holds for players in arbitrary infinite two-player games. *Weak Determinacy* says that, either one of the players has a winning strategy, or the other player has a strategy preventing the first player from ever reaching a winning position. Now *strategies* are objects that admit of logical description (van Benthem, 2014a) as programs where conditions on actions can depend on what players know about the world or about other players.[29] But as an illustration here, we stay at a high zoom level, that of the earlier-mentioned powers of players. At that level, Weak Determinacy is a typical law of 'temporally forcing' histories:

$$\{G,i\}\varphi \vee \{G,j\}Always\neg\{G,i\}\varphi$$

---

[27] A more general point here is that, in this way, global limit considerations about learning methods may be brought to bear on the issue of which local update rules one should choose.

[28] van Benthem, 2015 is a first exploration of logical patterns for such long-term group behavior, ranging from generalized fixed-point logics to extended dynamic-epistemic logics. A wide variety of logics for well-known informational social group phenomena is found in Christoff, 2016.

[29] For much more on logic and strategies, see Ghosh, van Benthem, and Verbrugge, 2015.

Dynamical systems Behind all this lies a challenge. How to interface dynamic or other logics of information-driven agency with the theory of dynamical systems that underlies evolutionary game theory and formal theories of social behavior?[30]

17 Theory of play

The general picture emerging from our considerations so far may be summarized as follows. When logic, game theory and computer science come together in the way that we have sketched, something new emerges that does not belong entirely to any of these fields, which may be called a *Theory of Play*. What is in focus here is 'play' instead of 'games': how agents engage in games and game-like activities, what information they absorb, how they keep this aligned with their preferences, and what actions result in the real world. This is a huge widening of the traditional focus on game trees as such (where one makes up stories about how they could have arisen without making their ingredients explicit)—but motivations and tools for this broader enterprise can be found in all three areas mentioned.[31]

Agent diversity While Theory of Play is attractive, this agenda expansion is not unproblematic. Many issues remain in charting the great variety of logical tasks involved, and finding their static and dynamic laws.

But a more difficult further issue is how much *diversity* we allow for the agents performing these tasks: agents could have any sort of abilities for picking up information or policies for changing beliefs. To tame the resulting explosion of options in analyzing social scenarios, we need a taxonomy of natural kinds: such as risk-seeking/risk-averse agents, competitive/ cooperative, and the like. We observed earlier that most logical systems assume uniformity of agents: allowing diversity seems a real challenge to which we will return below, where we may need to relocate what is logical to the level of interfacing agents of different types.

Instead of engaging in further general soul-searching, let us end with some consequences of Theory of Play for the interface of logic and game theory. We have seen several concrete instances of this dynamic perspective already, such as our new information-dynamic scenarios for Backward Induction via iterated public announcement or plausibility upgrade. We add a few more instances, starting with a fundamental theme discussed in earlier sections.

Game equivalence revisited In an agent-oriented view, the very notion of game identity is at stake: it becomes player-dependent. It no longer makes much sense to ask, as

---

[30]Some first explorations of interfacing dynamic logics with dynamical systems can be found in van Benthem, 2011, van Lee, Rendsvig, and van Wijk, 2016.

[31]For instance, the players in a Theory of Play can be modeled as automata, and then an extensive literature from computational logic becomes available, cf. Grädel et al., 2002.

before, when two given games are the same: rather, we need to ask if they are equivalent *for what players*, given their preferences, beliefs, and general modus operandi. We will only give one very small illustration to show what has to change then in our earlier style of analysis of actions and powers.

Logical equivalence for rational players Let us assume that players are rational in the sense of Backward Induction. When are two games the same for players like that? It makes sense to demand that the equilibria be correlated in some way. But then, at once, earlier logical laws will fail. Consider again propositional Distribution, but now apply it to our earlier running example of a simple game with preferences.

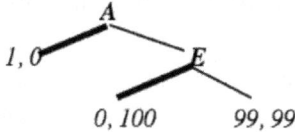

Figure 9.

The game depicted here is not Backward Induction-equivalent to its distributed form

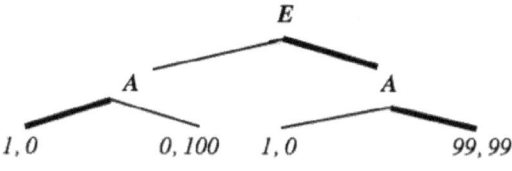

Figure 10.

as can be seen by comparing the Backward Induction strategies and outcomes in both cases. Thus, applying the logical law of distribution blindly would turn a competitive solution into a cooperative one!

It is an open problem to determine what weaker propositional algebra of game equivalence arises when we demand that the Backward Induction equilibrium outcomes are the same on both sides. For some simple, but suggestive results, see van Benthem et al., 2016.

Play equivalence But Backward Induction is only one style of reasoning-based play, and there are many others, depending on what we take the agents' modus operandi to be, perhaps including their computational and inferential limitations. The more general issue that arises then is analyzing equivalence between combinations of games plus play styles. The best way of doing this, preferably again tied to introducing matching languages, seems a serious open problem.

Bounded agency As a final theme, we mention the interaction of players that may have bounded resources of various sorts. One way of modeling these resources is in terms of 'awareness' in the recent game-theoretical literature, or of 'short sight' in the computational literature (cf. Grossi and Turrini, 2012, and the game-theoretic literature cited there). Another agent model would use various sorts of automata, and then fundamental results like the Positional Determinacy Theorem from computational logic (cf. Venema, 2015 for this and many related results) start telling us something about the reach of bounded agency.

Theory of Play has many further strands, but we leave matters here.

18 Logic, games, and general agency

In the remainder of this paper, we present a general programmatic discussion taking a still broader view. Not all intelligent interaction is game-like, and in fact, much of what we have been after in this paper has in fact been the broader arena of Logic and Agency. In the coming sections, we will show how this general shift in perspective has repercussions all around. The agency view can be taken to virtually any topic at this congress or beyond to find new issues or connect old ones.

We merely give a few examples of agency-oriented threads, many further illustrations for this perspective can be found in van Benthem, 2011, 2014a. Given the short compass of this paper, many of our claims will be somewhat apodictic, but even so, we hope that they will open some windows for the reader.

19 Epistemology: From foundation to correction

The traditional emphasis in epistemology has been on secure, cumulative knowledge claims. Even in contemporary work at the interface of logic and epistemology, finding the surplus of knowledge over belief is a focus, though there are further themes of interest to logicians (cf. Arló-Costa, Hendricks, and van Benthem, 2016; Baltag, van Benthem, and Smets, to appear).

Epistemic action In the perspective of this paper, two deviant viewpoints arose. One is that we should not focus on static attitudes like knowledge (or even belief), but in tandem with these, on the *epistemic actions* that create and modify such attitudes. Such actions involve the whole spectrum of informational events that we have discussed earlier on, from inferring to observing and communicating. But, especially in epistemology, they may well include other actions, that, say, generate doubts, or raise objections. Nothing is sacroscant.

From foundations to correction One of the most striking aspects of the agent repertoire studied here is that it does not presuppose unfailing correctness. To the contrary, mistakes and recoveries show logical ability at its best! Thus, the old foundational ideal of a safe haven once and for all for our theorizing largely disappears. It is both unreachable and not ambitious enough. *Correction* is the more exciting goal, not just correctness. Mistakes are natural, learning from mistakes is intelligent, and the real focus for epistemology is how we correct ourselves, repair our theories, and make the next creative leap to a fallible theory.

Correspondingly, our view of the role of logic changes. It is not the guardian of eternal correctness, and the key to a sterile world where nobody ever gets sick. It rather unleashes its powers in a world full of error and uncertainty, and it acts there as what one might call *the immune system of the mind*.

20 Natural language: From meaning to action

Here is an empirical angle on our program. It concerns *natural language*, the medium with which we describe the world, but also, and perhaps primarily, communicate with each other. It is illuminating to take some earlier themes to this setting.

'Of' and 'as' Recall our two directions connecting logic and games. Seeing language *as* agency, we get a dynamic view of what natural language use consists of, who uses it, and what it achieves, and all earlier topics apply. But also, looking at the language *of* agency, natural language provides a rich repertoire of expressions for driving agency, but also for discussing it, evaluating it, and reflecting on it.

However, common sense is in order. Natural language tends to blur methodological distinctions. As has often been observed, natural language is a 'universal medium' where one can discuss everything, including language itself. Our two directions not only occur in natural language, but they also allow for smooth switches. An agent can use language to communicate in a first-person 'participating stance', but also step out of this process, and comment on it in a third-person 'reporting stance'.

Natural language as agency To make this general agency perspective a bit more concrete, we quote some themes from van Benthem, 2014b.

Action/product duality Natural languages have a pervasive duality of static and dynamic vocabulary. "Dance" is both an activity verb and a noun denoting a product of that activity, and likewise for, say, "argument"—some languages do not even have a grammatical distinction here. Moreover, language has a general co-existence of static and dynamic verbs. This mirrors our approach to agency: the logic of static 'knowing' needs to be studied in tandem with the logic of the dynamic 'learning'.

Natural logic and inferential zoom A second foothold is the program of 'natural logic' from the 1980s that sought simple fast inference mechanisms that humans employ, living inside the more cumbersome full machinery of first-order or higher-order proof theory. A typical example were monotonicity inferences (performing upward or downward predicate replacement), computed by just understanding the parse trees of logical formulas (van Benthem, 1986). Inference is a family of modules, some less, some more complex—and the same might hold for our logical calculi of information dynamics. Inside their elaborate mechanisms, there may be natural language fragments that provide much simpler high-zoom reasoning about social action.

Translation as action Our final example is an agency perspective on the crucial linguistic notion of translation. Instead of a mere mapping between the syntax and semantics of two languages, translation seems correlation of behavior. This is more ambitious in that acts of communication and reasoning need to be 'translated' as well, but it is at the same time less demanding, since—in line with our earlier remarks about correction—mistakes and misunderstandings are not problematic in cross-language communication, as long as they are eventually detected and repaired.

21 Agents inside logic itself

Next, let us take agency to logic itself. Logical systems do not carry a description of their users, and presumably, these are taken to be idealized all-seeing agents that all have the same abilities. Can we tease out a parameter, and introduce agent types inside logic in a meaningful way—going to a theory of logical systems 'as used'? We need to give up hidden uniformity assumptions, finding meaningful parameters for different agents using the same system. There are some examples in the literature. Authors have looked at differences in memory, which can be modeled, for instance, with different levels in the automata hierarchy (van Benthem, 1986). Also, agents with different inferential resources have been used to model 'bounded agency', when agents may be unable to employ the full power of a standard proof system.

Fragments versus agents This perspective may change our view of many standard topics. Consider the earlier topic of 'natural logic', i.e., simple logical inference inside a complex total system. We can think of this in a standard way as a search for fragments of the full system that are decidable, or otherwise especially well-behaved. But we could equally well think that simplicity does not arise from simple fragments that guarantee success, but from simplicity of agents using complex systems. For instance, consider first-order logic as used by a finite automaton: how much correct model checking or inference could it do?

Pebble games and memory Memory modulation in standard logic occurs in 'pebble games', where access to objects in model games is restricted by a fixed supply of pebbles that are used to mark current objects of attention. As a result one parametrizes standard logical games to those that can be played successfully by players with a certain finite amount k of pebbles. However, it is typical for the state of the art that pebbling is only used in a few specialized domains, whereas it is clearly a general device for introducing memory in many logical settings. Also, players are given the same amount of pebbles, whereas again, there seems to be no need for this.

Agent diversity revisited But scenarios can be made still more exciting. What if agents are taken to be very different—as in the many scenarios of 'humans versus machines' that pervade the literature from the Turing Test onward? Our earlier logic games might now be played between different agent types, such as competing versus cooperative agents, giving up the usual uniformity of reasoners. What logical notions would then correspond to the resulting equilibria in a Theory of Play?

But this diversity also extends to other issues in logic. Consider the well-known diversity or plurality of logical systems, offering us options from classical logic to systems like intuitionistic or linear logic. What if we reinterpret this diversity, not as some sort of momentous dogmatic system choice, but as a reflection of agent variety?

These separate themes add up to a general challenge. Logic now resides, not in one ideal prescribed rationality, but at a higher level of rational interaction between different agents. How will this work precisely?

22 Logic meets reality

The preceding thoughts are still about repercussions inside theoretical academic fields. But many people read the focus on agency advocated in this paper as a step toward a more empirical account of logic as analyzing concrete reasoning and communication styles. There is certainly an influence from reality in all of the above, in that our themes are not chosen out of the blue, but in accordance with what we perceive as basic features of actual agents. This turn does not stand on its own. In recent years, logic as a discipline has been exposed to major outside influences, shaking up the cozy corners of the a priori mind. It is not the aim of this paper to chart all these influences, but the following facts are readily observable.

Logic is a source of computational devices that are transforming our world. At the same time, logic meets the empirical facts of human behavior in encounters with cognitive science. It may be hard to find logicians today who really feel at ease with splendid isolation behind the barrier of Frege's 'anti-psychologism'. Leaving the glass bead game of 'intuitions', learning instead what people really do, and how the human brain really works, is proving an irresistible challenge even to many theorists. And finally, human society itself creates an ever-growing fund of challenges such as the many whirlpools and cliffs in the world of public opinion that seems to be spinning out of control at an alarming rate.

I am not saying that our logic, games and agency view is a panacea for all the ills of our current society (or the source of all that is good), but it may become part of a reconceptualization that will equip us better to deal with major practical challenges.

In the final sections of this paper, I discuss a few themes arising in encounters with practice. I start with a simple concrete illustration, to show how the above flights of the imagination can quickly land on solid ground.

23 Designing new games and human cognition

In the real world, things go differently from theoretical scenarios. Here is an example: what happens to our standard algorithms when the world (or other agents) engage in sabotage? Rational agents should be able to cope with such changes, but let us first see what drastic effects they can have on situations we thought we understood.

From algorithms to games Under adverse circumstances computational tasks for single agents can turn quite easily into more-agent games. Consider the ubiquitous practical search problem of *Graph Reachability*:

> "Given two nodes s, t in a graph, is there a sequence of successive arrows from s to t?"

There are fast *Ptime* algorithms finding such a path if one exists (Papadimitriou, 1994). But what if there is a disturbance—a reality in travel?

Example Sabotage Games.

The following network links two European centers of logic and computation:

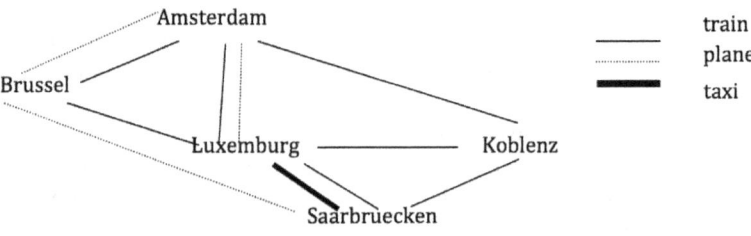

Figure 11.

Let us focus on the two nodes Amsterdam and Saarbruecken. It is easy to plan trips either way. But what if transportation breaks down, and a malevolent demon starts canceling connections, anywhere in the network? Let us say that, at each stage, the demon first takes out one connection. Now we have a two-player game, and the question is who can win it where.

Here is a Zermelo solution: the sabotage game satisfies the conditions Zermelo's theorem. From Saarbruecken to Amsterdam, a German colleague has a winning strategy. Demon's opening move may block Brussel or Koblenz, but she gets to Luxemburg in the first round, and to Amsterdam in the next. Demon may also cut a link between Amsterdam and a city in the middle—but she can then go to at least one place with two intact roads. But with a traveler starting from the Dutch side, it is the Demon who has the winning strategy. It first cuts a link between Saarbruecken and Luxemburg. If the traveler then goes to any city in the middle, Demon has time in the next rounds to cut the last intact link to Saarbruecken.

Logic in play

By now, sabotage scenarios have been used for various tasks, and sabotage is a general strategy for changing giving algorithms or games.[32] Here is an illustration.

Example Teaching, the grim realities.

A Student located at S in the next diagram wants to reach the *escape* E below, the Teacher wants to prevent him from getting there. Each line segment is a path that can be traveled. In each round of the game, the Teacher first cuts one connection, anywhere, the Student must then travel one link still open at his current position:

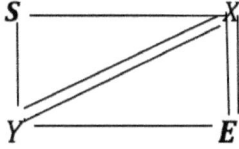

Figure 12.

Again Zermelo's Theorem applies. In this particular game, Teacher has a winning strategy: first cut a line to the right between X and E, and then wait for Student's moves, cutting lines appropriately. General games like this arise on any graph with single or multiple lines, and they have been used in real teaching scenarios.

New logics of model change There is also an interesting theoretical angle here. Sabotage games suggest a language that changes the models on which it is evaluated. Thus, their logical study involves dynamic modalities referring to a new model after some change has been made to its structure, in the spirit of the dynamic update logics considered earlier, but also some genuine extensions.[33]

Gamification Like our earlier logic games, sabotage is an instance of a general phenomenon of 'gamification' that can be observed in many fields today, designing new games for pleasure or even for serious social purposes. Given the striking human penchant for interactive game play, games are a way of changing the world. At the same time, newly designed games are a free cognitive lab where we academics can observe and even manipulate how people behave in specific informational scenarios.

In all this, games constitute 'hybrid forms' of natural and designed behavior. This is one of the most intriguing features about us humans: theory can influence design qua behavior, resulting in a society where natural and designed behavior have to live together. Of course, there are many examples of this phenomenon, starting with the

---

[32] It can be shown that the computational complexity of solving sabotage games jumps up from *NP*-complete for graph search to *Pspace*-complete, a typical complexity class for games.

[33] Our earlier dynamic-epistemic logics are about definable model changes, and hence their complexity tends to stay low, thanks to matching recursion axioms. Sabotage dynamic logics are about arbitrary stepwise model changes, and as it happens, even sabotaged basic modal logic is already undecidable. For the latest on logics of model change, cf. Aucher, van Benthem, and Grossi, to appear.

introduction in Antiquity of specialized reasoning disciplines with specialized hybrid languages. such as mathematics or the law. It seems fair to say that we do not really understand very well how such societies function, but clearly, it will involve the agent diversity of our Theory of Play in a major way.

A universal model for cognition? We conclude this section with a grand question. We saw how computational devices become games when more users are involved, competitive or adversarial. Now consider Turing's celebrated analysis of computation, using a stylized model of a human agent performing a single task: calculation. As suggested in van Benthem, 1990, could games with agent diversity, analyzed in the same sparse style, be a universal model for human cognition?[34]

24 Logic one last time

Our final question is what the topics in this paper mean for our understanding of logic. Given the (despite all our caveats) greater empirical and practical flavor of our agency perspective, what becomes of the status of logical theory? I merely mention a few themes, leaving other valid concerns for other occasions.[35]

Normative versus descriptive? Logic and cognitive psychology or neuroscience have often been worlds apart, with logic taken to be a normative source of valid laws, while actual human behavior may fail to follow these norms for various reasons. But this dividing line seems thin. Good logical theory lets itself be informed by the best available facts about human cognition, if only to see which topics for research are most urgent or relevant. The latter point even holds if we take our logical theories of agency to be normative, uncovering the correct laws of information flow, belief revision, strategic interaction, and so on.

But the topics in this paper call for an even deeper entanglement of normative and descriptive views. Consider our emphasis on correction and learning as key aspects of agency. This was presented as crucially going beyond the statics of what is and what ought to be the case to the dynamics of *improvement*. But working toward improvement and judging something to be an improvement requires a norm, and even though these need not be absolute norms, it is a fact that learning theory is replete with norms that allow us to judge whether we are going in the right direction. So, facts and norms can and should work in tandem.

---

[34]The original Turing Test itself concerned a hybrid scenario with possibly different agent types.

[35]For instance, the models for bringing out dynamic actions in this paper are extensional, whereas one might think that action is a typical intensional notion, involving 'how' (that is, ways of doing things) as much as 'what' state changes are achieved. While this is true, the program outlined here for making dynamic actions explicit seems orthogonal to the choice of a level of intensionality for the logic employed.

Limitations? The high point of classical logic are its famous limitation results saying that some things are unachievable, such as decidability for logics that are strong enough, or completeness for mathematical theories that are expressive enough. It is this balance of positive and negative results that makes us logicians feel we truly understand subjects such as proof, definability or computation. But in our program so far, anything seemed to work. What would be limitations to a logical approach to games and agency? Could there be new social paradoxes that show boundaries to what the current program can achieve? I believe that there are such boundaries, and as a logician, I even fervently hope there are—but this paper has nothing substantial to offer on this score, except for the obvious fact that some logics of agency are undecidable, and can even have quite high non-arithmetical complexity.

Object and metalevel Our final point is a new philosophical worry to be resolved. We have advocated a broad view of the logic of information-driven agency, where inference is just one source of information among others, such as making observations or asking questions. So, is there no privileged role left for reasoning and its laws? One might think there still is, since at the meta-level of formulating dynamical agent systems we studied information update or belief revision in terms of their logical laws and what follows from them. So, inference would still rule the meta-level.

But if one were to be consistent with our general program, the broader informational activity perspective might also have to enter the metalevel. Perhaps the answer is this. Just as in science, we do not just *have* logical theories as sets of laws and proof rules, we also find them, develop them, and *live* them[36]—and perhaps it is that dynamic activity that constitutes the proper content of the meta-level. Theorizing about activities is itself a many-faceted activity.

25 Conclusion

The main thrust of this light programmatic paper is easy to state. To the open-minded observer, there is a great deal of logical content to agency, with games as a striking example and as a natural laboratory where theory meets practice. And looking in the converse direction, there is also a lot of agency to logic. This dual interface is a pleasure to explore, especially in the compass of games—and it suggests new perspectives on past and future interfaces of logic with a wide range of disciplines.

---

[36]In that spirit, the various formalisms mentioned in this paper, such as the dynamic logics of hard and soft information or various current logics of games, would not be the final laws and the last word of representation, but just one stage in formulating the meta-thory of agency.

Bibliography

Abramsky, S. (2008). Information, processes and games. In P. Adriaans, & J. van Benthem (Eds.), *Handbook of the philosophy of information* (pp. 483–549). Amsterdam, Netherlands: Elsevier.

Arló-Costa, H., Hendricks, V. F., & van Benthem, J. (Eds.). (2016). *Readings in formal epistemology*. Dordrecht: Springer.

Aucher, G., van Benthem, J., & Grossi, D. (to appear). Modal logics of sabotage revisited. *Journal of Logic and Computation*.

Baltag, A., & Smets, S. (2011). Keep changing your beliefs, aiming for the truth. *Erkenntnis*, *75*(2), 255–270.

Baltag, A., Smets, S., & Gierasimczuk, N. (2011). Belief revision as a truth-tracking process. In K. Apt (Ed.), *TARK XIII proceedings of the 13th conference on theoretical aspects of rationality and knowledge* (pp. 187–190). New York, NY: ACM.

Baltag, A., van Benthem, J., & Smets, S. (to appear). *The music of knowledge*. Netherlands: ILLC, The University of Amsterdam.

van Benthem, J. (1996). *Exploring logical dynamics*. Stanford,CA: CSLI Publications.

van Benthem, J. (2011). *Logical dynamics of information and interaction*. Cambridge, UK: Cambridge University Press.

van Benthem, J. (1986). *Essays on logical semantics*. Dordrecht, Netherlands: D. Reidel.

van Benthem, J. (1990). Computation versus play as a paradigm for cognition. *Acta Philosophica Fennica*, *49*, 236–251.

van Benthem, J. (2014a). *Logic in games*. Cambridge, MA: The MIT Press.

van Benthem, J. (2014b). Natural language and logic of agency. *Journal of Logic, Language and Information*, *23*(3), 367–382.

van Benthem, J. (2015). Oscillations, logic, and dynamical systems. In S. Ghosh, & J. Szymanik (Eds.), *The facts matter: Essays on logic and cognition in honour of Rineke Verbrugge* (pp. 9–22). London, UK: College Publications.

van Benthem, J. (2016). Tracking information. In K. Bimbó (Ed.), *J. Michael Dunn on information based logics* (pp. 363–389). Dordrecht, Netherlands: Springer.

van Benthem, J., Bezhanishvili, G., & Enqvist, S. (2016). *Instantial game logic*. Amsterdam, Netherlands: ILLC.

van Benthem, J., & Gheerbrant, A. (2010). Game solution, epistemic dynamics and fixed-point logics. *Fundamenta Informaticae*, *100*(1-4), 19–41.

van Benthem, J., & Smets, S. (2015). Dynamic logics of belief change. In J. Halpern, W. van der Hoek, & B. Kooi (Eds.), *Handbook of epistemic logic* (pp. 313–393). London, UK: College Publication.

Branderburger, B. (2014). *The language of game theory: Putting epistemics into the mathematics of games*. Singapore: World Scientific.

Carnap, R. (1928). *Der Logische Aufbau der Welt*. Berlin, Germany: Weltkreis.

Christoff, Z. (2016). *Dynamic logic of networks: Information flow and the spread of opinion* (Doctoral dissertation, ILLC, University of Amsterdam).

Gärdenfors, P., & Rott, H. (1995). Belief revision. In D. M. Gabbay, C. J. Hogger, & J. A. Robinson (Eds.), *Handbook of logic in artificial intelligence and logic programming* (Vol. 5: Epistemic and temporal reasoning, pp. 35–132). Oxford, UK: Oxford University Press.

Ghosh, S., van Benthem, J., & Verbrugge, R. (Eds.). (2015). *Models of strategic reasoning: Logics, games and communities*. Lecture Notes in Computer Science. Berlin, Germany: Springer.

Grädel, E., Thomas, W., & Wilke, T. (Eds.). (2002). *Automata, logics, and infinite games: A guide to current research*. Lecture Notes in Computer Science. Berlin, Germany: Springer-Verlag.

Grossi, D., & Turrini, P. (2012). Short sight in extensive games. In V. Conitzer, & M. Winikoff (Eds.), *Proceedings of the 11th international conference on autonomous agents and multiagent systems - volume 2* (pp. 805–812). AAMAS '12. Valencia, Spain: International Foundation for Autonomous Agents and Multiagent Systems.

van Lee, H., Rendsvig, R., & van Wijk, S. (2016). *Merging frameworks for information dynamics*. Copenhagen: Center for Information and Bubble Studies.

Liu, F., Seligman, J., & Girard, P. (2014). Logical dynamics of belief change in the community. *Synthese*, *191*(11), 2403–2431.

Osborne, M., & Rubinstein, A. (1994). *A course in game theory*. Cambridge, MA: The MIT Press.

Papadimitriou, C. H. (1994). *Computational complexity*. Reading MA: Addison-Wesley.

Perea, A. (2012). *Epistemic game theory: Reasoning and choice*. Cambridge, UK: Cambridge University Press.

Shoham, Y., & Leyton-Brown, K. (2009). *Multiagent systems: Algorithmic, game-theoretic, and logical foundations*. New York, NY: Cambridge University Press.

Venema, Y. (2015). *Lectures on the modal μ-Calculus*. Netherlands: ILLC, University of Amsterdam.

Author biography. Johan van Benthem is University Professor, emeritus, of logic at the University of Amsterdam, Henry Waldgrave Stuart Professor of philosophy at Stanford University, and Changjiang Professor of humanities at Tsinghua University. He is a member of the Academia Europaea, the Dutch Royal Academy of Arts and Sciences, and the American Academy of Arts and Sciences. He has received the national Spinoza Award of the Dutch Organization for Scientific Research. His books include *The Logic of Time*, *Modal Logic and Classical Logic*, *Essays in Logical Semantics*, *Language in Action*, *Exploring Logical Dynamics*, *Modal Logic for Open Minds*, *Logical Dynamics of Information and Interaction*, and *Logic in Games*.

2 Models in geometry and logic: 1870-1920

PATRICIA BLANCHETTE [*]

Abstract. Questions concerning the consistency of theories, and of the independence of axioms and theorems one from another, have been at the heart of logical investigation since the beginning of logic. It is well known that our own contemporary methods for proving independence and consistency owe a good deal to developments in geometry in the middle of the nineteenth century, and especially to the role of models in establishing the consistency of non-Euclidean geometries. What is less well understood, and is the topic of this essay, is the extent to which the concepts of consistency and of independence that we take for granted today, and for which we have clear techniques of proof, have been shaped in the intervening years by the gradual development of those techniques. It is argued here that this technical development has brought about significant conceptual change, and that the kinds of consistency- and independence-questions we can decisively answer today are not those that first motivated metatheoretical work. The purpose of this investigation is to help clarify what it is, and importantly what it is not, that we can claim to have shown with a modern demonstration of consistency or independence.

Keywords: independence, consistency, Hilbert, Beltrami, history of logic, non-Euclidean geometry.

---

[*]University of Notre Dame.

1 Introduction

In 1939, Ernest Nagel wrote:

> The sort of questions considered by meta-mathematics, e.g. the consistency and independence of axioms, have been discussed in antiquity and have been cultivated ever since by mathematicians in every age; but such problems have been clearly formulated and systematically explored only after pure geometry had been freed from its traditional associations with space, and only after its character as a calculus had been isolated from its applications. (Nagel, 1939, §74, 202-3.)

While there is no doubt a good deal of truth to this claim, the purpose of this paper is to suggest a way in which the characterization it gives us of the role of models in meta-mathematics is misleading. The aspect of Nagel's view that I want to question is the idea that with the increasing formalization of mathematical theories in the late 19th and early 20th centuries, we achieved at last a method, and tools, for the rigorous treatment of ancient questions. What I will suggest in what follows is that the questions we can now answer with our modern rigorous tools are not the same questions as those that arise for mathematical theories prior to the modern era. The questions we can now raise and answer, I will argue, have been shaped significantly by the tools we have developed. One result of this is that the notions of independence and consistency that we now take for granted are not those that went by these names prior to about 1900, and the clean methods we now have for demonstrating independence and consistency do not answer what in e.g. 1700, or even in 1850, would have been called by these names.[1]

2 1900

We begin by looking at the state of the art of independence proofs in the penultimate year of the nineteenth century. Here our example is David Hilbert's 1899 *Foundations of Geometry*, in which we find a clear and systematic application of the technique that was becoming, at this point, the standard approach to demonstrations of independence.[2] As Felix Klein puts it in 1908, describing what he calls the "modern theory" of geometric axioms:

> In it, we determine what parts of geometry can be set up without using

---

[1] On the role of geometry in the early development of modern logic, see Webb (1985).

[2] See Hilbert (1899). Hilbert was not alone in using the technique to significant effect at this time. Its application can be seen as well in the work of the Italian school (see e.g. Peano, 1894, Padoa, 1901, Pieri, 1898; for discussion of Pieri's work see Marchisotto and Smith, 2007) and in closely-related work by e.g. Veblen (1904) and Dedekind (1888).

certain axioms ...
As the most important work belonging here, I should mention Hilbert's [1899].
(Klein, 1908, as quoted in Birkhoff and Bennett, 1988, p. 185.)

Hilbert's work in the *Foundations of Geometry* monograph involves a clear and careful axiomatization of Euclidean geometry, together with a consistency proof for the whole, and a series of independence proofs that demonstrate the connections of logical entailment holding between the various parts of the edifice. Hilbert characterizes both consistency and independence here in terms of the relation of logical deducibility: a set of axioms is *consistent*, he tells us, if "[I]t is impossible to deduce from them by logical inference a result that contradicts one of them" [§9], and a geometric axiom or theorem is *independent* of a collection thereof if it cannot be deduced from that collection.

Hilbert's method of demonstrating non-deducibility is as follows: Given a set AX of axioms, and a statement (perhaps another axiom) A, we begin by uniformly re-interpreting the geometric terms ("point," "line," "lies-on," etc.) in AX and in A, in terms of objects and relations given by a different theory, in this case a theory R of real numbers and collections thereof. We then note that, as re-interpreted, each of the sentences AX, together with any sentence deducible from them, expresses a theorem of R. Finally, the negation $\neg A$ of the target sentence A also expresses a theorem of R, which, assuming the consistency of R, guarantees that A itself does not express a theorem of R. Still assuming the consistency of R, then, we have a guarantee that A is not deducible from AX.

The consistency of AX, in the sense of the non-deducibility of a contradiction from it, is demonstrable similarly, again assuming the consistency of the background theory R. As Hilbert says,

> From these considerations, it follows that every contradiction resulting from our system of axioms must also appear in the arithmetic related to the domain [of the background theory]. (Hilbert, 1899, §9.)

An important point to note about the interpretation-theoretic technique used by Hilbert here is that it presupposes that the relation of deducibility in question is "formal" in the sense that it is unaffected by the reinterpretation of geometric terms. It is this that guarantees that the sentences deducible from AX will express theorems of R under the reinterpretation, given only that the members of AX do. But the deducibility relation is not, for Hilbert in 1899, "formal" in the sense of "syntactically specified;" there is no formal language at this point, and no explicit specification of logical principles. We will use the term "semi-formal" for such a relation. The first thing, then, that Hilbert's interpretations (or models) shows is that a given sentence is not semi-formally deducible from a collection of sentences.

Hilbert's models also show, importantly, the *satisfiability* of the conditions implicitly defined by the collections of sentences in question. Given a collection $AX \cup \neg A$ of sentences whose geometric terms appear schematically, a Hilbert-style reinterpretation on which each member of that collection expresses a truth about constructions on the real numbers demonstrates the satisfiability of the condition defined by the collection. Equivalently, it demonstrates that the condition defined by *AX* can be satisfied without satisfying the condition defined by A.[3]

Hilbert's technique, then, demonstrates the *independence* of a given sentence from a collection of sentences in two different senses. Taking as an example the question of the independence of Euclid's parallel postulate (PP) from the remainder of the Euclidean axioms (EU) for the plane, the two senses, with our labels introduced, are:

- *Independence$_D$*: (PP) is not (semi-formally) deducible from (EU);
- *Independence$_S$*: The condition defined by $(EU) \cup \neg(PP)$ is satisfiable.

Independence$_S$ is the stronger of the two notions, though they are extensionally equivalent in the setting of an ordinary first-order language.[4]

3 Frege

Gottlob Frege's work, in the same period, focuses on a notion of independence that's distinct from both of the relations demonstrable via Hilbert's technique. For Frege, *independence* is a relation not between sentences but between *thoughts*, i.e. between the kinds of things expressible by fully-interpreted sentences. Each thought, as Frege understands it, has a determinate subject-matter: thoughts about geometric objects and relations are entirely distinct from thoughts about collections of real numbers. Hence the re-interpretation of sentences along Hilbert's lines will result in the assignment to those sentences of different thoughts. Finally, logical connections between thoughts, connections like dependence and independence, provability and consistency, are sensitive as Frege sees it to the contents of the simple terms in the sentences used to express those thoughts. Hence Hilbert's re-interpretation strategy amounts, from Frege's point of view, to shifting attention from the geometric thoughts in which one was originally interested to an entirely different collection of thoughts, a collection whose logical properties are no guide to those of the original target thoughts. As a result, Frege takes it that Hilbert's technique is unsuccessful in demonstrating consistency and independence. In Frege's words,

---

[3]This understanding of consistency in terms of satisfiability was the more central concern for e.g. Dedekind and Veblen.

[4]The equivalence is given by the completeness of first-order logic. Hilbert's setting is that of natural language without strictly-defined relations of deducibility or satisfiability, so the question of the extensional relationship between the two independence relations is imprecise. The expressive richness of that language, however, is well beyond that of (what was to become) first-order logic, giving the second relation, in that setting, a narrower extension than the first.

> Mr. Hilbert appears to transfer the independence putatively proved of his pseudo-axioms to the axioms proper ... This would seem to constitute a considerable fallacy. And all mathematicians who think that Mr. Hilbert has proved the independence of the real axioms from one another have surely fallen into the same error.[5] (Frege, 1906, p. 402.)

Or, as we might more calmly put it, the relation that Frege calls "independence" is neither of the relations, also reasonably known by that name, demonstrable via Hilbert's technique. For Frege, the parallels postulate is independent of the other axioms of Euclid iff it isn't provable from those axioms (and in this he agrees with Hilbert), but the notion of *proof* that Frege works with is a very rich one: the question of whether a given thought is provable from others can turn on non-trivial conceptual analyses of the components of those thoughts.[6] Hence a sentence A can be Independent$_D$ and even Independent$_S$ of a set AX of sentences while the thought $\tau(A)$ expressed by A fails to be independent in Frege's sense of the set $\tau(AX)$ of thoughts expressed by the members of AX. We introduce a term for this third kind of independence:

- *Frege – independent*: The thought $\tau(A)$ is not provable, in Frege's rich sense, from the set $\tau(AX)$ of thoughts.

To fix ideas with a vivid example: consider the sentences

(BET$_A$) Point B lies on a line between points A and C.

(BET$_C$) Point B lies on a line between points C and A.

For Hilbert, a model can immediately show that (BET$_C$) is independent of (BET$_A$), in both of the relevant senses: (BET$_C$) is not semi-formally deducible from (BET$_A$), and the condition defined by (BET$_A$) $\cup$ $\neg$(BET$_C$) is satisfiable.[7] For Frege on the other hand, a model can show no such thing. Though Frege does not discuss this example, it is compatible with his views that the thoughts expressed by sentences (BET$_A$) and (BET$_C$) are the same thought, and hence provable immediately from one another. For an alternative example close to Frege's heart: each of the Dedekind-Peano axioms for number theory is Independent$_D$ and Independent$_S$ from the others, but this straightforwardly-demonstrable fact is in no tension with Frege's logicist thesis, according to which the thoughts expressed by those axiom-sentences are not Frege-independent of one another.

---

[5]The "pseudo-axioms" as Frege calls them are Hilbert's partially-interpreted sentences; the "real axioms" are thoughts about points, lines, and planes.

[6]Rigorous deduction, for Frege as for Hilbert, cannot make reference to the meanings of non-logical terms. But the demonstration that a given thought τ is provable from a collection Σ of thoughts can (and, in the logicist project, regularly does) involve non-trivial analysis of τ and/or of Σ *en route* to the expression of those thoughts in the sentences that will appear in the rigorous deduction. See Blanchette (1996, 2012).

[7]This is a simplification of Hilbert's more-interesting result that the biconditional [(BET$_A$) iff (BET$_C$)] is independent of the other axioms of order. See Hilbert (1899, §10).

4 The Parallels Postulate

The paradigm independence question prior to the twentieth century was the question of the provability of Euclid's parallels postulate from the remainder of Euclid's axioms. The firm conviction, by the end of the nineteenth century, that the parallels postulate is not provable from the rest of Euclid rested in large part on the construction of - as we now put it - "models of non-Euclidean geometry." But the canonical early models, and the lessons drawn from them, were in interesting ways different both from the later Hilbert-style models, and from their descendants, the models that we take for granted today. We begin with a sketch of some aspects of the well-known history, in order to make some observations about the early use of models.

In the middle of the 18th century, J. H. Lambert famously examined the independence question by working out some of the fundamental implications of $(\neg PP)$, the negation of the parallels postulate.[8] Working in the paradigm set by Saccheri, who divided the alternatives to the parallels postulate (itself called the "first hypothesis") into the *second hypothesis*, in accordance with which the internal angles of a triangle would sum to more than two right angles, and the *third hypothesis*, according to which the angle-sum would be less than that of two right angles, Lambert notes that the second hypothesis holds of the triangles drawn on the surface of an ordinary sphere.[9] He also, intriguingly, suggests that the third hypothesis would hold of triangles drawn on the surface of a sphere with imaginary radius (i.e. a radius whose length is a multiple of i), if such a figure were possible. Lambert does not, however, treat the possibility of such instances as a reason to suspect that the parallels postulate is independent, but continues to search for a proof of (PP) from the remainder of Euclid.

That Lambert does not see in the behavior of arcs on a (real or imaginary) sphere any indication that the parallels postulate might be independent of the rest of Euclid is not entirely surprising: arcs on a real sphere are not infinite; they also violate the Euclidean principle that two points determine a unique line. Most importantly for our purposes, there is also a relatively clear sense in which the sides of the triangle-like figures drawn on a sphere are not "straight:" the fact that the internal angles of such a figure sum to more than two right angles is in no tension with the principle that *real* triangles will have an angle sum equal to that of two right angles. As Katherine Dunlop has put it, regarding Lambert's view of the question "whether the principles that hold of figures on [the surface of a sphere] constitute a theory that is genuinely comparable to Euclid's":

> Lambert appears to share the consensus view that they do not. It was not news, in the second half of the 18th century, that Euclid's parallel

---

[8] See Lambert (1786). For discussion, see Ewald (1996b).

[9] Saccheri's division was actually couched in terms of the angle-sums of quadrilaterals rather than of triangles; the two issues are equivalent.

axiom did not hold of arcs on a sphere. But Lambert's contemporaries did not regard the arcs as lines. ...He clearly does not take the fact that the second hypothesis is satisfied on a spherical surface to show that it could belong to geometry after all. (Dunlop, 2009, p. 47.)

In short: the sides of spherical triangles are not lines. Therefore the fact that *they* behave "non-Euclideanly" gives no reason to suppose that *lines* might.

Similar concerns arguably attach to realizations of the third hypothesis, according to which two straight lines in a plane can converge without intersecting. Lambert's view is that this kind of asymptotic approach, discussed here with respect to two posited lines CD and AD, is contrary to the idea of straight lines:

> Whoever at this point objects that CD could perhaps approach AD asymptotically (like, for instance, the hyperbola and other asymptotic bent lines) in my opinion changes what the logicians call the *statum quaestionis*, or he deviates from Euclid .... I do not see how in the representation of *straight* lines objections about hyperbolas can be made. (Lambert, 1786, §3.)

Similarly, "[T]he *idea* that AD, CF are *straight* lines ... cannot coexist with the idea of an *asymptotic approach*. (Lambert, 1786, §10.)

So a surface whose geodesics approach asymptotically, of the kind now familiar as a model of hyperbolic geometry, could not have been seen from Lambert's point of view as a surface truly described by the negation of the parallels postulate: the only way to describe such a surface from this viewpoint is as a perfectly well-behaved Euclidean object whose geodesics are not lines.

Neither Lambert nor his contemporaries has the idea of a mathematical theory as providing an implicit definition of a structure-type, or of the mathematical terms in a theory as place-holders for the elements of satisfying structures. So the idea of curved lines as appropriate candidates for filling such a place, i.e. as objects satisfying implicitly-defined conditions, can make no sense from this perspective. Also lacking at this point is the idea that the proof of sentences one from another is unaffected by the reinterpretation of the geometric terms in those sentences. This idea, as natural as it seems now, requires an understanding of the mathematical language as peculiarly well-behaved in various ways, including a stratification into defined and primitive terms in such a way that none of the mathematical content resides in implicit connections between the contents of those terms. In short, neither of the views essential to the idea of reinterpretation as a method for proving independence, i.e. the idea of axioms as implicit definitions and the idea of rigorous proof as surviving reinterpretation, is part of the standard conception of mathematical theories in the middle of the eighteenth century.

The idea of geometry as a reinterpretable theory rather than a doctrine of space, and the idea that good principles of proof should survive re-interpretation, were both helped along by the success of projective geometry and its duality principles in the early

19th century: it is irresistible to view the counterpart theorems obtained by switching "point" with "line" etc. in the projective setting as also obtained by coordinated re-interpretations of those terms. The idea that an axiom and its dual are in some sense "the same" axiom under different interpretations is the beginning of a certain broadening in the understanding of the role of the term "line" within axioms. Also critical in this progression toward the loosening of the conception of geometric axiom was Riemann's reconceptualization of geometry as a theory of arbitrary n-dimensional manifolds, of which space itself forms merely a particular instance.[10]

The final piece of background to mention before looking at the role of models in the middle of the nineteenth century is the work of Lobachevsky. Following up on Saccheri's third hypothesis, Lobachevsky had provided in 1840 an elegant and deep theory involving the denial of the parallels postulate, and according to which the angle-sum of triangles is less than two right angles (Lobachevsky, 1840). The demonstration that one can go as far as Lobachevsky goes in developing the non-Euclidean theory, without encountering contradiction, provided compelling reason to think that the behavior of parallel lines might not, after all, be dictated by the rest of Euclid.

Against this backdrop, the middle of the nineteenth century saw the construction of surfaces deliberately understood as "models" of non-Euclidean geometry in the sense of containing line-like entities which, often under an alternative understanding of such central notions as that of distance along the surface, satisfied the postulates of two-dimensional Lobachevskian geometry, which we will call "L." Beltrami provides, for example, a metric on the open unit circle that satisfies L when its open chords are taken as lines, and another for that circle when selected arcs play the role of lines (Beltrami, 1868a, 1868b). In each case, the line-like entities are finite on the Euclidean measure, but not on the imposed metric; triangles formed by these entities are just what one would expect (enclosed, in the ordinary Euclidean sense, by three chords or arcs), and the triangles' angle-measure is on the imposed metric essentially what it 'actually' is on the Euclidean metric, summing in each case to less than two right angles. The simple geometric relations between points and lines (the intersection of lines, the lying of a point on a line) are, similarly, understood in the ordinary Euclidean way.

The most significant of the constructions at this point is Beltrami's celebrated pseudosphere and the cover he constructs on it, a surface of constant negative curvature whose geodesics play the role of lines in L.[11] In this case, the line-like entities are very much like lines: they are "really" infinite, i.e. infinite on the ordinary Euclidean metric, and as a result behave in the way that lines on an infinite Euclidean plane would, given an appropriate warping of that plane. Because the curvature of the surface is constant, it

---

[10]See Riemann (1873). For discussion of the mathematical background to the broadening of the concept of line, and especially the central role of the use of analysis in this reconceptualization of geometry, see Gray (1979).

[11]For discussion of the construction, see Stillwell (1996a). For discussion of the place of Beltrami's work in the history of the independence claim, see Scanlan (1988), Stump (2007).

satisfies the principle of the free mobility of figures on the plane, allowing the important proof-technique of superposition. The surface is, in short, just what one needs in order to provide a vivid depiction of what lines would be like if space had a constant negative curvature: they would be just as described by Lobachevsky.

The close analogy between the geodesics on the constructed surface and the lines of Euclidean geometry is crucial to the role played by the surface in Beltrami's thought. As he puts it,

> The most essential figure in elementary geometry is the straight line. Its specific character is that of being completely determined by two points, so that two lines which pass through the same points coincide throughout their extension. ...
>
> Now this characteristic ... is not peculiar to straight lines in the plane; it also holds (in general) for geodesics on a surface of constant curvature. ... [T]he surfaces of constant curvature, and only these, have the property analogous to that of the plane, namely: given two surfaces of constant and equal curvatures, in each of which there is a geodesic, the superposition of the two surfaces at two points of the geodesic causes them to coincide (in general) along its whole extension.
>
> It follows that, except in the case where this property is subject to exceptions, the theorems of planimetry proved by means of the principle of superposition and the straight line postulate, for plane rectilinear figures, also hold for figures formed analogously on a surface of constant curvature by means of geodesics. (Beltrami, 1868a, pp 8-9 of Stillwell, 1996b.)[12]

That is to say: the figures on this surface analogous to triangles will have all of those properties provable of Euclidean triangles with the exception of those that depend on the parallels postulate; the geodesics will share the corresponding portion of the Euclidean properties of lines, and so on. And that the geodesics are described by L establishes Beltrami's central point, that the surface is a "substrate for" Lobachevsky's geometry, so that this theory is not idle. (Beltrami, 1868a.)

We might ask what, exactly, constructions like Beltrami's were understood to show in this period about the parallels postulate and its connection to the rest of Euclidean geometry. The answer, arguably, is that there is no single precise lesson that was drawn from these constructions, aside from the general view that they showed, in some sense, the coherence of L and hence the independence (in some sense) of the parallels postulate. In 1868, axiomatic theories were still not understood as providing implicit definitions, and there was no well-developed sense of proof as semi-formal, or of the mathematical language as usefully subject to arbitrary reinterpretation. The

---

[12] The "straight line postulate" is the postulate that two points completely and uniquely determine a line.

constructed "models" of L were consequently not tools for demonstrating the modern notions of Independence$_D$ or of Independence$_S$. To get a sense of what the models were in fact taken to show, we look briefly here at the lessons drawn by Helmholtz and Poincaré.

To begin with, Helmholtz continues in Beltrami's mold to emphasize the importance of the analogy between Euclidean lines and the geodesics on curved surfaces, and the corresponding analogies between the figures constructed in each domain. As he puts it, speaking specifically of Gauss's surfaces:

> The difference between plane and spherical geometry has long been evident, but the meaning of the axiom of parallels could not be understood till Gauss had developed the notion of surfaces flexible without dilatation and consequently that of the possibly infinite continuation of pseudospherical surfaces. Inhabiting a space of three dimensions ... we can represent to ourselves the various cases in which beings on a surface might have to develop their perception of space ... (von Helmholtz, 1876, p. 308.)

and

> These remarks will suffice to show the way in which we can infer from the known laws of our sensible perception the series of sensible impressions which a spherical or pseudospherical world would give us, if it existed. In doing so we nowhere meet with inconsistency or impossibility ... We can represent to ourselves the look of a pseudospherical world in all directions ... Therefore it cannot be allowed that the axioms of our geometry depend on the native form of our perceptive faculty, or are in any way connected with it. (von Helmholtz, 1876, p. 319.)

In short, as Helmholtz sees it, these geometric constructions show that Kant is wrong. The pseudospherical models demonstrate that a non-Euclidean "world" is representable to us, with the consequence that Euclidean space is not uniquely determined by our representational capacities.

Despite having a radically different picture of the nature of geometrical truth, Poincaré shares essentially this reaction to the demonstrative value of geodesics on curved surfaces. Having claimed that Beltrami has shown via his pseudosphere that no contradiction is deducible from Lobachevsky's geometry, Poincaré continues:

> This he has done in the following manner: [I]magine beings without thickness living on [a surface of constant curvature] ... These surfaces ... are of two kinds: - Some are of *positive curvature*, and can be so deformed as to be laid on a sphere. ... Others are of negative curvature. M. Beltrami has shown that the geometry of these surfaces is none other than that of Lobachevsky. (Poincaré, 1892, p. 405.)

The result of this demonstration is, as Poincaré sees it, similarly anti-Kantian:

> [W]e ought ... to inquire into the nature of geometrical axioms. Are they synthetic conclusions a priori, as Kant used to say? They would appeal to us then with such force, that we could not conceive the contrary proposition, nor construct on it a theoretical edifice. There could not be a non-Euclidean geometry. (Poincaré, 1892, pp. 406-7.)

That there *is* in fact a non-Euclidean geometry, as demonstrated by the combination of Lobachevsky and Beltrami, shows in Poincaré's view the conceivability of that contrary science, and hence the falsehood of the claim that Euclid's geometry is synthetic a priori.

It is worth noting that this late-19th-century inference from the existence of Beltrami-style surfaces to the possibility or the conceivability of non-Euclidean space is by no means a trivial one. There is nothing about those surfaces that conflicts with Euclid: they are constructed within a purely Euclidean framework. And unless the actual behavior of the geodesics on a pseudosphere is taken to indicate the conceivable or possible behavior of lines, their satisfaction of L fails to tell us anything about the coherence of non-Euclidean surfaces or spaces. The question whether geodesics should be taken as such representatives is not a technical question but a conceptual one: Helmholtz and Poincaré, in keeping with the emerging confidence of the late 19th century in the richness and safety of the non-Euclidean framework, take it this way; those with a more conservative understanding of the nature of a line could in principle reject the inference, just as Lambert did with the sides of spherical triangles. That the geodesics on the pseudosphere did successfully play this representative role is due both to a certain loosening of the concept of *line* in the intervening century, and to the recognizably line-like character of the geodesics themselves.

The importance of the representative capacity of the curved surface, i.e. the representability of lines on a plane by geodesics on that surface, is emphasized by Poincaré in his commentary on just how such a surface undermines the claim that Euclid's axioms are synthetic a priori. In a truly synthetic a priori science like arithmetic, claims Poincaré, there can be no such representation of an alternative possibility:

> [L]et us take a true synthetical a priori conclusion; for example, the following: - If an infinite series of positive whole numbers be taken, ... there will always be one number that is smaller than all the others. ... Let us next try to free ourselves from this conclusion, and, denying [this proposition], to invent a false arithmetic analogous to the non-Euclidean geometry. We will find that we cannot .... (Poincaré, 1892, p. 406.)

There is of course no difficulty in providing a model, in a modern sense, for the negation of the principle Poincaré describes here, the least-number principle. Simply interpret "less-than" via the greater-than relation; alternatively, take "positive whole number" to be interpreted by the negative integers. Poincaré's claim is that we "cannot" represent *the positive whole numbers* as failing to satisfy the least-number principle; his point is that a collection that fails this principle is not the positive whole numbers. Lines, on the other hand, *can* be represented faithfully as failing to satisfy the parallels postulate. What we get from Beltrami's surface, on this account, is a demonstration that Lobachevsky's geometry is just as coherent as is Euclid's, not in the sense of (semi-) formal deductive consistency (which holds just as well in the case of the "false arithmetic"), but in the sense that each provides a coherent description of space.[13]

In general: the construction of Beltrami-style models of L demonstrated the independence of the parallels postulate in a sense quite different from those later notions of Independence$_D$ and Independence$_S$, which notions could not in any event have been made sense of in the setting of a traditional understanding of geometry and its language. The independence-claim in question was instead the less rigorously-demarcated idea of the coherence of a space whose lines behave non-Euclideanly. The role of the constructed surface in this project is to depict a genuine possibility for space itself. Hence the kind of independence shown is very strong: from the claim that it's possible for lines to satisfy most of Euclid without satisfying the parallels postulate, we conclude that one cannot prove the latter from the former, even if one employs proof procedures much richer and more content-sensitive than are the formal or semi-formal deductive principles favored after the turn of the twentieth century. The result is strong enough to establish not just Independence$_D$ and Independence$_S$, but arguably also the Frege-independence of the parallels postulate.

But while the representational strategy just described proves a strong result, an important weakness of the strategy is that its scope of applicability is severely limited. There is no way via appeal to representative surfaces to demonstrate, for example, the independence of (BET$_A$) from (BET$_C$), since no construction of points on a line-like element will satisfy one of the between-ness claims without satisfying the other. And as Poincaré points out above, no such representative strategy can successfully depict a situation in which the natural numbers fail the least-number principle. More generally, there is no straightforward way to extend the strategy beyond the scope of a small part of geometry, that part in which the statement to be demonstrated independent is one whose negation can be represented via recognizable analogues of lines and figures.

5   The turn of the century

In addition to the developments mentioned above concerning the recognition of duality principles in projective geometry and the Riemannian generalization of the scope of geometry, the path from Beltrami to Hilbert turns on closely-related developments concerning mathematical languages and their subject-matter. Here the developments are of both a mathematical and a logical kind. In the first category is the development

---

[13] For Poincaré, of course, unlike Helmholtz, the further lesson to be drawn is that the choice between the two is purely conventional: they describe the same empirical possibility for space.

of the view of mathematical theories not as statements about determinate collections of entities, but instead as collections of sentences that characterize multiply-instantiable structural properties. This development opens the way for a conception of consistency as satisfiability, and hence of independence as Independence$_S$. On the more purely-logical side is the increasingly-important idea near the end of the twentieth century of proof as characterizable in terms of semi-formal (or later purely formal) principles of sentential deduction, an idea that made Independence$_D$ a natural and central concern.

The modern strategy in place by the turn of the twentieth century, the Hilbert-style method described above, is both much more broadly applicable (since it can be applied to any relatively well-regimented axiomatic theory) and more rigorous than is the kind of appeal to constructed surfaces made by Helmholtz and Poincaré. There is no need for a Hilbert-style model to contain recognizable analogues of elements of an original subject-matter, and there is no conceptual question to be raised about whether the model depicts a conceivable arrangement of the target objects and properties. As long as the sentences in question are true as re-interpreted, and the rules of proof preserve truth under the reinterpretation, Independence$_S$ and Independence$_D$ are immediate. Poincaré's objections about the impossibility of representing false synthetic a priori theories have no purchase in this setting, and Lambert's concerns about the representability of lines by non-lines are irrelevant both to the newly-conceived notions of independence, and to their modern demonstration.

6 The formalization of logic

In the period of roughly 1880 to 1905, it became standard to apply to mathematical axiom-systems a handful of fundamental "meta-theoretical" questions, including those of completeness (in various senses), consistency, and mutual independence. Independence demonstrations for axioms of number theory, geometry, and analysis were familiar by the end of this period, and proceeded in essentially the Hilbert-style way.[14]

Between 1905 and 1915, the Hilbert school became increasingly interested in the development of systems of axioms for logic itself, and specifically in the formal, i.e. syntactic, specification of such systems. The natural question to raise at this point is whether the independence-proving techniques that are clearly useful in mathematical settings can be applied to systems of pure logic. Can one, for example, demonstrate the independence of logical axioms one from another using the modern technique of interpretation-style models?

---

[14] For discussion of these developments, see Awodey and Reck (2002), Mancosu, Zach, and Badesa (2009), Zach (1999).

Bertrand Russell, asked this question in 1909, responds as follows:

> I do not prove the independence of primitive propositions in logic by the recognized methods; this is impossible as regards principles of inference, because you can't tell what follows from supposing them false: if they are true, they must be used in deducing consequences from the hypothesis that they are false, and altogether they are too fundamental to be treated by the recognized methods. (Russell to Jourdain, April 1909, as reported in Grattan-Guinness, 1977, p. 117.)

A similar sentiment appears in *Principles of Mathematics*, and again in *Principia Mathematica*. (Russell, 1903, §17; Russell and Whitehead, 1910-1913, *1.)

It is not difficult to see the problem as Russell sees it. Suppose (with Russell) that the basic idea of an independence proof is to assume some collection of claims to be true, assume a further claim to be false, and then check to see whether any contradiction follows. As applied to the independence of the parallels postulate, the idea is that we are assured by the coherence of the non-Euclidean surface that no contradiction follows from supposing that the parallels postulate is false while the rest of Euclid is true. The role of the model on this picture is not that of re-interpretation to show Independence$_D$ or Independence$_S$; it is the earlier idea of a representation of a situation in which the axioms in question are true and false respectively. Given this understanding of an independence-proof and of the role of models, it is clear that one cannot, just as Russell says, apply the technique to principles of logic. To assume a handful of logical axioms true and a target logical axiom false is already to engage in contradiction. And there is no sense in which such a collection of sentences, i.e. one including the negation of a principle of propositional logic, can be taken to describe a coherent situation. In short, and as Russell notes, the principles of logic - to which we must appeal when demonstrating independence in the arenas of geometry or analysis - are "too fundamental to be treated by" such a method.

Nevertheless, there is clearly something wrong with Russell's idea that the "recognized method" at the time of his writing cannot be applied to systems of logic. By as early as 1905, Hilbert had already been lecturing on the use of arithmetical interpretations to show the mutual independence of axioms for propositional logic. And by 1918, Bernays uses just such interpretations to demonstrate the mutual independence of some of Russell's own propositional axioms (Bernays, 1918).[15]

Bernays' technique for demonstrating the independence of an axiom A from axioms $A_1 \ldots A_n$ is essentially as follows: We give a systematic assignment of values (e.g. the numbers 1, 2, 3) to every sentence of the language, in such a way that, for example: each of $A_1 \ldots A_n$, together with every sentence deducible from these via the specified inference rules, is assigned the value 1; A itself is not assigned 1. The immediate conclusion is that A is not deducible from $A_1 \ldots A_n$.

---

[15]For discussion of Bernays 1918, see Mancosu et al. (2009).

The first thing to note about this strategy is that, *contra* Russell, it does not involve supposing that A is false. It also doesn't involve the representation of a state of affairs that in any sense satisfies or exemplifies $A_1 \ldots A_n$; there is no need to try to make sense of the presumably-incoherent idea of a state of affairs in which an instance of an axiom of propositional logic is false. What is demonstrated by Bernays' method is simply the non-deducibility, now in an entirely *formal* - i.e. syntactic - sense, of the formula A from the collection $A_1 \ldots A_n$ of formulas.

The connection between the method of Hilbert 1899 and Bernays 1918 can be brought out as follows. First of all, both methods are instances of what we can call the *arbitrary valuation strategy*: That strategy, applied to a formula A and a collection $A_1 \ldots A_n$ of formulas, is to assign values to formulas in such a way that, for a designated value V, V is assigned to each of $A_1 \ldots A_n$ and to each formula *deducible* from them, but is not assigned to A. The two coordinated differences between Hilbert 1899 and Bernays 1918 are (i) the kind of values employed, and (ii) the nature of the deducibility relation.

For Hilbert 1899, the value V is in each case the value "expresses a theorem of B under interpretation I," where B is a background theory (typically a theory of constructions on the real numbers) and I is the interpretation of the formulas in question via the subject-matter of B. Deducibility is a relation not explicitly specified, but understood in terms of self-evident principles of inference, subject to the constraint that the principles be semi-formal, holding independently of the interpretations of the geometric terms appearing in those formulas. The critical fact about deducibility assumed throughout is that anything deducible from a formula that expresses a theorem of B is also a formula that expresses a theorem of B. This assumed feature of deducibility is the guarantee that V is preserved by deducibility. The guarantee that the target formula A *lacks* value V, i.e that A does not express a theorem of B under interpretation I, is given by the facts that (a) by design, $\neg A$ expresses a theorem of B under I, and (b) B is, by assumption, itself consistent.

No such valuation, and no such account of deducibility, can work in the setting of independence proofs for principles of pure logic. First of all, the designated value V must in this setting be one that is not automatically had by all truths of logic, since it is precisely a truth of logic that we will want to demonstrate lacks V. So V cannot be the value "expresses a theorem of theory B under interpretation I," for any B or I. The relation of deducibility, in addition, cannot be merely a generally-understood notion of (semi-formal) provability, since any such notion will count the formulas of pure logic as deducible from everything. Bernays' method rests on the existence of a syntactically-specified relation of deducibility, with respect to which it is not trivially true that each principle of logic is deducible from everything. It also rests on the choice of a targeted value V that has nothing to do with the "interpretation," in any ordinary sense, of the formulas in question: i.e. nothing to do with the idea of those formulas as expressing truths and falsehoods about either the intended or an alternate subject-matter.

With respect to the question, then, of whether Russell is right that the "recognized method" is not applicable to questions of independence in systems of logic, the answer will turn on what exactly we take to fall under the scope of "recognized method." Taking that method to be the very broad strategy we've called the "arbitrary valuation" strategy, Russell is wrong: the method works, as Bernays shows. Taking, on the other hand, that method to be the more narrowly-construed instance of that technique in which valuations are understood in terms of re-interpretations into an assumed-consistent background theory, then Russell is right: we cannot interpret the language in such a way that axiom-sentences of propositional logic express the negations of theorems of a consistent background theory. Finally: given Russell's own, old-fashioned way of understanding the "recognized method," as involving the supposition of the falsehood of the target axiom and a subsequent question about the consistency of the result, the method is clearly not applicable to principles of logic, for the reasons Russell himself gives.

A further important difference between the method as employed by Hilbert 1899 and its refinement in Bernays 1918 involves the kind and the strength of the independence claims thereby demonstrated. As above, Hilbert's technique shows Independence$_D$: it shows that A is not deducible from $A_1 \ldots A_n$, where "deducible from" is the semi-formal relation described above. Similarly, Bernays' technique shows that A is not deducible from $A_1 \ldots A_n$, where "deducible from" is now the rigorously-specified relation specific to a particular formal system. But Hilbert's technique also demonstrates, as we've discussed, the stronger result of "Independence$_S$,": it demonstrates that the condition implicitly defined by $\{A_1 \ldots A_n, \neg A\}$ is satisfiable. But Bernays' method provides no such further result: no domain is exhibited that satisfies conditions implicity defined by the formulas in question. And indeed, the satisfiability claim in question makes no sense as applied to the formulas to which Bernays applies it: the axioms are not implicit definitions of structural conditions, and there is no sense to be made of a domain with respect to which some of those axioms express falsehoods.

The importance of Independence$_S$ is most vivid in the setting of the kind of structuralist approach to mathematical theories and axioms that was beginning to take hold at the end of the nineteenth century, and remains of central importance today. In Dedekind's work, for example, the theory of natural-number arithmetic is the theory of any and all ordered collections of objects that satisfy the natural-number axioms, or equivalently the theory of any and all ω-sequences.[16] The role of each axiom on this conception is to provide a partial characterization of the type of structure in question. Given a collection of axioms $A_1 \ldots A_n$, the addition of a further axiom A would be redundant if every structured domain satisfying the former collection already satisfied the latter. The important independence relation, from this point of view, is the relation of non-redundancy in this sense, which is to say that it is a matter of the satisfiability of $\{A_1 \ldots A_n, \neg A\}$, i.e. of Independence$_S$.

Independence$_D$ is the relevant kind of independence if instead the goal of the axioms

---

[16] See Dedekind (1888).

Models in geometry and logic: 1870-1920   57

is the deductive characterization of a body of truths. In this setting, A is redundant with respect to $A_1 \ldots A_n$ if A is deducible from $A_1 \ldots A_n$, and hence independent in the relevant sense if Independent$_D$ of those axioms. In a setting in which axioms are intended to provide both a deductive characterization of a theory and a definition of the structural characteristics of its domain, both kinds of independence are relevant; and as above, both are demonstrable via the construction of a model in the mode of Hilbert 1899. Once we move to Bernays 1918, however, the goal of the axioms is purely deductive (there being no sense in which the axioms of the propositional calculus define properties of structures), and the relevant kind of independence is just Independence$_D$.

7 Summing up

The idea with which we began was the traditional idea that the development of model-theoretic methods around the turn of the 20th century provided, at last, a rigorous way of answering old independence questions. The contrary proposal suggested here is that this is not quite the right way to view the developments of the period 1870 - 1920. Instead of a single notion of independence that's given increasingly-rigorous treatment, we have a handful of different independence questions, some of which are susceptible to rigorous treatment, and some of which are not. As our methods have changed, so too have the questions that we are in a position to ask (and answer).

Passing from 1870 to 1899 to 1918, we see the following three lines of development.

First, we see a gradual increase in rigor. By 1899, questions of independence are divorced from questions about the representability of the subject-matter (e.g. of lines by geodesics, of positive numbers by negative numbers), and linked to the more tractable notion of reinterpretation. By 1918, the appeal to an informal notion of provability is replaced by appeal to an explicitly-defined relation of formal deducibility.

Secondly, we see an increase in the scope of the methods. In 1870, the canonical independence-proof technique applies to the parallels postulate and to that small collection of geometric propositions whose negation can be represented as holding on something recognizably like a surface. By 1899, we have a technique that applies to all of geometry and arithmetic. And by 1918, the standard technique allows us to prove independence even of the axioms for formalized systems of logic.

But, thirdly, we see in this period a gradual decrease in the strength of the independence claims demonstrable by the emerging methods. In 1870, as understood e.g. by Helmholtz, a model establishes a strong modal claim, i.e. the claim that space might really be a certain way, and that we can conceive of its being that way. By 1899, the method exhibited in Hilbert's *Foundations of Geometry* makes no claim to establishing such a strong claim about the possible configuration of space or about our conception of it; the claims made via the new method are claims of non-deducibility and of the satisfiability of implicitly-defined conditions. Finally, the method employed by Bernays in 1918 provides us with clear demonstrations of non-deducibility; it is neither intended to provide, nor is it capable of providing, any modal results about the subject-matter of the theory or any conclusions about the satisfiability of structural conditions.

If the oldest versions of the question of the independence of the parallels postulate are questions whose positive answer would establish the possibility or conceivability of non-Euclidean space, then they are not the questions answerable via either the method of 1899 or the method of 1918. And if the questions asked of arithmetic and geometry by Deekind and Veblen have to do with the satisfiability of conditions implicitly defined by the axioms in question, then those questions, decisively answerable via the method of 1899, are not answerable via the method of 1918.[17] The weakest of our independence relations, that of pure non-deducibility in a rigorously-specified formal system, is also the cleanest, and the most crisply demonstrable. It is natural to take it to be, in some sense, a refinement of older and more inchoate questions about independence. But if the suggestions made here are accurate, then the gap between the independence-claims cleanly demonstrable via the modern methods and the independence-claims that originally motivated much geometric work prior to the end of the 19th century is quite large; and it is sufficiently large that we cannot take the newer methods to be merely cleaned-up ways of answering the old questions.

Acknowledgements. Many thanks to the organizers of CLMPS 2015 in Helsinki, and also to audience members for helpful comments. This essay includes material previously presented at Logica 2015 in Hejnice, Czech Republic and published in *Logica 2015*, as well as material presented earlier at the Ohio State University and at the University of Notre Dame; thanks also to all three audiences and to the Logica 2015 organizers.

---

[17] See Dedekind (1888), Veblen (1904).

Bibliography

Awodey, S., & Reck, E. (2002). Completeness and categoricity part I: Nineteenth-century axiomatics to twentieth-century metalogic. *History and Philosophy of Logic*, *23*(1), 1–30.

Beltrami, E. (1868a). Saggio di interpretazione della geometria non euclidea. *Giornale di mathematiche*, *6*, 284–312. Translated by J. Stillwell as "Essay on the interpretation of noneuclidean geometry" in (Stillwell, 1996b, pp. 7-34).

Beltrami, E. (1868b). Teoria fondamentale degle spazii di curvatura costante. *Ann. Mat. pura appl*, *2*, 232–255. Translated by J. Stillwell as "Fundamental theory of spaces of constant curvature" in (Stillwell, 1996b, pp. 7-34).

Bernays, P. (1918). *Beiträge zur axiomatischen Behandlung des Logik-Kalküls. (Habilitationsschrift, Universität Göttingen)*. Bernays Nachlass, WHS, ETH Zurich Archive, Hs 973.192.

Birkhoff, G., & Bennett, M. K. (1988). Felix Klein and his 'Erlanger Programm'. In Aspray, & Kitcher (Eds.), *History and philosophy of modern mathematics* (pp. 145–176). University of Minnesota Press.

Blanchette, P. (1996, July). Frege and Hilbert on consistency. *The Journal of Philosophy*, *93*, 317–336.

Blanchette, P. (2012). *Frege's conception of logic*. Oxford University Press.

Dedekind, R. (1888). *Was Sind und was sollen die Zahlen*. English translation as "The Nature and Meaning of Numbers" in (Dedekind, 1901, pp. 29-115).

Dedekind, R. (1901). *Essays on the theory of numbers*. Edited and translated by W. W. Beman. Chicago: Open Court.

Dunlop, K. (2009). Why Euclid's geometry brooked no doubt: J. H. Lambert on certainty and the existence of models. *Synthese*, *167*, 33–65.

Ewald, W. (Ed.). (1996a). *From Kant to Hilbert*. Oxford University Press.

Ewald, W. (1996b). Johann Heinrich Lambert (1728-1777). In *Ewald, 1996a* (pp. 152–158). Oxford University Press.

Frege, G. (1906). Über die Grudlagen der Geometrie. *Jahresberichte der Deutschen Mathematiker-Vereinigung*, *15*, 293–309, 377–403, 423–430. English translation as "Foundations of geometry: Second series" in (Frege, 1984, pp. 293-340).

Frege, G. (1984). *Collected papers on mathematics, logic, and philosophy*. This is a translation of most of (Frege, 1990). Edited by B. McGuinness; translated by M. Black, V. H. Dudman, P. Geach, H. Kaal, E-H. W. Kluge, B. McGuinness, and R. H. Stoothoff. Blackwell.

Frege, G. (1990). *Kleine Schriften (2nd ed.)* Edited by I. Angelelli. Hildesheim: Georg Olms.

Grattan-Guinness, I. (Ed.). (1977). *Dear Russell - Dear Jourdain*. Columbia University Press.

Gray, J. (1979). Non-Euclidean geometry - A re-interpretation. *Historia Mathematica*, *6*, 236–258.

van Heijenoort, J. (Ed.). (1967). *From Frege to Gödel*. Harvard University Press.

von Helmholtz, H. (1876). On the origin and meaning of geometrical axioms. *Mind*, *3*, 302–321. Reprinted in (Pesic, 2007, pp. 53-70).

Hilbert, D. (1899). *Grundlegung der Geometrie*. English translation of the 10th edition: *Foundations of geometry*, L. Unger (trans), P. Bernays (ed), Open Court 1971. Stuttgart: Teubner.

Klein, F. (1908). *Elementarmathematik vom Höhere Standpunkt aus*. Teubner.

Lambert, J. H. (1786). Theorie der parallellinien. *Magazin für reine und angewandte Mathematik für 1786*, 137–64, 325–58. Partial translation by William Ewald in (Ewald, 1996a, pp. 158-167). Citations are to this translation.

Lobachevsky, N. (1840). *Geometrische Untersuchungen zur Theorie der Parallellinien*. Berlin: Funcke.

Mancosu, Zach, & Badesa. (2009). The development of mathematical logic from Russell to Tarski: 1900-1935. In L. Haaparanta (Ed.), *The Development of Modern Logic*. Oxford University Press.

Marchisotto, E. A., & Smith, J. T. (2007). *The legacy of Mario Pieri in geometry and arithmeic*. Springer.

Nagel, E. (1939). The formation of modern conceptions of formal logic in the development of geometry. *Osiris*, *7*, 142–224.

Padoa, A. (1901). Essai d'une théorie algébrique des nombres entiers, précédé d'une introduction logique à une théorie déductive quelconque. *International Congress of Philosophy 1900-1903*, *3*, 309–365. Partial translaation by Jean van Heijenoort in (van Heijenoort, 1967, pp. 118-123).

Peano, G. (1894). Sue fondamenti della geometria. *Rivista di matematica*, *4*(4), 51–90.

Pesic, P. (Ed.). (2007). *Beyond geometry classic papers from Riemann to Einstein*. Dover.

Pieri, M. (1898). I principii della geometria di posizione composti in sistema logico deduttivo. *Memorie della Reale Accademia delle Scienze di Torino (series 2)*, *48*, 1–62.

Poincaré, H. (1892). Non-Euclidean geometry. *Nature*, *45*.

Riemann, B. (1873). On the hypotheses which lie at the bases of geometry. *Nature*, *8*, 114–117, 136–136. Lecture delivered 1854. Translated by W K Clifford. Reprinted in (Pesic, 2007, pp. 23-40).

Russell, B. (1903). *Principles of mathematics*. Cambridge University Press.

Russell, B., & Whitehead, A. N. (1910-1913). *Principia mathematica*. Cambridge University Press.

Scanlan, M. (1988). Beltrami's model and the independence of the parallel Ppostulate. *History and Philosophy of Logic*, *9*(1), 13–34.

Stillwell, J. (1996a). Introduction to Beltrami's essay on the interpretation of noneuclidean geometry. In *Stillwell, 1996b* (pp. 1–5).

Stillwell, J. (Ed.). (1996b). *Sources of hyperbolic geometry*. American Mathematical Society.

Stump, D. (2007). The independence of the parallel postulate and the development of rigorous consistency proofs. *History and Philosophy of Logic*, *29*, 19–30.

Veblen, O. (1904). A system of axioms for geometry. *Transactions of the American Mathematical Society*, 5(3), 343–384.

Webb, J. (1985). Tracking contradictions in geometry: The idea of a model from Kant to Hilbert. In J. Hintikka (Ed.), *Essays on the development of the foundations of mathematics* (pp. 1–20). Springer.

Zach, R. (1999). Completeness before post: Bernays, Hilbert, and the development of propositional logic. *Bulletin of Symbolic Logic*, 5(3), 331–366.

Author biography. Patricia Blanchette is Professor of Philosophy at the University of Notre Dame. Her recent work includes *Frege's Conception of Logic* (Oxford University Press 2012), and a number of articles in the history and philosophy of logic, the philosophy of mathematics, and the history of analytic philosophy.

3 Squeezing arguments and strong logics

JULIETTE KENNEDY AND JOUKO VÄÄNÄNEN

Abstract. G. Kreisel has suggested that squeezing arguments, originally formulated for the informal concept of first order validity, should be extendable to second order logic, although he points out obvious obstacles. We develop this idea in the light of more recent advances and delineate the difficulties across the spectrum of extensions of first order logics by generalised quantifiers and infinitary logics. In particular we argue that if the relevant informal concept is read as informal in the precise sense of being untethered to a particular semantics, then the squeezing argument goes through in the second order case. Consideration of weak forms of Kreisel's squeezing argument leads naturally to reflection principles of set theory.

Keywords: generalized quantifiers, logical validity, reflection principles, second-order logic, squeezing principles.

1 Introduction

The foundational project is driven by the idea of modeling mathematical discourse, more or less globally, by giving an *adequate* formal reconstruction of it. Adequacy here, as in the phrase "adequate formal system," is delivered by the following: the formalism should be sound, complete, and at least to the degree possible, effective and syntactically complete. The formalism should also be *meaning-preserving*, relative to a given semantics (ideally). The idea of foundational formalism, as we called it in (Kennedy, 2013), is that with such a system in hand one could reasonably claim that the formalism has "captured" the informal discourse—whichever way one wishes to express this idea of "capturing."

At the same time the idea of considering just the informal mathematical discourse on its own, so to say *in situ*, has also attracted interest. This is implicit in so-called "practice-based" philosophy of mathematics—that practice is situated, after all, in natural language—while other philosophers take a more direct interest in natural language. M. Glanzberg, for example, in his (Glanzberg, 2015), argues that the notion of, e.g., consequence at work in natural language is to be distinguished from a genuinely logical consequence relation:

> The success of applying logical methods to natural language has led some to see the connection between the two as extremely close. To put the idea somewhat roughly, logic studies various languages, and the only special feature of the study of natural language is its focus on the languages humans happen to speak.
>
> This idea, I shall argue, is too much of a good thing.

Glanzberg is not only alerting us to the pitfalls of conceiving of natural language, at least in its logical aspects, as a kind of thinly disguised formal language, a matter of cleaning up the relevant definitions, concepts and so on. For Glanzberg, natural language and formal discourse are instead, with respect to their logical scaffolding anyway, two autonomous domains—"though the processes of identification, abstraction, and idealization can forge some connections between them." "Natural language has no logic" is the paper's central claim.[1]

What does it mean to say that natural language has no logic? For Glanzberg this is to deny *the logic in natural language thesis* (LNL thesis henceforth), namely the thesis that:

> A natural language, as a structure with a syntax and a semantics, thereby determines a logical consequence relation.

This is as opposed to the *logics in formal languages thesis* (LFL thesis henceforth):

> Logical consequence relations are determined by formal languages, with syntactic and semantic structures appropriate to isolate those relations.... Thus, the logics in formal languages thesis holds that consequence relations are *in* formal languages, in the sense that they are definable from them.

The LFL thesis is uncontroversial, if not trivial. The argument against the LNL thesis is subtle and turns partly on a critique of the model-theoretic account of logical consequence.[2] We will not consider Glanzberg's critique here, but will do so below, as the

---

[1] Glanzberg construes the term "logic" rather narrowly in the paper. The quote in the central claim is a deliberate reference to Strawson's (Strawson, 1950):

> Neither Aristotelian nor Russellian rules give the exact logic of any expression of ordinary language; for ordinary language has no exact logic.

[2] Glanzberg's arguments against the LNL thesis generally read "logic" as logical consequence.

issue becomes relevant to Kreisel's so-called squeezing arguments, to which we now turn.

2 Squeezing arguments and their critics

Squeezing arguments may be thought of as falling on the other side of the spectrum of belief in the LNL thesis (albeit tacitly). Introduced by Kreisel in his 1967 "Informal Rigour and Completeness Proofs" (Kreisel, 1967), and since taken up by W. Dean (Dean, 2016), H. Field (Field, 2008), V. Halbach (Halbach, 2016), P. Smith (Smith, 2011) and others, the arguments go as follows:

> Consider an informally defined mathematical concept I. Formally define two concepts A and B such that falling under the concept of A is a sufficient condition for falling under the concept of I, and falling under the concept of I suffices for falling under the concept of B. Thus $A \subseteq I \subseteq B$, where the inclusions are understood as applying to the extensions of the concepts A, B, I.
>
> Now suppose the formal notions A and B have the same extension. Then by the inclusions $A \subseteq I \subseteq B$ the informal concept I must coincide, again extensionally, with that of A and B.

For the informal concept I Kreisel took intuitive validity, denoted Val, understood as truth in all possible structures. This includes set and class-sized structures, as well as, in principle at least, structures that have no set-theoretical definition. Taking formal first order provability, denoted D_F, on the left, and taking truth in all set-theoretical structures,[3] denoted V, on the right, Kreisel argued as follows: By soundness, $D_F \subseteq$ Val. By the fact that truth in all structures entails truth in all set-theoretical structures, Val $\subseteq$ V. Thus

$$D_F \subseteq \text{Val} \subseteq \text{V}. \tag{3.1}$$

Invoking the completeness theorem for first order logic Kreisel concludes the following *theorem*, as he calls it, for α any first order statement:

$$\text{Val } \alpha \leftrightarrow \text{V}\alpha \text{ and Val } \alpha \leftrightarrow D_F \alpha.$$

Kreisel's presentation of the argument has been criticised in the literature. Smith (Smith, 2011) objects that Kreisel's somewhat model-theoretic construal of Val does not obviously capture the pre-theoretic notion in question, validity-in-virtue-of-form, as Smith prefers to think of Val.

---

[3] A set-theoretical structure is one whose domain, relations and functions are sets in the usual sense.

Field's criticism of the argument in (Field, 2008) involves the soundness claim, namely the first inclusion $\forall \alpha (D_F \alpha \to \text{Val}\,\alpha)$:

> In chapter 2 I argued that ...there is no way to prove the soundness of classical logic within classical set theory (even by a rule-circular proof): we can only prove a weak surrogate. This is in large part because we cannot even *state* a genuine soundness claim: doing so would require a truth predicate. And a definable truth predicate, or a stratified truth predicate, is inadequate for stating the soundness of classical logic, and even less adequate for proving it.[4]

Field goes on to prove soundness—or a weak surrogate of soundness—by means of a formal truth predicate, applying in restricted cases. Taken as a repair of Kreisel's argument one might argue that it ignores the methodology of the paper, which is heavily semantic (see below). Kreisel's argument does not depend on a proof of soundness in classical set theory. Kreisel is asking us to take soundness for granted, on the basis of *historical experience*—or as Kreisel puts it, *intuitive notions standing the test of time*. Instead, Field takes D_F as "primary", in Kreisel's terminology.[5]

As for Kreisel's own "proof" of soundness, extending to α^i (interpreting α as an ith order sentence) for all i, it amounts to arguing that the universal recognition of the validity of Frege's rules (D_F) at the time, together with the "facts of actual intellectual experience" accumulated subsequently, should amount to no more and no less than the acceptance of

$$\forall i \forall \alpha (D_F \alpha^i \to \text{Val}\,\alpha^i)$$

for us. And though a century of logical history has taught the logician nothing if not to be extremely suspicious of inclusions such as (3.1)—suspicions that Kreisel himself airs at the end of the paper—surely what Kreisel has in mind here is the idea that D_F was formulated *ex post facto*, that is precisely so as to *guarantee* soundness. D_F stood for Frege and his contemporaries, Kreisel claims, and stands for the contemporary mathematician too, as a completely adequate formalisation of Val.

Returning to our discussion of (Kreisel, 1967)'s critics, Halbach (Halbach, 2016) offers a repair of squeezing arguments in the form of a formal, syntactic substitutional notion of logical validity, to be substituted in for Kreisel's somewhat model-theoretic notion.[6] Such a concept of validity is, in Halbach's view, "closer to rough and less rig-

---

[4](Field, 2008, p. 191)

[5]From Kreisel (1967, p. 153):

> First (e.g. Bourbaki) 'ultimately' inference is nothing else but following formal rules, in other words D is primary (though now D must not be regarded as defined set-theoretically, but combinatorially). This is a specially peculiar idea, because 99 per cent of the readers, and 90 per cent of the writers of Bourbaki, don't have the rules in their heads at all!

[6]Halbach's conceptual analysis applies more widely, that is, it is an analysis of the "natural" concept of logical consequence in terms of substitutional validity überhaupt, i.e. not just in connection with squeezing

orous definitions of validity as they are given in introductory logic courses":

> I put forward the substitutional analysis as a direct, explicit, formal, and rigorous analysis of logical consequence. The substitutional definition of logical validity, if correctly spelled out, slots directly into the place of 'intuitive validity' in Kreisel's squeezing argument, as will be shown below.[7]

This is as opposed to the model-theoretic account of consequence, with its many (in Halbach's view) drawbacks:

> ...on a substitutional account it is obvious why logical truth implies truth *simpliciter* and why logical consequence is truth preserving. On the model-theoretic account, valid arguments preserve truth in a given (set-sized) model. But it's not clear why it should also preserve simple ('absolute') truth or truth in the elusive 'intended model'. Truth-preservation is at the heart of logical validity. Any analysis of logical consequence that doesn't capture this feature in a direct way can hardly count as an adequate analysis.[8]

Under a substitutional account, the connection with set theory is severed, or such is the claim; and interpretations of logical formulae are now syntactic objects:

> On the model-theoretic account, interpretations are specific sets; on the substitutional account they are merely syntactic and (under certain natural assumptions) computable functions replacing expressions.[9]

We do not wish to address the question here whether Halbach's is a reasonable conceptual analysis of the intuitive notion of logical consequence, the notion of consequence in itself. What seems clear to us is that the *informal* concept of consequence at work in natural mathematical languages is often plainly semantic, and moreover model-theoretic. That when the mathematician draws inferences in natural language, s/he imagines a situation in which the hypothesis is true—i.e. one has a model for the hypothesis in view—then s/he argues that the conclusion must hold in that model.

Kreisel states the point rather colorfully—"they don't have the rules in their heads at all!"—but what he means is that, e.g. group theorists do not derive theorems directly from the group axioms *in practice*, rather they employ the semantic method, i.e. they imagine a group and then show that the group has the property claimed for it. Analysts do not derive formal theorems from the axioms of real numbers, the real numbers are

---

arguments.

[7](Halbach, 2016)

[8](Halbach, 2016)

[9](Halbach, 2016). It is on account of this passage that we characterized Halbach's solution as syntactic. On the other hand, Halbach takes truth as a primitive in the paper, so in that sense his account is at the same time semantic.

taken as a *structure* which satisfies the axioms, and then theorems are proved about that structure. In fact this is what mathematicians are trained to do, as a cursory look at most standard introductory texts demonstrates. For example, Walter Rudin begins his classic text *Principles of Mathematical Analysis* (Rudin, 1976) with the existence of the field of reals:

> Theorem 1.19. There exists an ordered field $\mathbb{R}$ which has the least-upper-bound property.

The rest of the book is the investigation of this field $\mathbb{R}$. The proof of the existence of $\mathbb{R}$, i.e. the construction of the reals from the rationals, is relegated to an appendix.

For another example, Halsey Royden introduces the real numbers in his *Real Analysis* (Royden, 1988) with the following statement:

> We thus assume as given the set $\mathbb{R}$ of real numbers, the set P of positive real numbers and the functions "+" and "·" on $\mathbb{R} \times \mathbb{R}$ to $\mathbb{R}$ and assume that these satisfy the following axioms, which we list in three groups.

As for the reasons behind the mathematicians' semantic mode of thought, this is in some sense the moral of Gödel's speed-up theorem:

> Thus, passing to the logic of the next higher order has the effect, not only of making provable certain propositions that were not provable before, but also of making it possible to shorten, by an extraordinary amount, infinitely many of the proofs already available.[10]

This can be interpreted as saying that the semantic method, the method of establishing logical consequence by considering models, enjoys a so-called "speed-up" over the method of formal proofs. This may explain why the model-theoretic notion of logical consequence seemed natural to Tarski and others.

The above-mentioned objections to (Kreisel, 1967) are certainly appropriate. Kreisel's notion of intuitive validity is clearly overly theorised, in the sense that the intuitive notion considered is not sufficiently intuitive or pre-theoretic, per Smith; and under theorised, per Field and Halbach respectively. At the same time though one might ask, if the intelligibility of Kreisel's squeezing argument depends on replacing the pretheoretic notion of intuitive consequence by some formal, syntactic counterpart notion (as in (Field, 2008) or (Halbach, 2016)), what is the point of squeezing arguments at all? Why not simply analyse formal consequence directly, as logicians have always done?

For it seems to us that the interest of squeezing arguments lies in their being carried through in such a way as to fulfill what was originally claimed for them in (Kreisel, 1967), namely to capture an informal, natural language mathematical notion by "squeezing" it between two formal ones. It can be argued whether Kreisel himself

---

[10] (Gödel, 1986, p. 397)

succeeded in this. The point is that if the aim was to capture (or "squeeze") informal notions used in practice—to provide a conceptual analysis of the *informal* notion of validity, as it were—then as we have argued above, Kreisel's model-theoretic construal of intuitive consequence was the correct and natural one. This is not to question the validity of the conceptual analyses of logical consequence which have been pursued so vigorously, especially in the period since Etchemendy's (Etchemendy, 1990), but rather to ask whether the concepts emerging from such analyses ought to replace the informal notions which serve as the object of Kreisel's analysis here, i.e. in the original squeezing argument. We will return to the issue of "genuine informality" below.

A final objection may concern what is surely the very unnatural restriction to the first order case, that is to propositions of the form α^1. The issue is addressed by Kreisel in the paper, who points to a partial result in this direction, an analogue of the squeezing argument derivable in the case of extensions of D_F to the ω-rule.[11]

Of interest to us here, and one of the topics of this paper, is the possible development of squeezing arguments in the direction of infinitary and second order logics. As it turns out somewhat more is known about completeness theorems for extensions of first order logic than was known in 1967. Going beyond the ω-rule, which Kreisel mentions, completeness theorems have been obtained for a number of infinitary logics, as well as for logics intermediate between first and second order. The second order logic perspective has also been developed a great deal since the publication of Kreisel's (Kreisel, 1967).

As a result of this logical work squeezing arguments of the kind Kreisel seems to be asking for in (Kreisel, 1967) may now be available. It is an oddity of the paper that while Kreisel mentions both the completeness of ω-logic and infinitary logics, he doesn't mention completeness theorems that would have been available already during the writing of (Kreisel, 1967), namely the Henkin Completeness Theorem (Henkin, 1950) for second order logic with the so-called Henkin semantics, and the completeness theorem for $L(Q_1)$, the extension of first order logic by the quantifier "there are uncountably many" due to Vaught (Vaught, 1964) and published in 1964. We will return to these logics below.

3 Squeezing arguments and the logic in natural language thesis

Before we consider the possibility of expanding squeezing arguments in the direction of second order and other strong logics, we ask, are the relevant natural language concepts available for this analysis at all, i.e. even in the first order case? Keeping

---

[11]The completeness of the ω-rule is due to Orey (1956).

Glanzberg's rejection of the LNL thesis in mind, can one simply extract what one thinks of as the notion of informal validity at work in the natural language mathematicians use, and devise a squeezing argument for that notion? In other words, do squeezing arguments require the LNL thesis?

Many researchers in the foundations of logic and mathematics may share a tacit belief in what one might call the "logic in natural *mathematical* language thesis". At the more formalist end of the spectrum of foundationalist views which have been pursued traditionally, one might even attribute a belief in the identity of the notion of logical consequence at work in natural mathematical languages with the notion defined by the relevant formal language. This is just the thought that the logical consequence relation defined by a suitable, maximally adequate formal language is the correct version of the logical consequence relation at work in natural language mathematics—what had been *meant* by the natural language concept all along. Tarski, though no formalist, seemed to argue for this or a similar view in his conversations on nominalism with Quine and Carnap at Harvard in 1940-41, when he remarked that "the difference between logic and mathematics" was that "Mathematics = logic + ϵ".[12]

Others, with Glanzberg, might see the notion of consequence at work in the mathematician's natural language as exact but fundamentally different from the formal notion. In that case one might ask, what separates formal entailment from its counterpart in (mathematical) natural language? We remarked above on Kreisel's observation that Frege's rules gained acceptance among mathematicians over time. This is to say, presumably, that if the relevant part of the discourse is formalized, then Frege's rules would be the mathematician's obvious choice of logical rules. Formalization also involves a choice of a semantics, but in contrast to the rules (D_F) it is not clear that a choice of semantics is determined by the informal practice in the second order case, which is our interest here. The view taken in this paper is that in the case of *intuitive*, informal second order validity, a choice of semantics is entirely irrelevant to the conceptual analysis of the notion of informal consequence, or so we will argue below. To be informal in the second order case necessitates prescinding from a choice of semantics.

We first consider the case of strong logics in general.

4 Squeezing arguments with completeness theorems

Kreisel's paper is entitled "Informal Rigour and Completeness Proofs," and indeed an apparently implicit assumption in (Kreisel, 1967) is that any time one has a completeness theorem in hand for a given logic, the associated squeezing argument should go through. More precisely, let $\mathscr{L}$ be a logic, and let $V_{\mathscr{L}}$ denote $\mathscr{L}$-validity understood

---

[12] Tarski is quoted in Carnap's notebooks. See Mancosu, (Mancosu, 2005).

Squeezing arguments and strong logics

in the standard semantic sense, that is set-theoretically. Let $\text{Val}_\mathscr{L}$ denote informal validity (via $\mathscr{L}$), understood in Kreisel's sense, that is as referring to truth in all structures. Finally, let $D_\mathscr{L}$ denote derivability in the formal system introduced for $\mathscr{L}$.[13] Then one would expect that a completeness theorem for the logic $\mathscr{L}$ together with the inclusions:

$$D_\mathscr{L} \subseteq \text{Val}_\mathscr{L} \subseteq V_\mathscr{L}$$

should underwrite the extensional equivalence of the concepts $\mathscr{L}$-provability, informal $\mathscr{L}$-validity and $\mathscr{L}$-validity construed semantically.

Is this plausible? That is, can new squeezing arguments be obtained from completeness theorems for strong logics? The following completeness theorems for strong logics are known: in ZFC, completeness theorems have been obtained for $L(Q_1)$, the extension of first order logic by the quantifier "there are uncountably many", due to Vaught (Vaught, 1964), as was mentioned; for $L_{\omega_1\omega}$, the logic which is otherwise first order, but allowing conjunctions and disjunctions of countably many formulae, due to Karp (Karp, 1964); for so-called cofinality logic, the extension of first order logic by the quantifier denoted $Q_\kappa^{cf} xy\varphi(x,y)$, meaning "$\varphi$ defines a linear order of cofinality κ" for κ a regular cardinal, due to Shelah (Shelah, 1975); and for so-called stationary logic, the extension of first order logic by the quantifier denoted $\mathrm{aa}s\varphi(s)$, meaning "a club of countable sets s satisfies $\varphi(s)$", due to Barwise, Kaufmann and Makkai (Barwisw, Kaufmann, & Makkai, 1978). The last two require the Axiom of Choice for their completeness theorems.

Going beyond ZFC there is the logic $L(Q_2)$, the extension of first order logic by the quantifier "there are at least $\aleph_2$ many" proved complete by C.C. Chang (Chang, 1965), as pointed out by G. Fuhrken (Fuhrken, 1965), using the continuum hypothesis CH.

Another interesting case going beyond ZFC is the extension of first order logic by the Magidor-Malitz quantifier (Magidor & Malitz, 1977), defined as follows: $\mathscr{M} \models Q_\alpha^{MM,n} x_1 \ldots x_n \varphi(x_1,\ldots,x_n) \iff \exists X \subseteq M(|X| \geq \alpha \wedge \forall a_1,\ldots,a_n \in X \mathscr{M} \models \varphi(a_1,\ldots a_n))$. The completeness theorem for this logic uses the set-theoretical principle $\Diamond$, which is stronger than CH, and it is consistent that completeness fails in the absence of $\Diamond$ (Abraham & Shelah, 1993).

The squeezing argument for the logic $L(Q_1)$, for example, would look like this: Let $D_{L(Q_1)}$ denote the concept of formal provability relative to this logic. Let $V_{L(Q_1)}$ denote the truth of $L(Q_1)$-statements in all set-theoretical structures. Finally let $\text{Val}_{L(Q_1)}$ stand for the validity of $L(Q_1)$-statements relative to all possible structures. Then if the inclusions

$$D_{L(Q_1)} \subseteq \text{Val}_{L(Q_1)} \subseteq V_{L(Q_1)}$$

---

[13] For many $\mathscr{L}$ it is obvious what $D_\mathscr{L}$ should be, but this is not always so.

hold, the squeezing argument relative to the logic $L(Q_1)$ must also hold, by the completeness theorem for $L(Q_1)$.

Strengthening the logic escalates one's set-theoretic commitments, clearly. The completeness theorem for first order logic is actually equivalent to Weak König's Lemma (WKL), which is also required to prove the completeness both of ω-logic and of $L_{\omega_1\omega}$. The Axiom of Choice is used for proving completeness theorems for the cofinality and stationary logics, corresponding to the generalised quantifiers $Q^{cf}_\kappa xy\varphi(x,y)$ and $\mathrm{aa}s\varphi(s)$. The *CH* is used for proving the completeness theorem for the logic $L(Q_2)$, and finally $\diamond$ is required for proving the completeness of the extension of first order logic obtained from the Magidor-Malitz quantifier. A comprehensive study of the exact nature of these commitments would seem to be in order, but is not our concern here.

Countenancing such a hierarchy of commitments is acceptable in some quarters and unacceptable in others—a matter of deciding whether the relevant completeness theorems "speak for themselves," to quote Kreisel.[14] What about the soundness claim in this advanced setting? In the first order case we claimed, in the spirit of informal rigour, that D_F was formulated so as to guarantee soundness—in fact Kreisel's soundness claim $\forall i \forall \alpha (D_F \alpha^i \to \mathrm{Val}\ \alpha^i)$ extends to all orders, as we saw. There is no obvious reason why Kreisel's argument could not be extended to strong and infinitary logics. In the case of ω-logic, the order is somewhat reversed. That is, formal validity (V) is considered with respect to ω-models, in which the positive integer part is standard. Thus in the case of ω-logic the *semantics* is designed so as to underwrite the soundness of the omega-rule.

The case of second order logic in this regard is also striking, in that the Henkin semantics is formulated specifically so as to guarantee not soundness but completeness. We will now take up the question of whether squeezing arguments can be obtained in the second order case.

5 Squeezing for second order logic

The mathematician's informal discourse very naturally includes second order concepts—quantifying over functions and relations and so forth—so it is reasonable to ask for a squeezing argument for informal second order validity. But if a logic has a completeness theorem, then if the proof system of the logic is effective in the sense that the set

---

[14]Kreisel used this phrase in discussing the independence of the *CH* in the paper (p. 140):

> The present conference showed beyond a shadow of doubt that several recent results in logic, particularly the independence results for set theory, have left logicians bewildered about what to do next: in other words, these results do not 'speak for themselves' (to these logicians).

of axioms and rules are recursive and proofs are finite, then the set of valid sentences is recursively enumerable. By (Väänänen, 2001, Theorem 1) the set of valid sentences of second order logic is actually Π_2-complete in the Levy hierarchy. Thus on simple grounds of complexity no reasonable completeness theorem can exist for second order logic.

Does this mean one shouldn't pursue a squeezing argument for second order α? Kreisel himself took the view that "For higher order formulae we do not have a convincing proof of $\forall \alpha^2 (V\alpha^2 \leftrightarrow \mathrm{Val}\,\alpha^2)$ though one would expect one." We will now argue that a squeezing argument for second order α is available, once one incorporates the concept of validity with respect to general models.

Before addressing this point, recall that the Henkin semantics is defined simply so that in the so-called general models, the second order variables of a given formula are thought of as ranging over a fixed subset of the power-set of the domain. This subset of the power set may be a proper subset but it has to satisfy the axioms of second order logic, including the full comprehension axioms. In case the domain of quantification is actually the full power set, one refers to the model as "full" or "standard", and the associated semantics as full or standard semantics.

Taking the definition of Henkin semantics into account, a squeezing argument for informal validity of second order α would be the following:

Let D_Γ denote the usual axiom system of second order logic, already given in (Hilbert & Ackermann, 1928). As above, let $\mathrm{Val}\,\alpha^2$ mean that α^2 is informally true in all structures, including class-sized structures and including, in principle at least, structures that have no set-theoretical definition. Now let $V\alpha^2$ mean that α^2 is true in all set-theoretical structures. The unproblematic implications are:

$$D_\Gamma \alpha^2 \to \mathrm{Val}\,\alpha^2 \to V\alpha^2.$$

Note that if the completeness theorem held for second order logic, we could conclude straightway that $\mathrm{Val}\,\alpha^2 \leftrightarrow V\alpha^2$ and $\mathrm{Val}\,\alpha^2 \leftrightarrow D_\Gamma \alpha^2$, as before. Now denote by $V'\alpha^2$ the statement that α^2 is valid with respect to set-theoretically defined general models (satisfying—as we have assumed—the full comprehension axioms). Consider the following implications:

$$D_\Gamma \alpha^2 \to \mathrm{Val}\,\alpha^2 \to V'\alpha^2 \to V\alpha^2. \tag{3.2}$$

Suppose we assume (3.2). Then by Henkin's (Henkin, 1950) proof of

$$V'\alpha^2 \to D_\Gamma \alpha^2$$

together with (3.2) we would obtain:

$$\mathrm{Val}\,\alpha^2 \leftrightarrow D_\Gamma \alpha^2 \leftrightarrow V'\alpha^2.$$

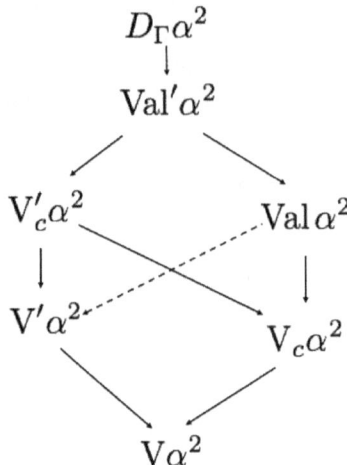

Figure 3.1: Varieties of validity in second order logic.

The first implication of (3.2) is clear, modulo the soundness claim, and so is the last, for trivial reasons. What about the middle implication Val $\alpha^2 \to$ V$'\alpha^2$? If α^2 is informally valid in all structures, why is it that general models should count as such structures? If the second order variables of α^2 are thought of as ranging over the full power set of the domain in question, why is it the case that these second order variables can be regarded as ranging over the subsets in a general model? Is there a principled way to distinguish non-standard general models from standard (full) structures, from the "informal practice" point of view?

To analyse the situation in more detail, let us write $V_c \models \alpha^2$ if α^2 is valid in both class-sized structures and set structures. Let us also write $V'_c \models \alpha^2$ if α^2 is valid in both class-sized general models and set-sized general models. Finally, let us write Val$'_c \, \alpha^2$ if α^2 is informally true in all structures, full or general, including class-sized structures. Figure 3.1 depicts the trivial implications. The possible implication Val $\alpha^2 \to$ V$'\alpha^2$ is the problematic one.

So what distinguishes Val α^2 from Val$'\alpha^2$? We claim that on the informal level it is impossible to see a difference between a standard model and a general model. It is true that if we consider a general model in isolation, from outside, so to speak, it is easy to imagine that something is missing from the model, in order for it to count as a standard model. For example, if we consider an infinite general model with a countable set of relations as the range of second order variables, we know that the model is not standard. There may be other ways of seeing the non-standardness from the outside. We may, for example "see" that a general model of arithmetic has an element with infinitely many predecessors. The position taken here is that it is contrary to the idea of informal validity that one should be able to look at the situation from outside.

One might still think[15] that there really is an *informal* concept of a general model, encapsulated by the thought: "All the sets I need are there and if some are missing, they do not change anything". This would seem to be different from the informal concept of a standard model, encapsulated by the thought: "All the sets are there and no set, whether I need it or not, is missing". If this is the case it is conceivable that for some α^2 we make the judgement that it is informally valid only in standard models, not in all general models. However, while it is crystal clear what the difference is between standard models and general models in the technical, logical sense, it is a different matter to see the difference on the informal level.

We go further and claim that on the informal level, the difference is not discernable. The reason for this is (essentially) that the general models "know" all the definable sets and relations (by the Comprehension Axioms) and they are the ones we refer to in mathematical practice.[16]

A similar line is articulated in (Väänänen, 2012), in which the second author has argued that from the point of view of mathematical practice, when we actually use second order logic we do not and in fact cannot see a difference between ordinary ("full", or "standard") models and general models.

> I will argue in this paper that if second-order logic is used in formalizing or axiomatizing mathematics, the choice of semantics is irrelevant: it cannot meaningfully be asked whether one should use Henkin semantics or full semantics. This question arises only if we formalize second-order logic after we have formalized basic mathematical concepts needed for semantics. *A choice between the Henkin second-order logic and the full second-order logic as a primary formalization of mathematics cannot be made*; they both come out the same.[17]

For example, let us consider Bolzano's Theorem:

Theorem 1. *(Bolzano) Every continuous real function on* $[0, 1]$ *which has a negative value at 0 and a positive value at 1 assumes the value 0 at some point of* $(0, 1)$.

For the proof, by the second order comprehension axiom one can instantiate a universal second order quantifier at $X = \{x | f(x) < 0\}$. The set X is even first order definable, with f as parameter. This is a paradigm example: we operate on sets definable from existing sets. Of course, principles such as the Axiom of Choice force us to introduce also non-definable sets, but they do not exist because "all" sets exists but because we assume—and the general models are assumed to satisfy—the Axiom of

---

[15]We are indebted to A. Blass for suggesting this line of thought.

[16]Note that the definable sets taken on their own are not sufficient as they do not satisfy the Comprehension Axiom. One needs a little "blurring" around the edges, otherwise one can diagonalise out of the class.

[17](Väänänen, 2012, p. 505, emphasis ours.)

Choice.

We now turn to the issue of set vs class-sized models. Consider the weaker claim that

$$V\alpha^2 \to V_c\alpha^2 \tag{3.3}$$

that is, the claim that second order formulas valid relative to set-theoretical structures are also valid relative to class-sized structures. In other words, we ask, is it true that if a second order sentence has a class-sized model, it also has a set-sized model? This cannot be proved from the axioms of von Neumann-Gödel-Bernays class theory (NGB), as the following "Zermelian" argument shows: Let α^2 be the second order sentence which says that the universe of the model is an inaccessible cardinal. Let κ be the least inaccessible and let M_κ denote the cumulative hierarchy up to κ. Then $\langle M_\kappa, \mathscr{P}(M_\kappa)\rangle$ is a model of NGB satisfying α^2. But no set-sized model, in the sense of $\langle M_\kappa, \mathscr{P}(M_\kappa)\rangle$, satisfies α^2.

Paul Bernays (Bernays, 1961) formulated more or less exactly (3.3), albeit in dual form, as a reflection principle, and observed that it implies the existence of inaccessible cardinals.[18] In fact, (3.3) implies a parameter-free version of so-called Levy's Schema (Lévy, 1960), which says that every definable closed unbounded class C of ordinals contains a regular cardinal. In the original Levy's Schema the definition of C is allowed to have parameters. Since the class of all cardinals is definable without parameters, we obtain from (3.3) a proper class of inaccessible cardinals. Bernays goes on to formulate (3.3) with second order parameters and arrives at what became later to be known as indescribable cardinals.[19] L. Tharp (Tharp, 1967) showed that the parametrized principle implies that for every n, the class of Π^1_n-indescribable cardinals is a proper class. This gave immediately a proper class of e.g. weakly compact cardinals. For an analysis and discussion of the situation we refer to Tait (Tait, 2005, Ch. 6).

Thus we cannot expect a proof of $V\alpha^2 \to V_c\alpha^2$, at least without additional axioms. On the other hand, the assumption (3.3) formulated in a reasonable class theory (such as NGB) seems plausible. By a result of Scott (Scott, 1961), it is true in the above $\langle M_\kappa, \mathscr{P}(M_\kappa)\rangle$, assuming that κ is not only weakly compact, but even measurable. In fact, it suffices to assume that κ is $\Pi^1_{<\omega}$-indescribable, hence (3.3) is consistent with $V = L$, assuming the consistency of a $\Pi^1_{<\omega}$-indescribable cardinal.

What about (3.3) for sentences α in other extensions of first order logic than second order logic? For first order logic this is an immediate consequence of the Levy Reflection Principle. For extended logics of the form $L(Q_\alpha)$ we can use translation to first order set theory and get the analogue of (3.3) as for first order logic. The same is true for $L(Q_\alpha^{MM,n})$, $L(Q_\kappa^{cf})$, and the extension of first order logic by the Härtig-quantifier

---

[18] We are indebted to A. Blass for pointing this out.

[19] In our model theoretic context second order parameters would correspond to adding generalised quantifiers to second order logic.

$Ixy\varphi(x)\psi(y)$, meaning: the cardinality of the set of elements x satisfying $\varphi(x)$ is the same as the cardinality of the set of elements y satisfying $\psi(y)$. For these powerful logics, unlike for second order logic, the analogue of the small part of the squeezing argument represented by (3.3) is simply provable in ZFC. The situation with stationary logic is more complicated. We leave the status of (3.3) open, if α^2 is taken to be a formula of so-called stationary logic rather than second order logic. The Open Question is, whether it is provable in ZFC or not.

Attempts to formulate higher order reflection with higher order parameters leading to larger large cardinals than $\kappa(\omega)$ have failed (see Koellner, 2009). However, a different approach, due to P. Welch, to a very strong reflection principle with second order parameters, called the *Global Reflection Principle*, gives a proper class of Woodin cardinals (Welch & Horsten, 2016; Welch, 2012).

6 Löwenheim-Skolem Theorems

Kreisel asks for a convincing proof of $\forall \alpha^2 (V\alpha^2 \leftrightarrow \text{Val}\, \alpha^2)$, on its face impossible as we saw. Short of such a proof, Kreisel then asks a more specific question, which *can* be answered. Stating the Löwenheim-Skolem Theorem for first order logic in the form $\forall \alpha \forall \sigma > \omega (V^{\omega+1} \alpha^1 \leftrightarrow V^\sigma \alpha^1)$, what is the analogue to ω for second order formulae?[20]

First we recall some definitions. Given a logic $\mathscr{L}$, we say that $\mathscr{L}$ has Löwenheim-Skolem number κ if κ is the least cardinal such that for all vocabularies τ such that the cardinality of τ is $\leq \kappa$, if a sentence φ in the vocabulary τ of the logic has a model $\mathfrak{M}$, then it has a model $\mathfrak{N}$ of size $\leq \kappa$. In case $\mathfrak{N}$ can be taken to be a submodel of $\mathfrak{M}$ then κ is called the Löwenheim-Skolem-Tarski (*LST*) number of the logic.

Let $\mathscr{L}^2$ denote second order logic. We can now state Magidor's result (Magidor, 1971), which answers Kreisel's question: κ is the the least supercompact cardinal if and only if $\kappa = LST(\mathscr{L}^2)$.

In fact there is now a whole range of logics calibrated by large cardinals, in the sense that the assumption of the cardinal is equivalent to or implies a Löwenheim-Skolem-Tarski theorem for the logic. For the cases already mentioned the results are as follows: for cofinality logic, corresponding to the generalised quantifier $Q_\kappa^{cf} xy\varphi(x,y)$, the *LST* number is $\aleph_1$.[21] For stationary logic, corresponding to the quantifier $\text{aa} s\varphi(s)$, the *LST* number is consistently $\aleph_1$, assuming the consistency of a supercompact cardinal,[22] but the *LST* number of stationary logic can also be the first supercompact

---

[20] In Kreisel's notation $V^\sigma \alpha^1$ denotes the assertion "α^1 is true is the cumulative hierarchy up to σ".
[21] (Shelah, 1975).
[22] (See Ben-David, 1978).

cardinal.[23]

Finally, the interesting case of the Härtig quantifier: It is now known that if the *LST* number $LST(I)$ of this logic exists, then there is a weakly inaccessible cardinal and $LST(I)$ is at least the least weakly inaccessible cardinal. It is consistent relative to the consistency of a supercompact cardinal that $LST(I)$ is the first weakly inaccessible, and also consistent that it is the first supercompact.[24]

A general approach to strong logics and the reflection principles they give rise to is presented in J. Bagaria et al. (Bagaria & Väänänen, 2016), where a close connection is established between *LST* numbers of strong logics and so-called *structural reflection principles* in set theory.

Just as in the completeness theorems, and the ensuing squeezing arguments, obtaining Löwenheim-Skolem type theorems may require principles that go beyond Weak König's Lemma (WKL), sufficient in the case of first order logic.

7 Squeezing very simple concepts

Consider the concept W of finite words in a given vocabulary X. Intuitively we construct a word by placing letters from X one after another a finite number of times. What does this mean? We can use a squeezing argument to shed light on this question. As an analogue of derivability consider the concept D of starting from the empty word and then adding one letter from X at a time to the end of any word we already have. As an analogue of set theoretic validity we take the concept C of being a member of every closed set, where a set A is called *closed* if the empty word is in A, all one-letter words are in A, and the concatenation ww' of any two words w, w' of A are in A. Clearly,

$$D \subseteq W \subseteq C.$$

The first "$\subseteq$" is intuitively obvious because adding one letter to the end of a word certainly yields another word. The second is less obvious but one can run an informal induction on the length of the word to see that if A is closed, then the word is in A. It is a mathematical fact that

$$C \subseteq D,$$

because D is one of the sets that C is the intersection of. Hence

$$D = W = C,$$

---

[23] Magidor, unpublished.
[24] (Magidor & Väänänen, 2011).

and the informal concept of a finite word is squeezed between two (extensionally) identical exact concepts. Although everything in this squeezing argument in on a very elementary level, it is noteworthy that strictly speaking the inclusion $C \subseteq D$ is based on an impredicative argument.

Similarly, we may consider the concept F of a finite set. Intuitively we call a set *finite* if we can use some natural number to list the elements of the set. On the other hand, natural numbers can be identified with finite ordinals. Thus there is a certain amount of circularity in the concept of finiteness. So what does "finite" exactly mean? Let us take as D the concept of starting from the empty set and then adding one element at a time to get more sets. Let C be the concept of belonging to every *ideal class* i.e. to every class which contains the empty set, all singletons and is closed under unions of any two elements of the class. Clearly,

$$D \subseteq W \subseteq C.$$

The first "$\subseteq$" is again intuitively obvious because adding one element to a finite set certainly preserves the set finite. For the second one can use informal induction on the finite size of the set to see that if A is an ideal class, then the set is in A. It is a mathematical fact that

$$C \subseteq D,$$

because D is one of the classes that C is the intersection of. Hence

$$D = W = C,$$

and the informal concept of a finite set is squeezed between two (extensionally) identical exact concepts of class theory.

8 Conclusion

Do squeezing arguments capture the mathematician's informal discourse, even as it strays beyond first order talk, quantifying over relations and functions, and making implicit use of infinitary rules? This is difficult enough to argue for in the first order case. Nevertheless, we hope to have reinforced Kreisel's original argument in (Kreisel, 1967) that squeezing arguments have a general role in the conceptual analysis of informal mathematical concepts. Moreover, we have pointed out and given evidence to the claim that the circumstance that the two sides of the squeeze (extensionally) agree is based in general on a non-trivial mathematical fact.

In particular, we hope to have shown for strong logics that if we refashion the relevant informal concepts appropriately (here validity), we can, so to say, filtrate the informal discourse involving those concepts through a hierarchy of set-theoretic commitments ranging from Weak König's Lemma (WKL) up to $\diamond$.

We also saw that various strategies present themselves in the second order case, that go beyond what Kreisel suggests, if intuitive second order validity is understood in the right way.

Acknowledgments. This paper was written while the authors were participants in the Intensive Research Program - IRP Large Cardinals and Strong Logics at the Centre de Recerca Matemàtica during the fall semester of 2016. The authors thank the CRM for their support. Jouko Väänänen thanks the Simons Foundation for their support.

Bibliography

Abraham, U., & Shelah, S. (1993). A Δ_2^2 well-order of the reals and incompactness of $L(Q^{MM})$. *Ann. Pure Appl. Logic*, *59*(1), 1–32.

Bagaria, J., & Väänänen, J. (2016). On the symbiosis between model-theoretic and set-theoretic properties of large cardinals. *J. Symb. Log. 81*(2), 584–604.

Barwisw, J., Kaufmann, M., & Makkai, M. (1978). Stationary logic. *Ann. Math. Logic*, *13*(2), 171–224.

Ben-David, S. (1978). On Shelah's compactness of cardinals. *Israel J. Math*, *31*(1), 34–56.

Bernays, P. (1961). Zur Frage der Unendlichkeitsschemata in der axiomatischen Mengenlehre. In *Essays on the foundations of mathematics* (pp. 3–49). Hebrew Univ., Jerusalem: Magnes Press.

Chang, C. C. (1965). A note on the two cardinal problem. *Proc. Amer. Math. Soc. 16*, 1148–1155.

Dean, W. (2016). Squeezing feasability. In *Pursuit of the universal (lecture notes in computer science)* (pp. 78–88). Springer.

Etchemendy, J. (1990). *The concept of logical consequence*. Cambridge, MA: Harvard University Press.

Field, H. (2008). *Saving truth from paradox*. Oxford: Oxford University Press.

Fuhrken, G. (1965). Languages with added quantifier "there exist at least $\aleph_\alpha$". In *Theory of models (Proc. 1963 Internat. Sympos. Berkeley)* (pp. 121–131). Amsterdam: North-Holland.

Glanzberg, M. (2015). Logical consequence and natural language. In *Foundations of logical consequence* (pp. 71–120). Oxford: Oxford University Press.

Gödel, K. (1986). *Collected works. Vol. I*. New York: The Clarendon Press, Oxford University Press.

Halbach, V. (2016). The substitutional analysis of logical consequence. Retrieved from https://mdetlefsen.nd.edu/assets/201030/halbach.consequence.pdf

Henkin, L. (1950). Completeness in the theory of types. *J. Symbolic Logic*, *15*, 81–91.

Hilbert, D., & Ackermann, W. (1928). *Gründzuge der theoretischen Logik*. Berlin: J. Springer (Die Grundlehren der mathematischen Wissenschaften Bd. 27). VIII, 120 S.

Karp, C. R. (1964). *Languages with expressions of infinite length*. Amsterdam: North-Holland Publishing Co.

Kennedy, J. (2013). On formalism freeness: Implementing Gödel's 1946 Princeton bicentennial lecture. *Bull. Symbolic Logic*.

Koellner, P. (2009). On reflection principles. *Ann. Pure Appl. Logic*, *157*(2-3), 206–219.

Kreisel, G. (1967). Informal rigour and completeness proofs. In I. Lakatos (Ed.), *Proceedings of the International Colloquium in the Philosophy of Science, London, 1965, Vol. 1.* (pp. 138–157). Amsterdam: North-Holland Publishing Co.

Lévy. (1960). Axiom schemata of strong infinity in axiomatic set theory. *Pacific J. Math*, *10*, 223–238.

Magidor, M. (1971). On the role of supercompact and extendible cardinals in logic. *Israel J. Math, 10*, 147–157.

Magidor, M., & Malitz, J. (1977). Compact extensions of $L(Q)$. Ia. *Ann. Math. Logic, 11*(2), 217–261.

Magidor, M., & Väänänen, J. (2011). On Löwenheim-Skolem-Tarski numbers for extensions of first order logic. *J. Math. Log. 11*(1), 87–113.

Mancosu, P. (2005). Harvard 1940–1941: Tarski, Carnap and Quine on a finitistic language of mathematics for science. *Hist. Philos. Logic, 26*(4), 327–357.

Orey, S. (1956). On ω-consistency and related properties. *J. Symb. Logic, 21*, 246–252.

Royden, H. L. (1988). *Real analysis* (3$^{\rm rd}$ ed.). New York: Macmillan Publishing Company.

Rudin, W. (1976). *Principles of mathematical analysis* (3$^{\rm rd}$ ed.). International Series in Pure and Applied Mathematics. New York-Auckland-Düsseldorf: McGraw-Hill Book Co.

Scott, D. (1961). Measurable cardinals and constructible sets. *Bull. Acad. Polon. Sci. Sér. Sci. Math. Astronom. Phys. 9*, 521–524.

Shelah, S. (1975). Generalized quantifiers and compact logic. *Trans. Amer. Math. Soc. 204*, 342–364.

Smith, P. (2011). Squeezing arguments. *Analysis (Oxford), 71*(1), 23–30.

Strawson, P. F. (1950). On referring. *Mind, 9*, 320–344.

Tait, W. (2005). The provenance of pure reason. In *Logic and computation in philosophy*. Essays in the philosophy of mathematics and its history. New York: Oxford University Press.

Tharp, L. H. (1967). On a set theory of Bernays. *J. Symbolic Logic, 3*, 319–321.

Väänänen, J. (2001). Second-order logic and foundations of mathematics. *Bull. Symbolic Logic, 7*(4), 504–520.

Väänänen, J. (2012). Second order logic or set theory? *Bull. Symbolic Logic, 18*(1), 91–121.

Vaught, R. L. (1964). The completeness of logic with the added quantifier "there are uncountably many". *Fund. Math. 54*, 303–304.

Welch, P. (2012). Global reflection principles. Isaac Newton Institute preprint series NI12051-SAS.

Welch, P., & Horsten, L. (2016). Reflecting on absolute infinity. *Journal of Philosophy, 113*(2), 89–111.

Author biography. Juliette Kennedy received her Ph.D. from the Graduate School and University Center of CUNY in 1996 with a thesis in mathematical logic. She is now University Lecturer (eq. Associate Professor) at the Department of Mathematics and Statistics at the University of Helsinki. She works in set-theoretic model theory and in the history and philosophy of mathematics.

Jouko Väänänen is a Professor of Mathematics of the University of Helsinki. He is also a Professor Emeritus of Mathematical Logic and Foundations of Mathematics of the University of Amsterdam. He works in several fields of logic, such as set theory, model theory, computer science logic, and foundations of mathematics. His current work is best described by his two books "Dependence logic", CUP 2007, and "Models and Games", CUP 2011.

4 Global reflection principles

PHILIP WELCH *

Abstract. We consider a structural *reflection principle* that seeks to go beyond the traditional reflection principles of Mahlo, Levy, Bernays, *et al.* The latter are all firmly *intra-constructible* that is they produce justifications for only small large cardinals consistent with Gödel's constructible universe L. Our *Global Reflection Principles* by contrast ensure that the universe of set theory has unboundedly many measurable Woodin cardinals. This is a hypothesis which (with variants) is used by current set theorists to establish many absoluteness (and consistency) results concerning the universe V. It is argued that the Principles are justified by appealing to a Cantorian Absolute perspective, together with a mereological view of its classes as *parts*: thus we may distinguish the mathematico-set theoretic part of the realm, the sets, from the classes. As there are no 'super-parts' a hierarchy over and above the universe V does not threaten.

Keywords: Set Theory, Reflection Principles, Large Cardinals, Strong Axioms of Infinity.

> *Das Absolute kann nur anerkannt, aber nie erkannt, auch nicht annähernd erkannt werden.*
>
> Cantor, *Über unendliche lineare Punktmannigfaltigkeiten.*
> Math. Ann., 1883

> *To say that the universe of all sets is an unfinished totality does not mean objective undeterminateness, but merely a subjective inability to finish it.*
>
> Gödel, in (Wang: "A Logical Journey: From Gödel to Philosophy")

---

*University of Bristol

1 Reflection principles in set theory: Introduction

Historically *reflection principles* in set theory have been associated with attempts to say that no one notion, idea, or statement can capture our whole view of the universe $V = \bigcup_{\alpha \in On} V_\alpha$; the motivating idea seemingly that the universe $(V, \in)$ is in some sense ineffable: all attempts to pin it down will fall short. In particular this will be so, once we have formalised our notions, for any formal linguistic expression.

Such reflection principles have usually been formulated in some language (first or higher order) as sentences φ (when interpreted in the appropriate way over V) that hold in $\langle V, \in, \ldots \rangle$, hold in some $\langle V_\beta, \in, \ldots \rangle$ - *sentential* reflection. Here we are initially allowing set objects to be substituted for variables of a purely first order formula to construct a sentence. For formulae of higher orders than one it then becomes a question as to how to reflect second order parameters, and indeed is a genuine problem at third and higher orders. In Section 3 we summarise the history of such principles. As is well known, such principles had been used to a validate the large cardinals known at that time, or at least the plausibility thereof. From a modern perspective such principles are disappointing as they never reveal any arguments for strong axioms that suggest anything beyond those consistent with Gödel's constructible universe L.

However, since Scott's Theorem (itself surprisingly late) that the existence of a measurable cardinal implies that V is not L, and the wide use of axioms implying the existence of much stronger cardinals than measurable, reflection principles have tended to be ignored, (at least by set theorists) as being part of the historical development of the subject and not much more.

This is not entirely true however. Perhaps the most well known attempt was that of Reinhardt (1974a) to give justifications of some ideas (that later were adopted to imply very strong axioms when once re-expressed in an appropriate "set form") using a notion of 'projecting V into idealised realms'.

In Section 4 We suggest a *Global Reflection Principle* to overcome the limitations that these principles are all *intra-constructible*. The concept of a *Woodin cardinal* has become central to the development of set theory in the last few decades. This is not just for the development of determinacy results, but more germane here, for Woodin's remarkable results on the absoluteness of various truths resisting any attempt by Cohen style forcing arguments to falsify them. Many of such theorems use as background assumptions the existence of a proper class of such Woodin cardinals, or even measurable Woodin cardinals. The Global Reflection Principle delivers just such hypotheses.

Thus: we must rise to the challenge to justify a set-theoretic reflection principle that will ensure the existence of large cardinals (or strong axioms of infinity) that are sufficient to deliver the hypotheses needed for modern set theoretical principles.

1.1 Argument

As we shall see that purely syntactic, or linguistic means only deliver intraconstructible principles, we need to go beyond them. We shall need to express the ineffability of V not purely in considerations about $(V, \in)$ alone, but in the general idea of a *set structure* which we view in any case, as extending Martin's concept of "set structure" (Martin, 2012). We thus outline our viewpoint for this section in terms of which we wish to argue for new principles. In summary:

> (I) (*cf.* Martin, 2012) We have no need to 'perceive' in a Gödelian sense, or otherwise 'locate' any mathematical objects in order to understand the concepts involved, and to communicate that understanding.
> Thus: *instantiation* of our concepts is not necessary, but what is needed is uniqueness of the concept up to isomorphism.

> (II) We then seek to extend arguments of Martin's along the above lines which he advocated for the 'concept of ω-sequence' and the 'concept of set of x's' to a 'concept of set with absolute infinities'.

> (III) We then formulate within this framework *reflection principles* that establish large cardinals beyond those consistent with Gödel's L.

2 Martin on set structures

2.1 The reality of concepts: Sets and classes

Martin (2005) has questioned the level of realism that Gödel, although on occasion expressing this with talk of 'perception of mathematical objects' *etc.*, needs in order to make some of his arguments work, *eg.* , of the analyticity of mathematical truths.

Martin identifies two kinds of sense to 'concept of set'.

> (i) That more nearly akin to pure platonism - the whatever it is that falls under the extension of 'concept of set (of x's)' - that is sets (of x's), and
> (ii) a more general sense of 'concept of set' under which falls concepts of sets, or additionally that of 'set-structures'.

It is this final 'additional version' that I shall want to mostly take here, and go beyond Martin.

His primary point is perhaps plainly put: the example of Axiom of Extensionality: it does not say what a set is, on this it is silent, it only prescribes what it means for any two sets to be equal. The concept of set alone does not determine what it is for an object to be a set (as he states in Martin, 2012). The notion is objective: we understand it, talk about it, as no doubt they do on some other planet with discretely individuated intelligences. (It is not for nothing that we engrave on steel plates pictures of Pythagoras' theorem and place them on the moon, or send them out on Voyager 23.)

However first:

> A concept of set expressed by axioms such as comprehension axioms cannot put any constraint on which objects count as sets and which do not. Such axioms put constraints on the isomorphism type of a set theoretic structure ... a concept of set could count as concept of set in my [indirect *(ii)*] sense even if it determined completely what objects count as sets and what counts as the membership relation. A concept of this sort would have at most one instance: it would allow at most one structure to count as a set-theoretic universe ...

What is ultimately at play here is the point Martin wishes to make that *instantiation* of a concept for mathematics (or set theory) is not needed: what we require is *uniqueness* (up to isomorphism) in order to make sense and understand concepts. At the end of the quotation he says that it *might* have only one structure instantiating it, but this need not be the case. It might not have any instantiations. (Who seriously believes there is a well ordering in our own physical manifold that there is well ordering order type ω_{23}?) He reads Gödel as primarily not needing instantiation in many crucial places: for example, he notes that neither it nor perception of objects plays any significant role in Gödel's justifications of **strong axioms of infinity.** In short instantiation is not needed either in mathematics or in set theory. Hence: this is closer to an eliminative structuralist viewpoint.

Basic concepts and their properties

He (Martin, 2012) discusses in the following terms two basic concepts: the *concept of an ω-sequence* and the *concept of set*. He identifies three properties a basic concept may have:

1. First order completeness: the concept determines truth values for all first order statements.
2. Full determinateness: the concept fully determines what any instantiation would be like.
3. Categoricity.

An analysis of the notion of $\mathbb{N}$, as to whether the notion is *first-order complete, categorical, or perhaps fully determinate*. He claims that instantiation of the natural numbers is not needed for number theory: what we require is *categoricity*, although mathematicians probably think that $\mathbb{N}$ is in fact fully determinate. He identifies *In-*

formal Axioms schemes of PA with which one can, informally, prove the categoricity of $\mathbb{N}$.

Turning to set theory, we have only glimmerings of what goes on when considering subsets of $V_{\omega+1}$: is the Continuum Hypothesis true? Is every definable subset of the plane definably uniformisable? So we are hopelessly far from first order completeness. (However, when considering subsets of V_ω we are in a better position. We now know that adding the assumption of Projective Determinacy to analysis, or to the theory of hereditarily countable sets give us as complete a picture of HC as PA does for $V_\omega = \mathrm{HF}$. (See the discussion in Woodin, 2001.))

Martin identifies four components of the modern iterative concept of set.

(1) the concept of natural number
(2) concept of 'set of x's'
(3) concept of transfinite iteration
(4) concept of absolute infinity.

Corresponding to the informal axioms for Peano Arithmetic, we may then ask: which informal axioms are implied by the concept of set?

(i) *If a and b have the same members, then $a = b$. (Extensionality)*
(ii) *For any property P, there is a set whose members are those x's that have P (Informal Comprehension).*

The Comprehension scheme is called informal since "property" is not specified in generality. However any worries can be dispelled since it will be clear that the few instances we shall use are clear cases of properties.

Martin seeks to further soothe any worries that we need to specify what objects sets are in order to 'fully understand' the concept. He will ignore whatever structural constraints one may put on what sets actually are, other than the structural constraints of (i) and (ii).

Theorem 2. *(Essentially Zermelo) Informal Axioms (i) and (ii) are categorical: if $(\mathfrak{V}_1, \in_1)$, $(\mathfrak{V}_2, \in_2)$ are two structures satisfying (1) and (2) with the same x's, then with each set of x's $b \in_1 \mathfrak{V}_1$ we associate a set of x's $\pi(b) \in_2 \mathfrak{V}_2$.*

Proof: Let P be the property of being an x such that $x \in_1 b$. By the Informal Comprehension Scheme there is a $c \in_2 \mathfrak{V}_2$ such that

$$\forall x [x \in_2 c \leftrightarrow P(x)]$$

Q.E.D.

This is the basis of Zermelo's proof that any two models of ZFC (without *urelemente*) of the same ordinal height are isomorphic. Similarly for any two $\mathfrak{V}_1 = (V_1, \in_1)$, $\mathfrak{V}_2 = (V_2, \in_2)$ obtained by iterating the V_α function throughout all the absolute infinity of ordinals, we have an isomorphism

$$\pi : (V_1, \in_1) \cong (V_2, \in_2).$$

Note for later that we see that $\pi \restriction \text{On}^{\mathfrak{V}_1} : \text{On}^{\mathfrak{V}_1} \cong \text{On}^{\mathfrak{V}_2}$.

2.2 From set structures to set structures with classes

We shall want to extend the application of Martin's analysis to a broader territory - no doubt broader than he might like. The components of the modern iterative concept of set (1)-(4) above lead, in our view inexorably, to classes associated to absolute infinities in the sense that Cantor recognised and spoke of. We continue our informal development of the theory of sets, just as any mathematician would develop her theory of some mathematical construct. The absolute infinity of *On* we regard as just that: it is determinate what constitutes an ordinal (*à la* von Neumann: a transitive set well ordered by $\in$), and *On* is the class of such, and is an absolute infinity as Cantor realised, and Burali-Forti showed (albeit half unknowingly it seems, see van Heijenoort, 1967). The notion is not regarded as indefinitely extensible: if we posit beyond *ZFC* stronger axioms of infinity, such as measurable cardinals, we enrich our view of *V*, but we do not somehow 'add more ordinals'.

V itself is, then, also one of many 'absolutely infinite' or 'inconsistent multiplicities' $V, \text{On}, \text{Card}, \ldots$, the class of all singletons, ... that Cantor recognised. Let us imagine that $\mathscr{C}$ is the collection of all such. The nature of $\mathscr{C}$ is admittedly somewhat ineffable, but we can agree with him that as absolute and final knowledge cannot be obtained by *V* alone, neither can we find such for $\langle V, \in, \mathscr{C} \rangle$. Just as Martin used the idea of 'property' in Informal Comprehension, without specifying exactly what that constituted, as he would only need to use this in a small number of uncontroversial cases, so we shall think about $\mathscr{C}$, and shall see that we only need to consider a small part of $\mathscr{C}$. Later we shall consider formalizations of this.

We may, if we wish, tell a *mereological* story about the whole mathematical universe: we think of sets as the sole representatives of *mathematical* objects. The class of sets thus forms the realm of, or the arena for, mathematical discourse. Thus strictly, the class of sets is not a mathematical object. However absolute infinities are *parts* of the whole realm of that discourse. (One may wish to add to this sets as well as parts of the realm, but we shall just identify such parts with the sets themselves.) We take $\mathscr{C}$ to be the collection of all absolute infinities (or non-set parts, or proper classes if you will). $\mathscr{C}$ then, contains the possible parts of *V*.

The usual danger is that we may risk having to build super-classes, and more such above those; in other words a hierarchy that looks too similar to the V_α hierarchy. However this is to extrapolate the use of *mathematical* methods of set formation - principally power set and recursion along the ordinals. The power set axiom is a mathematical operation and not a warrant for a power class operation, even if we could form a definite domain of quantification. In our pre-formalised thinking, we bite the Cantorian bullet and admit two types of objects to our notion of set-structure: *V* together with its absolute infinities.

In Welch and Horsten, 2016 we discuss further the status of such classes. In particular the possible use of plural quantification as a means of establishing the 'existence' of such classes. However we reject that view, principally for reasons to come: we need actual entities of some kind in $\mathscr{C}$ not just something that is ultimately a linguistic construct: "some sets such that ..."

2.3 Isomorphism again

Theorem 3. *If we have two structures of sets $\mathfrak{V}_i = (V_i, \in_i)$ ($i = 1, 2$) satisfying Martin's (i) and (ii) above, let $\pi : (V_1, \in_1) \to (V_2, \in_2)$ as above be an isomorphism. Now assume we have collections of classes $\mathscr{C}_i$. Then π extends to an isomorphism:*

$$\pi : (V_1, \in_1, \mathscr{C}_1) \cong (V_2, \in_2, \mathscr{C}_2).$$

Proof: Let $(V_1)_\alpha$ be the set of $\mathfrak{V}_1$-sets of rank α in the sense of $\mathfrak{V}_1$ etc. It suffices to show for every class $X \subseteq V_1$ (thus $X \in \mathscr{C}_1$) there is a $Y \subseteq V_2$ with $\pi(X \cap (V_1)_\alpha) = Y \cap (V_2)_\beta$ where $\alpha \in_1 \text{On}^{\mathfrak{V}_1}$ and $\beta \in_2 \text{On}^{\mathfrak{V}_2}$ with $\pi(\alpha) = \beta$, and conversely - since then we may define $\pi(X) = \bigcup_{\alpha \in_1 \text{On}^{\mathfrak{V}_1}} \pi(X \cap (V_1)_\alpha)$.

Q.E.D.

Here we are taking the 'informal union' of the sets of the form $\pi(X \cap (V_1)_\alpha)$. However we are not declaring this union to be a 'set' or any such, so no formal axiom is needed. This is unproblematic as it is simply taking a union (or fusion if you will) of the classes $\pi(X \cap (V_1)_\alpha)$ and thus is a class of V_2.

3 Reflection

We now very briefly survey some of the history of reflection principles, before discussing some more recent work.

3.1 Historical reflection

First order reflection is unproblematic: any one syntactic instant of it for a formula in $\mathscr{L}_{\dot\in}$, the language of first order set theory alone, is provable in ZF:

(R_0) : *For any $\varphi(v_0, \ldots, v_n) \in \mathscr{L}_{\dot\in}$*

$$ZF \vdash \forall \alpha \exists \beta > \alpha \forall \vec{x} \in V_\beta[\varphi(\vec{x}) \leftrightarrow \varphi(\vec{x})^{V_\beta}].$$

Indeed by formalising a Σ_n-Satisfaction predicate we have (still in ZF, here "c.u.b." abbreviates "closed and unbounded"):

For each n there is a term c_n for a c.u.b. class of ordinals, so that so that for any $\varphi \in \text{Fml}_{\Sigma_n}$:

$$\text{ZF} \vdash \forall \beta \in c_n \forall \vec{x} \in V_\beta [\varphi(\vec{x}) \leftrightarrow \varphi(\vec{x})^{V_\beta}].$$

Informally we write for each n: $\forall \beta \in c_n : (V_\beta, \in) \prec_{\Sigma_n} (V, \in)$. More formally we can define first order satisfaction for Σ_n formulae as a two place predicate $\text{Sat}_n(\ulcorner \varphi \urcorner, x)$ and have that for $\beta \in c_{n+1}$ that

$$\forall \ulcorner \varphi \urcorner \in \Sigma_n, \forall x \in V_\beta (V_\beta, \in) \models \text{Sat}_n(\ulcorner \varphi \urcorner, x) \leftrightarrow (V, \in) \models \text{Sat}_n(\ulcorner \varphi \urcorner, x)).$$

We thus have reflection of $(V, \in, \text{Sat}_n)$ to $(V_\beta, \in, \text{Sat}_n)$ for such β. As is well known, we can not do this for all n simultaneously.

Reaching back to ideas of P. Mahlo, using *normal functions* and their fixed points we have:

(F) *Any (definable) normal function $F : \text{On} \to \text{On}$ has a regular fixed point.*

Such fixed points are inaccessible cardinals (and indeed strong limits of limits of such). A reflection principle is related to this. We let Inacc denote the class of strongly inaccessible cardinals.

(R_1) : For any $\varphi(v_0, \ldots, v_n) \in \mathscr{L}_{\dot\in}$

$$\forall \alpha \exists \beta > \alpha (\beta \in \text{Inacc} \wedge \forall \vec{x} \in V_\beta [\varphi(\vec{x}) \leftrightarrow \varphi(\vec{x})^{V_\beta}]).$$

Then it easy to see that $(F) \iff (R_1)$. One then may ask for (F) itself to reflect down from normal functions definable over V to *all* normal functions $f : \kappa \longrightarrow \kappa$ for $\kappa \in \text{Inacc}$. Then, if $(V_\kappa, \in)$ witnesses (F), in this extended sense for all such functions f, κ is called a *Mahlo* cardinal. Being Mahlo is a second order property; Bernays investigated such in more generality, and one may then end up with *indescribability properties* that in general a Π^1_m sentence about second order parameters may reflect down to V_κ. Here φ is a first order statement about the illustrated upper case second order variables F_i and parameter P, and possible further parameters $\vec{x} \in V_\kappa$.

$$\forall F_1 \exists F_2 \cdots QF_m(\varphi(\vec{F}, \vec{x}, P)) \leftrightarrow (\forall F_1 \exists F_2 \cdots QF_m(\varphi(\vec{F}, \vec{x}, P \cap V_\kappa)))^{V_\kappa}.$$

One observation at this point is that first order satisfaction is second order, indeed Π^1_1-definable, consequently if we have Π^1_1-reflection in the above sense, we may reflect $(V, \in, \text{Sat})$ down to arbitrarily large $(V_\beta, \in, \text{Sat})$ (where Sat codes all of first order satisfaction simultaneously). Reflection of first order satisfaction is not strong.

The localisation of this idea to a cardinal λ in place of V itself (just as a Mahlo cardinal is the localised version of a concept considering class functions over V) then yields the notion of a Π^1_m-indescribable cardinal. We thus arrive at indescribable cardinals, indeed not just second order, but of any n'th order: Π^{n-1}_m-*indescribable cardinals*.

This can be extended to transfinite orders. There are difficulties with 3rd order parameters as Reinhardt observed. This again is tracing the lines of the following thought of Gödel:

> *The Universe of sets cannot be uniquely characterized (i.e. distinguished from all of its initial segments) by any internal structural property of the membership relation in it, which is expressible in any logic of finite or transfinite type, including infinitary logics of any cardinal number.*
> (Wang - 1996)

In a different direction Reinhardt studied Ackermann's set theory, Ackermann, 1956 and his thesis Reinhardt, 1967 (published in Reinhardt, 1970) established its conservativity over ZF.

From Ackermann, 1956 (p.337): "We must require from already defined sets that they are determined and well-differentiated, thus the conditions of a collection [to be a set] only turn on that it must be sufficiently sharply delimited what belongs to a collection and what does not belong to it. However now the concept of set is thoroughly open." For Ackermann a well-defined and differentiated condition for an object to be a set cannot include in its definition assertions of the form "such and such a set is in V" since the latter is not a sufficiently sharply determined concept. Thus the definition of a set x cannot include clauses of the form '$y \in V$' although the latter is a *bona fide* expression of the formal language.

Reinhardt interpreted this as realising in some form ideas about the *projection* of the universe V into some virtual realm. The language for this theory includes a symbol V to denote the 'constructed' sets, but objects such as $\{V\}, \{\{V\}\}$ *etc.* are permissible (and such are classes for Ackermann). The Cantorian ordinals form Ω and V is V_Ω, but we imagine further set-like objects. We posit ZF for V and the axiom $V = V_\Omega$. The following (S2) schema asserts that any first order sentence θ true in V of sets in V, is true in the 'projected universe' and conversely.

$$(\forall x, y \in V)(\theta^V(x,y) \leftrightarrow \theta(x,y)) \qquad (S2)$$

This system was called "ZA". The extension of a formula θ in V is thus $\{x \in V \mid \theta(x)\}$ and is merely the restriction of that in the projected or virtual universe where it is of course $\{x \mid \theta(x)\}$.

Reinhardt (1974b) also studied and formalised a view of Shoenfield (1967) that thought of V as the actually existing sets, but that there is some 'formal extension' of the ordinals, and a class of 'imaginable sets' U. This came about by studying a thesis of Shoenfield that had shades of Ackermann set theory (Reinhardt, 1974b p.6):

If C is a collection of stages [ranks], and if we can imagine *a situation in which all stages in P have been completed, then there is to be a stage s after all the stages in C.*

After some further considerations Reinhardt formalises this as having "imaginable sets" and at the same time real existing sets, with the latter elements of V. (Ackermann would have called the imagined sets 'classes' and Shoenfield's V is Ackermann's class of all sets.) V cannot be a real existing set, and the property P such that $x \in V \leftrightarrow P(x)$ cannot be said to exist either. However as a consequence one has the slogan:

If $X \subseteq Y$ and Y is imaginable, and suppose X is definable using only $\vec{x} \in V$ as parameters then $X \in V$.

Reinhardt in Reinhardt, 1974a argued for a formally projectible realm V_λ for some '$\lambda >>$ On' with a corresponding $V_{\lambda'}$ related to V_λ via an embedding j:

$$j : V_\lambda \longrightarrow V_{\lambda'} \; ; \; j(\text{On}) = \text{On}', \text{crit}(j) = \text{On}.$$

We finish this subsection with just one more quote of Gödel that is often referred to when discussing reflection in the set/class theoretic setting, and can be seen as motivating reflection properties in general.

> *All the principles for setting up the axioms of set theory should be reducible to Ackermann's principle: The Absolute is unknowable. The strength of this principle increases as we get stronger and stronger systems of set theory. The other principles are only heuristic principles. Hence, the central principle is the reflection principle, which presumably will be understood better as our experience increases. Meanwhile, it helps to separate out more specific principles which either give some additional information or are not yet seen clearly to be derivable from the reflection principle as we understand it now."* (Wang - 1996).

3.2 Strengthening reflection principles

Tait considered in Tait, 2005 some strengthened reflection principles that were of order higher than two. As already alluded, Reinhardt had observed that third order with third order parameters would be false. Tait considered how to use the higher order languages with some syntactical restrictions. Koellner showed (Koellner, 2009) that those that were consistent were all weaker than $\kappa(\omega)$ (an ω-*Erdős* cardinal). The latter cardinal is of only modest size, and is also consistent with V being L.

We are thus left with the thesis that all *syntactically based* Reflection Principles to date are all consistent with a view of the universe as being L the constructible one: they are *intra-constructible*.

The moral that may be adduced is that if we need stronger reflection principles: those that generalise Montague-Levy in terms of enlarging our set/class-theoretic language are not up to the task of providing any justification for the large cardinals needed for modern set theory.

Why ask for stronger reflection principles?

Theorem 4 (Woodin). *Suppose there is a proper class of Woodin cardinals. Then $Th(L(\mathbb{R}))$ is immune to change by set forcing.*

This supposition is now ubiquitous in modern set theory. The above is just singled out as a dramatic instance of how large cardinals affect the universe V. One might have cited the equally dramatic results of Martin-Steel that infinitely many Woodin cardinals implies determinacy for projectively defined sets of reals. If there is a proper class of Woodin cardinals, then again Cohen's set forcing techniques cannot change that fact. Projective determinacy implies a host of results about the projective sets: their Lebesgue measurability, their having the *perfect subset property*, implying that every uncountable such set has size the full continuum; that projectively defined subsets of the plane $\mathbb{R} \times \mathbb{R}$ can be *uniformized* by a function with a projective graph; the non-existence of a Banach-Tarskian paradoxical decomposition of the sphere into projective pieces. The list can be extended, but the point is made: large cardinals at the level of Woodin cardinals *prove* such results to hold. These are not relative consistency proofs of one theory relative to another given by the forcing method, they are facts about the universe of mathematical discourse, V, which must hold if those requisite large cardinals exist.

In order to obtain such large cardinals, we need therefore a means of thinking about, or justifying, or obtaining *extra-constructible* reflection principles. We therefore define a *Global Reflection Principle* (GRP).

4 III Global Reflection Principles

Instead of 'formally projecting V' *à la Reinhardt* let us turn it around and generalise to obtain a strong reflection principle. Let us take a naive Cantorian (and non-Zermelian) stance: we bite the bullet of the necessity of the seeming existence of two types of objects: the *sets* of mathematical discourse and the *absolute infinities* (or *proper classes*) that Cantor, Burali-Forti, Russell and others early on saw the need for. We let $\mathscr{C}$ denote the informal collection of such absolute infinities, without worrying about a formal definition delineating these objects. We shall see that we shall only be concerned with principles that require $\mathscr{C}$ to contain a small number of such proper classes, where a 'small number' only means set-many.

We then consider the constellation $(V, \in, \mathscr{C})$ of the universe of sets V obtained in the usual mathematical fashion of iterating the power set operation as above, and writing membership $\in$ in the usual sense: $x \in y$ and $x, y \in Y$ for a Y in $\mathscr{C}$. As this indicates here, we shall be considering the usual first order language for set theory $\mathscr{L}$, but augmented by second order variables $X, Y, Z, \ldots, X_n, \ldots$ which appear atomically as statements of the "$x \in Y$", "$x \notin y$". Equality between classes X, Y is given extensionally: $X = Y \leftrightarrow \forall z(z \in X \leftrightarrow z \in Y)$ etc. However there are no second order quantifiers. (In any case, we have left the domain $\mathscr{C}$ of any such putative quantification vague.) But we shall allow *bounded quantifiers* as $\exists x \in Y(\cdots), \forall x \in Y(\cdots)$ into our Δ_0 classified formulae of our extended language $\mathscr{L}^+$.

Just as earlier reflection principles expressed the ineffability of V by asserting principles that reflect the whole realm to a small part of itself, we shall do the same, whilst reflecting at the same time some satisfaction. One may note that the truth or otherwise of $\varphi(x, X)^{(V, \in, \mathscr{C})}$ for an $X \in \mathscr{C}$ and $\Delta_0 \varphi$ is only dependent on a limited number of other classes in $\mathscr{C}$, depending on a few other class parameters derived from sub-formulae, and rudimentary combinations of such. The pathway is thus open to consider reflection from the whole universe of $(V, \in, \mathscr{C})$ with its classes to some smaller set-sized domain. We shall want to reflect ordinary first order expression of $\mathscr{L}$ about objects x which are members of the domain, but also some collection of, or possibly all, statements from $\mathscr{L}^+$ involving classes. Clearly 'class many' statements are not going to be reflectable.

The domain, the "V" of this small realm, should be set sized, and indeed we should speak of a small number of "classes" for it, as possible interpretations for the variables $X_i \ldots$

So motivated by earlier reflection principles, we choose the small realm to be a 'typical normal domain' V_κ, and we seek some form of satisfaction preserving reflection of $(V, \in, \mathscr{C})$ together with some collection of its classes, to some $(V_\kappa, \in, D)$ with $D \subseteq V_{\kappa+1}$, which we regard as a collection of classes D over V_κ. We may obtain such smaller structures by considering *set-sized substructures* of $(V, \in, \mathscr{C})$, $(V', \in, \mathscr{C}') \prec (V, \in, \mathscr{C})$ with V' equal to some V_κ and consider its isomorphic transitive image $(V', \in, \mathscr{C}') \cong (V_\kappa, \in, D)$ say. We shall want that $D = V_{\kappa+1}$ but first we discuss the kind of elementarity we are pursuing.

A class $X \subset V_\kappa$, with X from D, is the reflection of something satisfied in the realm $(V, \in, \mathscr{C})$ by one of the parts, or classes, of the whole realm, some $\tilde{X}$ say. What should $\tilde{X}$ be? If even truth of atomic statements of the form "$y \in X$" are to be faithfully reflected then it should be a part of the whole (V) that extends X, i.e. $\tilde{X} \cap V_\kappa$ should be the same as X. This then is no different from the reflection of second order parameters in the earlier principles. In order then to have simultaneous preservation of Δ_0 formulae of our language between $(V_\kappa, \in, D)$ and $(V, \in, \mathscr{C})$, we only need that $\mathscr{C}$ contain extensions $\tilde{X}$ for every $X \in D$. We are not yet requiring it (although we shall do shortly), but to have *full* or 'global' downwards reflection, we ask that $D = V_{\kappa+1}$. The implicit posit here is that there is some choice of some κ to fix a domain, and, for every X in the given D, there is some choice of $\tilde{X} \in \mathscr{C}$ with $\tilde{X} \cap V_\kappa = X$. If these choices can be made then we have a Δ_0-reflection property between $(V, \in, \mathscr{C})$ down to $(V_\kappa, \in, D)$. (Indeed making such choices is just another way of stating that we shall take a substructure $(V', \in, \mathscr{C}') \prec (V, \in, \mathscr{C})$ as above.) Some elementary observations can be made: for example, as $On \in \mathscr{C}$ we'd better have $On \cap \kappa = \kappa$ in D, and thus $\tilde{X} = On$ has to be our choice of extension for $X = (On \cap \kappa)$ as $\tilde{\kappa}$, if we are to have Δ_0

reflection.

Having sorted out this version of reflection for Δ_0-Satisfaction we may naturally require that our choices of $\tilde{X} \in \mathscr{C}$ with $\tilde{X} \cap V_\kappa = X$, for $X \in D$, can further sustain reflection of satisfaction for the Σ_1 formulae, the Σ_n formulae, or indeed the whole language $\mathscr{L}^+$. These desiderata are fulfilled in the following definition.

Definition 5 (GRP). *The* Global Reflection Property *holds if the universe $(V, \in, C)$ admits of a set-sized $\mathscr{L}^+$-elementary substructure, $(V', \in, \mathscr{C}') \prec (V, \in, \mathscr{C})$ with $V' = V_\kappa$ for some κ and which is isomorphic to a structure of exactly the same kind, namely $(V_\kappa, \in, V_{\kappa+1})$.*

As discussed, by 'elementary' we mean first order elementary in the usual language of set theory $\mathscr{L}_\in$ but we allow the addition of second order predicate symbols $P_0, P_1, \ldots, P_n, \ldots$.

4.1 Reflection increased: A spectrum

We step back somewhat and consider properties in a spectrum of increasing strength leading up to GRP.

Definition 6. *[Partial Global Reflection] The* Partial Global Reflection Property *holds if the universe $(V, \in, \mathscr{C})$ admits of a set-sized $\mathscr{L}^+$-elementary substructure, $(V', \in, \mathscr{C}') \prec (V, \in, \mathscr{C})$ with $V' = V_\kappa$ for some κ and $(V', \in, \mathscr{C}') \cong (V_\kappa, \in, D)$ with $D \subseteq V_{\kappa+1}$.*

We can immediately rewrite this in a way familiar to set theorists[1].

Definition 7. *Let $(V', \in, \mathscr{C}'), D, V_\kappa$ satisfy the last definition. We define j to be the inverse of the isomorphism $(V', \in, \mathscr{C}') \cong (V_\kappa, \in, D)$ and write:*

$$j : (V_\kappa, \in, D) \longrightarrow_{\mathscr{L}^+} (V, \in, \mathscr{C}).$$

Then, to be specific, note that $j \restriction V_\kappa$ is simply the identity function: $j(x) = x$ for any $x \in V_\kappa$.

---

[1] Readers of Welch and Horsten, 2016 will note that the primary definition of Global Reflection (Axiom 3) there is based on Def. 6 (with $D = V_{\kappa+1}$). Conversations with Sam Roberts incline us to the view, that, although making no difference mathematically, for presentational purposes it is perhaps nicer to define the *range* of j first (as in Def. 6) and then j^{-1} as its transitivisation collapse map. We warmly thank him for these discussions. He has defined (Roberts, preprint, 2016) wide ranging principles extending *GRP* by defining certain particular substructures of V with its classes, and certain satisfaction predicates.

To spell it out further: j must preserve as follows for φ in $\mathscr{L}^+$, $x \in V_\kappa$, $X \subseteq V_\kappa$, $X \in D$:

$$\varphi(x,X)^{(V_\kappa,\in,D)} \leftrightarrow \varphi(j(x),j(X))^{(V,\in,\mathscr{C})}$$

but $j(x) = x$ so
$$\leftrightarrow \varphi(x,j(X))^{(V,\in,\mathscr{C})}.$$

Note that we shall have $(V_\kappa,\in) \prec_\mathscr{L} (V,\in)$, but of course the latter is not a strong property.

The strength of these reflection principles is entirely signified by the richness of D. We may impose closure conditions on D. The intra-constructible reflection principles such as those from indescribability, could be obtained from principles such as these, that is when $D \subsetneq \mathscr{P}(\kappa)^L$.

If $D \supseteq \mathscr{P}(\kappa)^L$ then this will deliver to us $0^\sharp$: define U on $\mathscr{P}(\kappa) \cap L$ by

$$X \in U \leftrightarrow X \in L \wedge \kappa \in j(X).$$

By the Σ_1 reflection property of j, standard arguments show that this is a normal L-ultrafilter and we may form $\tilde{j}: (L,\in) \to_e Ult(L,U) \cong (L,\in)$ with $\tilde{j}(\kappa) > \kappa$. The existence of such an embedding from L to L contradicts $V = L$ and is often referred to as the 'least' large cardinal property, or strong axiom of infinity. We thus should have our first extra-constructible principle.

Similarly for other definable inner models, M say in place of L, then requiring $D \supseteq \mathscr{P}(\kappa)^M$ will reveal that there is a $j: M \to_e M$, and so forth.

The natural limit of the above is when we require $D = V_{\kappa+1}$. There is a step difference here: $(V_\kappa, \in, V_{\kappa+1})$ we consider as a normal domain with *all* of its parts, and we assert the property that anything satisfied in it by a set $x \in V_\kappa$ and a class $X \subset V_\kappa$, is the reflection of something satisfied in the realm $(V, \in, \mathscr{C})$ by x and by one of the parts of V, $\tilde{X} \in \mathscr{C}$ with $\tilde{X} \cap V_\kappa = X$. We thus have that $(V_\kappa, \in, V_{\kappa+1})$ is a simulacrum of $(V, \in, \mathscr{C})$. The j above then maps as follows:

$$j: (V_\kappa, \in, V_{\kappa+1}) \longrightarrow_e (V, \in, \mathscr{C})$$

with j fully $\mathscr{L}^+$-elementary.

If we assume the axiom of choice, AC, then $(V, \in, \mathscr{C})$ will be a model of Global Choice (since if $R \subseteq V_\kappa^2$ is a well ordering of V_κ then $j(R)$ is a wellorder of V). (It would be entirely within keeping of other reflection principles if we were to ask that there be unboundedly many such κ in On for which there is such a j, but we don't pursue that here.) Just as above for L, define a field of classes U on $\mathscr{P}(\kappa)$ by

$$X \in U \leftrightarrow \kappa \in j(X)$$

As $\mathscr{P}(\kappa) \subseteq V_{\kappa+1} \subseteq \text{dom}(j)$ by Σ_1-elementarity (in j), this is an ultrafilter on $\mathscr{P}(\kappa)$. Standard arguments show that U is a normal measure on κ, and thus κ is a measurable cardinal. This again shows that $V \neq L$ (by an old theorem of Scott, 1961).

But then:

$\forall \alpha < \kappa \langle V, \in \rangle \models$ "$\exists \kappa > \alpha(\kappa$ a measurable cardinal)" $\implies$
$\langle V_\kappa, \in \rangle \models$ "$\forall \alpha \exists \lambda > \alpha(\lambda$ a measurable cardinal)" $\implies$
$\langle V, \in \rangle \models$ "There are unboundedly many measurable cardinals".

Known methods improve this to:

Theorem 8 (GRP). $(V, \in) \models \forall \alpha \exists \lambda > \alpha(\lambda$ a measurable Woodin cardinal).

Consequently by results of Martin-Steel, Woodin GRP this implies:

a) Projective Determinacy (PD) and $(AD)^{L(\mathbb{R})}$.

b) (Woodin) $\mathrm{Th}(L(\mathbb{R}))$ is fixed: no set forcing notion can change $\mathrm{Th}(L(\mathbb{R}))$, and in particular the truth value of any sentence about reals in the language of analysis, thereby including PD.

One may extend the above and reason that there is a proper class of Shelah cardinals. Although it is easy to see that

Lemma 9. *If $Con(ZFC + \exists \lambda(\lambda$ a 1-extendible cardinal$))$ then $Con(NBG + GRP)$*

originally GRP was thought of as obtained from a weakening of the notion of another downward reflecting cardinal, a *sub-compact cardinal*:

Definition 10. μ *is a* subcompact cardinal *iff*
$\forall A \subseteq H_{\mu^+} \exists \kappa < \mu \exists B \subseteq H_{\kappa^+}, \exists j \neq id, j \restriction H_\kappa = id$, *so that:*

$$j : (H_{\kappa^+}, \in, B) \longrightarrow_e (H_{\mu^+}, \in, A).$$

This is a third order principle (over H_μ). By dropping the A, B component, we arrive at the notion of $\mu = On$ being subcompact, or in other words, something like (GRP).

5 Philosophical reflections

In Welch and Horsten, 2016 we discuss the possible interpretations in the light of current philosophical positions regarding classes. We regard mathematics as taking place within $(V, \in)$ - using sets as representations of mathematical conceptual objects. ('Large' categories indeed need some special consideration as definable classes. However we might point out that for any *theorem* that depends on, or builds on such large categories we may invoke the Montague-Levy Reflection theorem and bring the result down to one within ZF. (Even in one's wildest moments, no set theorist seriously thought that Wiles's theorem really required a proper class of Grothendieck universes.)

It is the author's view that fully second (or higher order) methods should be eschewed as these require a domain of quantification: the class of all classes at the very least. We do not wish to quantify over the collection $\mathscr{C}$ in any universal fashion (as we are unconvinced of the coherence of such an act); although we do wish to say that certain elements of $\mathscr{C}$ stand in certain relations to each other (and to a set-sized domain of sets). We just want the ability to say that such and such is a sufficiently rich substructure of $(V, \in, \mathscr{C})$. The richness is expressed in the form of elementarity in $\mathscr{L}^+$. Thus we do not wish to collect the classes together to form a new level V_{On+1}, and then regard them simply as some sets we forgot to add on to the top of V. We do not wish to form $\{V\}$.

One might think of the members of the class $\mathscr{C}$ as *pluralities*. The plural reading may allow us to talk of the such-and-such things without committing ourselves to the existence of entities. We rejected that view as it ties the notion too closely to linguistic concerns. We wish to have entities to go with V to be able to assert a substructure of V together with (some of) those entities in $\mathscr{C}$. Those entities must be in the range of our j.

We can however use the language of mereology to talk about classes as the *parts* of V. This is similar to David Lewis's interpretation of sets (Lewis, 1991). For Lewis the parts of sets are obtained from the singleton function and the process of mereological fusion. Here classes have subclasses as their parts. A set-sized part of a class we identify with the corresponding set. Indeed an instance of Comprehension can be interpreted as declaring that part of set through a defining clause, is a set. Unlike Lewis, we take the sets and so V as given, and use the part/whole relation to discuss the absolute infinities and V's parts. An absolute infinity then becomes the fusion of all of its parts.

There are thus either the set-sized parts which we identify with the sets themselves, or the parts that are proper-class sized. (As noted GRP implies Global Choice and so for any latter part there is a bijection with a proper class of ordinals.)

The reflection principle embodied in GRP is one involving quantification over sets, our mathematical objects (although classes are allowed unquantified). It is possible to have a strengthened GRP requiring the set-sized substructure of $(V', \in, \mathscr{C}') \prec (V, \in, \mathscr{C})$ to be Σ^1_∞-elementary in the full second order language $\mathscr{L}^2_{\in}$. This would indeed be taking quantification over classes and stepping out into higher orders. In this case after taking an isomorphism to a set structure $(V_\kappa, \in, V_{\kappa+1})$ we shall have: *There is* $\kappa \in \text{On}$, *there is* $j \neq \text{id}$, $\text{crit}(j) = \kappa$ *with*

$$j: (V_\kappa, \in, V_{\kappa+1}) \longrightarrow_{\Sigma^1_\infty} (V, \in, \mathscr{C}).$$

This we may call *mereological reflection*: we fully reflect on the parts of V by allowing now quantification over parts. By sticking to "mathematical reflection" we have a principle consistent relative to *ZFC* with a 1-extendible cardinal.

We can go further and using Global Choice to define Σ^1_n-Skolem Functions, define full second order satisfaction, *Sat*. We might in turn add that to our language, call it $\mathscr{L}^2_{Sat}$ and get a yet stronger principle, GRP$^+$ say, with

$$(V', \in, \mathscr{C}', Sat') \prec (V, \in, \mathscr{C}, Sat)$$

now required to be an $\mathscr{L}^2_{Sat}$-elementary substructure.

Just as a sample of what one could obtain then, we see that V can be filtrated as an *On*-length system of embeddings of GRP type:

Proposition 11 (GRP$^+$). *There is a commuting system $\langle \kappa_\alpha, j_{\alpha\beta} \rangle_{\alpha \leq \beta \in On}$ of embeddings $j_{\alpha\beta} : (V_{\kappa_\alpha}, \in, V_{\kappa_\alpha+1}) \longrightarrow_{\Sigma^0_1} (V_{\kappa_\beta}, \in, V_{\kappa_\beta+1})$ with, each $j_{\alpha\beta}$, $\alpha < \beta$, witnessing the simple GRP at the 'universe' V_{κ_β} with its parts $V_{\kappa_\beta+1}$. Thus each $j_\alpha \upharpoonright \kappa_\alpha = \mathrm{id} \upharpoonright \kappa_\alpha$, and $j(\kappa_\alpha) = \kappa_\beta$. Moreover for $\alpha \in On$, there are maps*

$$j_\alpha : (V_{\kappa_\alpha}, \in, V_{\kappa_\alpha+1}) \longrightarrow_{\Sigma^0_1} (V, \in, \mathscr{C})$$

also witnessing GRP *in the universe.*

It is not hard to derive the consistency of such a system from a subcompact cardinal (Welch, 2016).

Bibliography

Ackermann, W. (1956). Zur Axiomatik der Mengenlehre. *Mathematische Annalen, 131*, 336–345.

Cantor, G. (1883). Über unendliche lineare Punktmannigfaltigkeiten. *Mathematische Annalen, 21*(188), 545–591.

van Heijenoort, J. (Ed.). (1967). *From Frege to Gödel: A source book in mathematical logic, 1879-1931*. Harvard University Press.

Koellner, P. (2009). On reflection principles. *Annals of Pure and Applied Logic, 157*(2), 206–219.

Lewis, D. (1991). *Parts of classes*. Basil Blackwell.

Martin, D. A. (2005). Gödel's conceptual realism. *Bulletin of Symbolic Logic*, 207–224.

Martin, D. A. (2012). Completeness or incompleteness of basic mathematical concepts. In *EFI Harvard Workshop Papers*.

Reinhardt, W. N. (1967). *Topics in the metamathematics of set theory* (Doctoral dissertation, University of California, Berkeley).

Reinhardt, W. N. (1970). Ackermann's set theory equals ZF. *Annals of Mathematical Logic, 2*(2), 189–249.

Reinhardt, W. N. (1974a). Remarks on reflection principles, large cardinals, and elementary embeddings. In T. Jech (Ed.), *Axiomatic set theory, Proceedings of Symposia in Pure Mathematics* (Vol. 13(2), pp. 189–205). Providence, Rhode Island: American Mathematical Society.

Reinhardt, W. N. (1974b). Set existence principles of Shoenfield, Ackermann, and Powell. *Fundamenta Mathematicae, 84*(1), 5–34.

Roberts, S. (preprint, 2016). A strong reflection principle.

Scott, D. (1961). Measurable cardinals and constructible sets. *Bulletin de l'Académie Polonaise des Sciences, Série des Sciences Mathématiques, Astrononomiques et Physiques, 9*, 521–524.

Shoenfield, J. R. (1967). *Mathematical logic*. Addison-Wesley.

Tait, W. W. (2005). Constructing cardinals from below. In W. W. Tait (Ed.), *The provenance of pure reason: Essays in the philosophy of mathematics and its history* (pp. 133–154). Oxford, UK: Oxford University Press.

Wang, H. (1996). *A logical journey: From Gödel to philosophy*. MIT Press.

Welch, P. D. (2016). Obtaining Woodin's cardinals. In *Proceedings of the Logic Conference Harvard in Honor of Woodin's 60th Birthday, Contemporary Mathematics*. American Mathematical Society.

Welch, P. D., & Horsten, L. (2016). Reflecting on absolute infinity. *The Journal of Philosophy, 113*(2), 89–111.

Woodin, W. H. (2001). The continuum hypothesis, Part I. *Notices of the American Mathematical Society, 48*(6), 567–576.

Author biography. Philip Welch is a set theorist and mathematical logician working in Bristol since 1986. In 1997-2000 he was at Kobe University Graduate School setting up a research group in set theory. He is the author of some 80 papers in set theory, logic, theories of truth, and transfinite models of computation. He is currently President of the British Logic Colloquium, on the Board of Trustees of the European Set Theory Society, and the Coordinating Editor of the *Journal of Symbolic Logic*. His doctoral 'grandfather' is Alan Turing, his supervisor at Oxford (1975-78) Robin Gandy, being Turing's only PhD student.

5 Logic revision: Some formal and semi-formal techniques for logic choice

EDWIN MARES [*]

Abstract. This paper sets out a probabilist theory of logic revision and discusses some of its consequences and the philosophical challenges that if faces. The probability theory is semantic. It is built upon a model for a very weak logic (the logic of bounded lattices) but that has sub-models for a wide variety of other logical systems – modal logics, relevant logics, linear logics, other substructural logics, fuzzy logic, and so on. The probability theory used is itself non-classical, although in those regions of the model in which classical logic holds probabilities act classically. Suggestions are made as to the treatments of debate and negotiation about logical rules and about the problems of logical omniscience and the apriority of logic.

Keywords: philosophy of probability, non-classical logic, non-classical probability, logic choice.

1 Introduction

Here is a problem. People can sensibly debate which logical system is right. They can take a purpose for logic, let's say its use in the evaluation of arguments that are presented as deductive, and argue which is the right logic for the job. These debates

---

[*]Philosophy Programme, Victoria University of Wellington, Wellington, New Zealand

appear reasonable. We don't know for certain which logic is the correct one for this purpose. Thus, we need a theory that makes sense of these debates and, more generally, about our reasoning about logic. But it would seem that such a theory needs to incorporate some sort of logic itself, because it needs to set rules of inference that can be used in such debates. Moreover, in order to allow that someone might hold (what turns out to be) the wrong logical views but be rational, it would seem that we want a theory that is neutral between all substantive logical positions.

The purpose of this paper is to outline one such theory. The theory that I construct is a form of probabilism. It uses a probabilistic mechanics. It treats ideal agents as having their degrees of beliefs, including their beliefs about logic, conform to the principles of a probability theory. This probability theory is formalised by a generalisation of the standard Kolmogorov axioms to fit with a wide range of classical and non-classical logics. The theory of belief revision presented employs a version of Jeffrey conditionalisation.

I present this theory to demonstrate how we can construe logic choice as a rational process. This is important, not only demonstrate that those who actually do choose logics – largely philosophers and mathematicians – can be construed to be rational, but also to show that the choice of such apparently foundational beliefs as those concerning logic can be understood as rational and revisable. Moreover, although I think that the probabilistic theory is defensible, the construction here also shows how other theories of belief revision, such as AGM based theories and dynamic doxastic logics can be used as bases for logic revision. In short, the probabilistic theory of logic revision is presented as a test case to show that many of our most fundamental beliefs can be treated as rational and revisable.

The plan of this paper is as follows: First, I set out the requirements on models used in the present theory and look at some logical systems for which there are semantics that can be used to construct models that fit these requirements. Second, construct a single *background model* that is used as a basis for the theory of belief revision. Third, I set out the the probability theory that corresponds to this class of models and define a Jeffrey update function. Fourth, I use a scorekeeping pragmatics as a framework for interpreting debates and negotiation about logic. Fifth, I look at two issues that arise from this theory of logic choice. One of these problems is the omniscience problem and the other concerns the apriority of logic.

I do not include any proofs of theorems in this paper. For these, see (Mares, 2014).

2 Logics and models

In order to place the logical systems in a single framework, I formulate them in a single language. This language has propositional variables $p, q, r, ...$, negation ($\neg$), conjunction ($\wedge$), disjunction ($\vee$), and implication ($\rightarrow$), as well as two propositional constants $\top$ and t. I define equivalence using implication and conjunction, as usual:

$A \leftrightarrow B =_{df} (A \to B) \wedge (B \to A)$. Standard formation rules apply. The logics that I can model using the techniques in this paper are ones that can be given an indexical semantics (a semantics in which truth is relativised to worlds in a frame) and for which propositions can be interpreted as sets of worlds in frames.

For each logic, L, there is a set of models, M_L. A model is a structure,

$$\mathfrak{M} = (W, N, [\![\cdot]\!], Prop),$$

where W is a non-empty set (of points, worlds, situations, ...), N is a non-empty subset of W, $[\![\cdot]\!]$ is a function from formulas into $\wp(W)$, and $Prop$ is a lattice of subsets of W that includes N, W, and $\emptyset$.

I also stipulate that the conjunction of each logic behaves semantically as set-theoretic intersection. I do not, however, assume that disjunction acts semantically as set-theoretic union. Instead, I merely require that disjunction behave as a least upper bound in the algebra of propositions. That is, in a model $\mathfrak{M}$, for any formulas A and B, there is no proposition in $\mathfrak{M}$ that is a proper superset of $[\![A]\!]$ and of $[\![B]\!]$ and also is a proper subset of $[\![A \vee B]\!]$. These requirements entail that the algebra of propositions on any model constitute a lattice, but this lattice need not be a distributive lattice. The avoidance of distributivity is because certain of the logics that I wish to model reject the rule of distribution of conjunction over disjunction, in particular, linear logic and orthologic.

A set of models M_L is the set of models for a logic L if and only if

1. A is a theorem of L if and only if for every $\mathfrak{M} \in M_L$, $N_\mathfrak{M} \subseteq [\![A]\!]_\mathfrak{M}$;

2. for $n \geq 1$, $A_1,...,A_n \vdash B$ is a derivable rule in L if and only if $[\![A_1]\!]_\mathfrak{M} \cap ... \cap [\![A_n]\!]_\mathfrak{M} \subseteq [\![B]\!]_\mathfrak{M}$ for every $\mathfrak{M} \in M_L$.

The first condition corresponds to weak semantic completeness and the second condition corresponds to strong completeness. In the present context, strong completeness does not entail weak completeness, so both need to be stated.

Here is a familiar example. The modal logic S4 (in which $\to$ is taken to be strict implication) has as its models triples, (W, R, V), where W is a non-empty set (of possible worlds), R is a reflexive and transitive binary relation, and V is an assignment from formulas into subsets of W that accords with the standard truth conditions for Kripke semantics. The truth condition for strict implication is

$$a \in V(A \to B) \text{ iff } \forall b(Rab \implies (b \notin V(A) \vee b \in V(B))).$$

The model that I extract from this is a quadruple, $\mathfrak{M} = (W, N, [\![\cdot]\!], Prop)$ such that for all formulas A, $[\![A]\!] = V(A)$ and $N = W$.

The set N is the set of worlds in its model in which the theorems of the logic are verified. For some logics, such as classical logic, classical modal logic, intuitionistic logic, first-degree entailments (FDE), and strong Kleene (K_3), this set of so-called normal worlds is just the whole set of worlds, W. For substructural logics such as relevant logics and linear logic, N may not be the same as W.

What happens when I integrate a modal logic into the present framework is that the modal accessibility relation is no longer part of the surface grammar of the model theory. What is assumed is that there are mechanisms in the individual models that determine the truth values of the formulas at each of the points in the model. In the current framework, the model in the sense it is being used here is abstracted from the model in the standard sense.

2.1 Some logics and their models

The model theory for different logics require different treatments to produce models in the required sense. In what follows, I look at a few logics to illustrate in more depth the method used. I start with lattice logic, which is the minimal logic that fits with probability theory.

Lattice logic

The weakest logic that has a semantics of this sort is *lattice logic* (Paoli & Restall, 2005). Lattice logic is just the logic that has as its algebraic semantics the class of lattices. Alasdair Urquhart's representation theorem for lattices (Urquhart, 1978) can be employed to show that lattice logic is complete over a set of models, which has an interesting topological frame theory. But I am interested here in a more general and more abstract characterisation of a set of models that characterises lattice logic.

A model for lattice logic is a $(W, Prop, [\![\cdot]\!])$ such that W is a non-empty set, $Prop \subseteq W^2$, and $[\![\cdot]\!]$ is a function from formulas into $Prop$ such that

- $[\![A \wedge B]\!] = [\![A]\!] \cap [\![B]\!]$;
- $[\![A \vee B]\!] = [\![A]\!] \sqcup [\![B]\!]$.

where $X \sqcup Y$ is the least upper bound of X and Y in $Prop$. In models for lattice logic, the least upper bound always exists. The treatments of implication and negation are arbitrary in models for lattice logic, since neither belongs to the language of lattices.

It is easy to show that, given any lattice, such a model can be constructed. The worlds in this model are the filters of the lattice. A member of $Prop$ is a set of filters X such that there is some point a in the lattice for which $a \in X$. Let $[a]$ be the set of filters X such that $a \in X$. It can be shown that $[a \vee b] = [a] \sqcup [b]$ and $[a \wedge b] = [a] \cap [b]$.

To construct a model in the sense used in this paper, we add a set N to the definition of a model for lattice logic such that $N = W$.

Logic revision

An easy way to formulate lattice logic is by means of sequent calculus in which the sequents are of the form $\Gamma \vdash \Delta$ where Γ and Δ are sets. The axioms of the system are all sequents of the form $A \vdash A$ and its one structural rule is K:

$$\frac{\Gamma \vdash \Delta}{\Gamma' \vdash \Delta'}$$

where $\Gamma \subseteq \Gamma'$ and $\Delta \subseteq \Delta'$. The other classical structural rules are obtained from the fact that antecedents and succedents are sets. Lattice logic has the following connective rules:

$$\frac{\Gamma \vdash A, \Delta \quad \Gamma \vdash B, \Delta}{\Gamma \vdash A \wedge B, \Delta} \qquad \frac{\Gamma \vdash A \wedge B, \Delta}{\Gamma \vdash A(B), \Delta}$$

$$\frac{\Gamma \vdash A, \Delta}{\Gamma \vdash A \vee B(B \vee A), \Delta} \qquad \frac{A \vdash \Delta \quad B \vdash \Delta}{A \vee B \vdash \Delta}$$

Apart from the requirement that the premises be singular on the left in the disjunction on the left rule, these are standard classical meaning rules for the connectives.

One variant of lattice logic in which I am interested is the logic of bounded lattices. To create this logic we add the following axioms:

$$\Gamma \vdash \top \quad \bot \vdash \Delta \quad \Gamma \vdash t \quad f \vdash \Delta$$

where $\bot = \neg \top$ and $f = \neg t$. I also add to the definition of a model for lattice logic the following conditions:

$$[\![\top]\!] = W = N = [\![t]\!] \quad [\![\bot]\!] = \emptyset = [\![f]\!]$$

The choice of lattice logic has to do with the theory of probability given below. It is the weakest logic that supports this theory of probability. It can be proven that any model for lattice logic can have a probability function that satisfies the generalised probability axioms imposed on it. And, as I argue, there is good reason for wanting probability functions in this sense.

Fuzzy logic

There are various fuzzy logics, but I only discuss here a simple fuzzy logic defined by a set of fuzzy models. A fuzzy model is a triple $([0,1], d, V)$ such that $[0,1]$ is the real interval between 0 and 1, d is some distinguished member of $[0,1]$, and V is a function from the formulas into $[0,1]$ such that the following clauses hold:

- $V(A \wedge B) = \min\{V(A), V(B)\}$;
- $V(A \vee B) = \max\{V(A), V(B)\}$;
- $V(\neg A) = 1 - V(A)$;
- $V(A \to B) = \max\{1 - V(A), V(B)\}$.

A formula A is said to be true on V if and only if $V(A) \geq d$.

A worlds model for fuzzy logic is a quadruple $(W, N, d, \mathbb{D}, \mathbb{V})$ such that W is a non-empty set of points, N is a non-empty subset of W, d is a distinguished point in $[0,1]$, $\mathbb{D}$ is a function that assigns to each world in W a real number in $[0,1]$ such that for all worlds $a \in N$, $\mathbb{D}(a) = d$, and $\mathbb{V}$ is a function that assigns to each world in W a fuzzy valuation.

To turn a worlds model for fuzzy logic into a model in the sense required by this paper, we define a new valuation function $[\![\cdot]\!]$ such that for all formulas A,

$$[\![A]\!] = \{a \in W : \mathbb{V}(a)(A) \geq \mathbb{D}(a)\}.$$

This says that the proposition assigned to A is the set of worlds in which A has a value that is at least as great as the designated value of the world. The reason that I include non-normal worlds in models is so that models will represent fuzzy consequence as it is intended. That is, for each fuzzy model, $\mathfrak{M}$, as usual, I set

$$A_1, \ldots, A_n \models_{\mathfrak{M}} B \text{ iff } [\![A_1]\!]_{\mathfrak{M}} \cap \ldots \cap [\![A_n]\!]_{\mathfrak{M}} \subseteq [\![B]\!]_{\mathfrak{M}}$$

where $A \models_{\mathfrak{M}} B$ if and only if, for all $a \in W$, if $a \in [\![A]\!]$ then $a \in [\![B]\!]$. Given the variation of designated values on non-normal worlds, this definition characterises the same consequence relation as one that is defined over a set of fuzzy valuations and sets $A \models B$ if and only if $V(A) \leq V(B)$ on every valuation in that set.

Supervaluations

The version of supervaluational frames that I present is a slightly simplified form of that given by Kit Fine in (Fine, 1975). I begin with a set of points W, a distinguished subset C of W, and a partial order $\leq$ on W. The set C is the set of *complete* points in the frame. We add the condition that if $c \in C$, then for any point a if $c \leq a$, $a = c$.

A valuation V on a supervaluational frame is a function from propositional variables to sets of worlds closed upwards under $\leq$. Thus, for example, if $a \leq b$ and $a \in V(p)$, then $b \in V(p)$. Moreover, we constrain valuations such that if for any propositional variable p and any point a, if all complete points $c \geq a$ are such that $c \in V(p)$, then $a \in V(p)$.

Each valuation determines a satisfaction relation, $\models$, according to the following recursive definition:

- $a \models p$ iff $a \in V(p)$;
- $a \models A \vee B$ iff $\forall c \in C(a \leq c \Rightarrow (c \models A \vee c \models B))$;
- $a \models A \wedge B$ iff $a \models A$ and $a \models B$;
- $a \models \neg A$ iff $\forall c \in C(a \leq c \Rightarrow c \not\models A)$;
- $a \models A \to B$ iff $\forall c \in C(a \leq c \Rightarrow (c \not\models A \vee c \models B))$.

We can prove easily that $a \models A$ if and only if, $\forall c \in C(a \leq c \Rightarrow c \models A)$, for all a and all formulas A. Truth and falsity at each of the complete points behaves in a completely classical manner. It is is also easy to show that $[\![A \vee B]\!]$ is the least upper bound of $[\![A]\!]$ and $[\![B]\!]$ in the set of propositions in a super valuational model, and hence these models can be turned into models in the sense required by the present project.

Putting together models for supervaluations, fuzzy logic, and epistemic modal logics, we can create a background model that can be used to represent reasoning about a choice between approaches to vagueness. The way in which models for various logics are to be combined to create background models is treated in the next section.

3 Background models

Given a set of models for the various different logics that one wishes to choose between, we can construct a single model to act as the basis for a belief revision structure. Crudely put, the single model is a fusion (largely a union) of the models.

Let's make this more precise. I assume that a set of models M has been selected. The background model based on M is a quadruple $(W, N, Prop, [\![\cdot]\!])$ that is defined as follows. The set of worlds is just the union of the sets of worlds of the models in M, that is, $W = \bigcup_{\mathfrak{M} \in M} W_\mathfrak{M}$ and similarly the set of normal worlds is just the collection of all the normal worlds in M, i.e. $N = \bigcup_{\mathfrak{M} \in M} N_\mathfrak{M}$. The construction of $Prop$ needs a bit more care. For each formula, A, $[\![A]\!] = \bigcup_{\mathfrak{M} \in M} [\![A]\!]_\mathfrak{M}$. $Prop$ is just the set of $[\![A]\!]$ for all formulas A.

For each logic L that is being considered, I also add two constants $\top_L$ and t_L. Let M_L be the set of all the models in M for the logic L. The interpretations of the new constants are as follows:

$$[\![\top_L]\!] = \bigcup_{\mathfrak{M} \in M_L} W_\mathfrak{M} \qquad [\![t_L]\!] = \bigcup_{\mathfrak{M} \in M_L} N_\mathfrak{M}$$

These two constants give us a means in the language to talk about the logic L. $\top_L$ represents all of the formulas and rules that are true or valid in every world in every model for L and t_L represents the theorems of the language and the rules under which the theorems are closed. (The sets of worlds that these constants represent are members of $Prop$.)

In the next section, I continue to examine the different ways in which we can understand formulas and rules of inference.

3.1 Representing theorems and rules of inference

In accepting or rejecting a formula, we may accept or reject it as a simple truth or we may accept or reject that it is a logical truth. To accept or reject a formula as a logical truth (as an axiom or theorem), what one accepts or rejects is not the proposition expressed by the formula but rather the normal sets that verify that proposition as a logical truth. In other words, a formula, taken as an axiom, is the union of the sets of normal worlds that make true that formula. In formal notation:

$$[\![A]\!]^{Th} = \bigcup \{N_\mathfrak{M} : \mathfrak{M} \in M \wedge N_\mathfrak{M} \subseteq [\![A]\!]\}$$

I add to *Prop*, for each formula A in the language, the union of sets of sets of normal worlds that verify A.

The treatment of rules is more complicated. There are various types of rules. Two sorts of rules that are, in my opinion, most important are the ones that Dana Scott calls "horizontal" and "vertical" rules of inference (Scott, 1971). A horizontal rule of inference, can be represented as a single conclusion sequent of the form

$$A_1, ..., A_n \vdash B$$

This is to be understood in the semantics as saying that if any world makes true all of the premises, $A_1, ..., A_n$, then it also makes true the conclusion B. A vertical rule is a rule that sets a closure condition for the set of theorems of a logic. For example, weaker relevant logics such as B and MC obey the following version of the rule of modus ponens:

$$\frac{\vdash A \rightarrow B \quad \vdash A}{\vdash B}$$

These logics, however, do not obey the horizontal version of this rule, i.e. $A \rightarrow B, A \vdash B$.

It also might be desirable to distinguish between types of vertical rule in order to provide the correct semantic interpretations of them. Consider the rule of universal generalisation for first order classical logic:

$$\frac{\vdash A(a)}{\vdash \forall x A(x)}$$

where a is free for x. This rule is quite different from the modus ponens rule presented above. There are models for classical first order logic in which the normal worlds, either individually or as a set, do not satisfy this rule. Consider a one world model of the natural numbers. The domain of quantification is the natural numbers and in it the formula $\neg 0 = 1$ is true. But the formula $\forall x \neg x = 1$ fails to be true. What is true, however, is that where a is an individual constant, if $A(a)$ is true in every world in every model for classical first order logic then $\forall x A(x)$ is also true in every world. We can represent a rule like this as the union of sets of normal worlds of all the logics that are closed under this rule.

In a more involved version of the theory, we might add a ternary relation R to models in order to represent *intensional sequents*. An intensional sequent is a sequent of the form $\Gamma \vdash A$ where Γ is a *structure* of formulas and A is a formula. The set of structures of formulas is the smallest set such that: if A is a formula, then A is a structure; if Γ and Δ are structures then so is $(\Gamma;\Delta)$ (Restall, 2000, ch 2). The semi-colon is used to express an intensional connection between premises, often formalised in the object language of substructural logics by fusion ($\circ$). An intensional sequent $A;B \vdash C$ is valid in a model $\mathfrak{M}$ if and only if for all worlds a,b,c in $\mathfrak{M}$, if $Rabc$, $a \models A$, and $b \models B$, then $c \models C$. This idea is extended to treat all intensional sequents using products of the ternary relation with itself. The notation $R^2(ab)cd$ means that $\exists x(Rabx \land Rxcd)$. And we can have products of R any finite power. Thus, for example, we have $(A;B);C \vdash D$ valid in $\mathfrak{M}$ if and only if for all a,b,c,d in $\mathfrak{M}$, if $R^2(ab)cd$, $a \models A$, $b \models B$, and $c \models C$, then $d \models D$. In models for substructural logics the ternary relation is also used to give a truth condition for implication: $a \models A \rightarrow B$ iff $\forall x \forall y((Raxy \land x \models A) \Rightarrow y \models B)$. This connection between the representation of intensional sequents and implication yields the following form of the deduction theorem:

$$\Gamma;A \models B \text{ iff } \Gamma \models A \rightarrow B$$

In this paper, however, I do not make any use of this ternary relation except in this digression.

4 Belief revision

Once a background model is constructed it can be used as the basis for a model of a theory of belief revision. For instance, we can impose on it a system of spheres in the sense of Grove's semantics for the AGM theory of belief revision (Grove, 1988; Gärdenfors, 1988). We could also place a binary preference relation on worlds and construct a model for some form of dynamic epistemic logic or dynamic doxastic logic (van Benthem, 2011; Segerberg, 1995). Instead of these, I use the background model as a basis for a model of probability, and develop a probabilist theory of belief revision.

4.1 Probability functions

In (Mares, 2014), the notion of a probability function is generalised to fit the range of logics treated here. In this section, I briefly state the axioms of the resulting class of functions and discuss some of their properties that are salient for the current project.[1]

A probability function P on a model $\mathfrak{M} = (W,N,V,Prop)$ is a function from the formulas into $[0,1]$ such that

---

[1] In (Mares, 2014), I do not have the axioms $P(W) = 1$ and $P(\emptyset) = 0$, but rather slightly more complicated axioms. The simpler axioms will do for my purposes here.

1. $P(W) = 1$ and $P(\emptyset) = 0$;
2. If $X \subseteq Y \cup Z$, then $P(X) \leq (P(Y) + P(Z)) - P(Y \cap Z)$;
3. If $Y \cup Z \subseteq X$, then $(P(Y) + P(Z)) - P(Y \cap Z) \leq P(X)$;
4. If $X \subseteq Y$, then $P(X) \leq P(Y)$.

The complex forms of additivity expressed by postulates 2 and 3 are needed because the standard form of additivity, viz. $P(A \cup B) = (P(A) + P(B)) - P(A \cap B)$, assumes that the probability of unions of members of *Prop* are always defined. This isn't the case in models for logics without distribution, such as linear logic, or in my background model.

I adopt the traditional definition of conditional probability:

$$P(Y/X) = \frac{P(X \cap Y)}{P(X)}$$

It can be easily shown that if in a model the propositions are closed under union and boolean complement, any probability function over this model also satisfies the standard finite additivity axiom:

$$P(X \cup Y) = (P(X) + P(Y)) - P(X \cap Y)$$

For all logics that have classical models (such as classical and modal logics), the associated probability functions are all traditional Komolgorov functions. This is just an example of how the probability functions adapt to be appropriate to the logics, given both the base definition of a probability function (given above) and the models for the logics. The logics incorporated into the present framework satisfy the following "equation":

Base Probability Theory + Logic L = Appropriate Probability Theory for L

The class of probability functions in which I am most interested in are those that are defined over the background model. It is clear, however, that for each probability function P defined over the background model and for each sub-model $\mathfrak{M}$ the function $P \upharpoonright \mathfrak{M}$ is a probability function meeting the postulates given above.

Using this notion of probability, I define a version of Jeffrey conditionalisation as the method of updating agents' probability functions in response to new information. Given a new piece of information $P_{New}(X) = r$, for any proposition Y,

$$P_{New}(Y) = [\frac{P(X \cap Y)}{P(X)} \times P_{New}(X)] + [\frac{P(Y) - P(X \cap Y)}{1 - P(X)} \times (1 - P_{New}(X))].$$

This is just Jeffrey's definition adapted to the present context. I replace the negations in the original definition because not all of the logics considered here have negations that express boolean complement on sets in their model theory. The problem with this form of updating is that it does not treat cases in which an agent has given a proposition a prior probability of zero and then later discovered that it should be given a positive value. To deal with this difficulty, I suggest that a move be made to a more complicated structure in which a class of probability functions on a model are placed in a linear order, using something like Wolfgang Spohn's ranking theory (Spohn, 2012).

4.2 Beliefs and rejections

In thinking about logic, we both decide which principles to accept and which to reject. Formulating rejection can be problematic. Some logics do not have a negation that can be understood as representing rejection. Paraconsistent negations, for example, allow one to accept both a proposition and its negation without commitment to every proposition. These logics have non-trivial models and are used to represent, among other things, inconsistent belief states. Thus I think of a cognitive state of an agents as including both a set of beliefs and a set of propositions that he or she rejects. Rejection is a primitive of the present theory (as in (Mares, 2002)). The implementation of the probability theory here is to explain both how an individual organises his or her beliefs and his or her rejections.

One's beliefs and rejections do not just require internal organisation, they need to bear certain relations to one another. Let's call the pair (Γ, Δ) of an agent's beliefs and rejections, respectively, her *content*. This content is said to be *pragmatically consistent* if and only if Γ does not entail Δ, that is,

$$\bigcap \Gamma \not\subseteq \bigsqcup \Delta.$$

Where the generalised join operation used here is defined as:

$$\bigsqcup \Delta = \bigcap \{X \in Prop : \forall Y (Y \in \Delta \Rightarrow Y \subseteq X)\}$$

Note that whereas the infinite meets (intersections, in this case) and joins are definable, an infinite join or meet of propositions need not belong to *Prop* itself.

This notion of pragmatic consistency can be modified to produce a probabilistic notion in the same way as the notion of classical consistency is altered by classical Bayesians to a notion of probabilistic consistency. On the classical notion, an agent's beliefs are probabilistically consistent if and only if the strength of her beliefs accord with the probability calculus. Similarly, an agent's content can be said to be *probabilistically pragmatically consistent in a context* if and only if the strength of beliefs and rejections accord with the probability calculus and, in the context there is a threshold of belief b and a threshold of rejection r that do not overlap. This requires some explanation.

On the present theory, an ideal agent's cognitive state in a context can be represented by a triple (P, b, r) where P is a probability function over a background model and $b, r \in [0, 1]$ such that $b - r > 0$. Her content in that context is (Γ, Δ) where Γ is the set of propositions X such that $P(X) \geq b$ and Δ is the set of propositions Y where

$P(Y) \leq r$. The requirement that $b - r$ be strictly greater than 0 entails that $\Gamma \cap \Delta = \emptyset$. This gives us a very weak notion of consistency.

The notion of pragmatic consistency requires that the intersection of the believed propositions is always non-empty. This, I argue below, is a reasonable constraint for debates and negotiations concerning logic, but not a reasonable principle to govern an individual's beliefs. The lottery paradox shows that there are cases in which one should believe a collection of propositions but not believe in their conjunction. A virtue of probabilism is that it allows one to hold pragmatically inconsistent beliefs, even about logic, and to have a very nuanced attitude towards logical principles.

5 Positions and scorekeeping

In this section, I use a version of score-keeping pragmatics to act as a foundation for a social-epistemological version of the theory belief revision that I have outlined above.

According to David Lewis's score-keeping theory, in any well-run conversation there is a running score. At each point in the conversation, the score represents various parameters that are essential to understanding the conversation, such as, the degree of vagueness that is acceptable, and what presuppositions are being made (D. Lewis, 1979). The element of a score that is of most interest in the present context is what David Ripley and Greg Restall call the common *position* of the participants of the conversation. The position at a particular point in a conversation is the social counterpart of what I call in the previous section an agent's content, that is, a position is a pair (Γ, Δ) such that Γ is the set of propositions that are accepted by all the participants in the conversation and Δ is the set of propositions that are jointly denied by all of them.

Ideally, positions should be pragmatically consistent. When our aim is to decide (by debate, negotiation, or other means) which logic to apply, we need to accept a class of models. In debates and negotiation, the aim to is to accept a set of rules as the ones to apply. In one's own content, one might hedge one's acceptance of a particular logic with less certain beliefs in other principles, perhaps in some cases with weak rejections of related principles, and so on. But if the aim is to find one logic to accept, these nuanced attitudes are to be avoided.

It is a necessary condition of the acceptance of a proposition by a participant in a conversation that her degree of belief in that proposition be above the threshold that she has set for belief. It cannot be a sufficient condition because of conflicts with others in the conversation and because of the rejection of the sorts of nuances that I have discussed above.

In thinking about debates concerning logic, I adopt some of the connections between probabilist reasoning and argumentation theory developed by researchers working in Bayesian argumentation theory (Betz, 2012, 2013; Zenker, 2013), and combine these with scorekeeping pragmatics. Probability theory can be used to analyse the strength of evidence and arguments used in debates and to give norms for when participants in the debate and members of the audience should accept or reject claims made in the debate. At a particular time in a debate, the logic that is acceptable is given by the region of the background model constructed from largest class of models that all contain all the propositions that are accepted at that time by the participants of the debate and fail to contain any of the propositions that are rejected at that time.

5.1 Circular justification

The key to understanding debate on this theory is that evidence is understood as supporting a contrast between logical systems. This is particularly useful because evidence for logical systems is often seen as circular. Paul Boghossian has us consider a logical system with modus ponens as its only primitive rule. Then any rule of inference that we use in its justification must either be modus ponens or derived from modus ponens (Boghossian, 2001, p. 10).

Boghossian thinks of the epistemic justification of logical rules in terms of deductive arguments. Probabilists, on the other hand, generally treat epistemic justification inductively or contrastively. Here is a Bayesian analysis of allegedly circular justification by Tomoji Shogenji:

> Here is a general procedure for avoiding epistemic circularity. Suppose the suspicion of epistemic circularity arises because the evidence that we hope to use for confirming the hypothesis is useful only if the truth of the hypothesis is assumed. To avoid epistemic circularity, assume the hypothesis only in the sense of envisioning its truth. ... Assume next the negation of the hypothesis – again in the sense of envisioning its truth. Then proceed to compare the degree of coherence between the evidence and the hypothesis and the degree of coherence between the evidence and the negation of the hypothesis, using the background assumptions when necessary. (Shogenji, 2013, p. 180)

Shogenji's analysis is available for justification of basic logical principles and rules, such as modus ponens, in the present framework. The only principles and rules that are excluded are those of lattice logic. The idea is that, given the position that the participants occupy in a debate, they take a principle and contrast the probabilities we give to the union of the class of models that accept the principle or rule with the union of the class of models that do not accept it.

Tonk

Let us consider Paul Boghossian's example of a circular justification of a logic that contains Prior's tonk rules. Boghossian shows that if we allow circular justification, then anyone who accepts the tonk rules can justify these very same rules (Boghossian, 2001). The probabilist solution to this problem is to look at the models for the tonk logic. The tonk logic only has models that are trivial in the sense that every world is in every proposition. Trivial models are excluded by the definition of 'model' that I am using. Every model, on this definition, contains at least one world. In addition the empty set is a proposition in every model. Hence in no model is there is a world is in every proposition. There are good reasons from the theory of probability for these structural constraints on models. If we have an empty model, then the probability of the set of worlds will be the same as the probability of the empty set, but the former is one and the latter zero. Thus, the background model cannot be a model for the tonk logic, nor can any submodel of the background model be a model for the tonk logic.

6 Negotiating logic

In actuality, the problem of alternative logics has been confined to conversations amongst philosophers, computer scientists, and mathematicians. But what if it broke out of these disciplines and became more widely discussed? In particular, what if in wider society there were a debate about what sorts of rules of inference we could use in political discussions and other social activities? C.I. Lewis posed this problem in 1921:

> Now whoever enters a discussion, pragmatically assumes that the logical sense of those engaged is the same with his. The pursuit of common enterprises, regarded as rational, rests at the bottom upon a similar assumption. But in making this assumption – as we are frequently aware – we take a certain risk. In the interest of our rational enterprise we must take this risk. ... The ideal of a universal logical sense is one strongly demanded by its importance to all social enterprises, and is more closely approximated in fact than most of our ideals. But sticking to the facts, in the spirit of the facts, we are obliged to admit that it does not completely exist and probably never will. It is easy to define "rationality" in one's own terms. But that can only lead to the familiar conclusion, "All the world is strange save thee and me – and thee's a little strange." With respect to our ideals, we all of us stand in the egocentric predicament; we can only assert our own and hope for agreement. (C. Lewis, 1921, p. 379)

Lewis thinks that there is *in fact* widespread agreement about which rules of reasoning should be used. I'm not sure that he is right about this, but even if there is agreement it is an important question to ask whether we could find rational support for the sort of agreement we have.

Different social enterprises may require different methods of reaching agreement. Consider, for example, engineers who wish to build a large and potentially dangerous structure, who are in disagreement about the logic should underly the mathematics that they use. Some wish to use calculus based on classical logic, others will accept only intuitionist mathematics. In such cases what is appropriate is to determine what the relative risks involved are in making one choice over the other. If the people involved are rational in the sense of the present paper, then they will be able to construct subjective assessments of the risks involved and then a discussion can begin between them in a rational manner.

With regard to political discussions, the values that need to be satisfied include political ideals such as fairness. In order to understand how to implement the theory in dealing with disagreement about the logical structure of political discourse, we would have to become clear about what is at stake for individuals with regard to different choices of logic. If we can determine what the gains and costs are of allowing certain rules of inference, we can then deal with the participants as Bayesian agents negotiating to gain advantage. Although there has been some work on the oppression of women and logic (Plumwood, 1993), there has been overall very little work on this, but it is an interesting avenue of research.

7 Problems concerning logical omnicience

Surely one sort of logical evidence concerns what can be proven given a particular logic. People do often compare logics in terms of their theorems and in terms of how much mathematics we can do using them. This sort of evidence is easily dealt with on the present theory. We compare two (or more) classes of models with one another, and set our preference for one or the other in terms of the probabilities we assign to the models.

A problem arises, however, when one discovers, against her prior beliefs, that a contested formula is or is not a theorem of a given logic. For example, suppose that we wish to model an agent who is trying to decide between two logics, L_1 and L_2. She would like it if A were a theorem of the logic chosen, but does not know whether it is. In fact, A is a theorem of L_1 but not of L_2. So, in a background model representing her space of alternative logics, $[\![A]\!]^{th} = [\![t_{L_1}]\!]$. So, $P([\![A]\!]^{th}) = P([\![t_{L_1}]\!])$. But,

$$P([\![t_{L_1}]\!]/[\![A]\!]^{th}) = \frac{P([\![t_L]\!] \cap [\![A]\!]^{th})}{P([\![A]\!]^{th})}.$$

Therefore,

$$P([\![t_{L_1}]\!]/[\![A]\!]^{th}) = \frac{P([\![t_{L_1}]\!])}{P([\![A]\!]^{th})} = 1.$$

So, representing the fact that the agent doesn't know that A is a theorem of L_1 and not of L_2 seems impossible on the present theory as does treating the discovery that A is a theorem of L_1 but not of L_2 in a non-trivial manner. This problem is a variant of the familiar problem of logical omniscience that is faced by most epistemic and doxastic logics.

In order to deal with this difficulty, I complicate the structure somewhat. I add a finite set of agent states AS each of which can be thought to be considering a background model with two logics, but logics that are perhaps slightly different from L_1 and L_1. In some of these background models, the logic labelled 'L_1' has A as a theorem and in some it does not and in some the logic called 'L_2' has A as a theorem and in some it does not. This range of states represents the ambiguity produced by an agent's thinking about a logic but attributing to it the wrong theorems or failing to attribute to it the right theorems.

I also add to each agent state a set of formulas that in that state the agent attributes as theorems to L_1 and a set of formulas that the agent in that state thinks are theorems of L_2 (the same can done with rules, of course). In addition, I assume a function $P_{AS} : AS \longrightarrow [0, 1]$ such that $\sum_{a_1 \in AS} a_1 = 1$. The P_{AS} function represents the weighting of the various states in terms of their approximation to the actual cognitive state of the agent who is being modelled. I write $[\![A]\!]_i$ to refer to the proposition expressed by A in the background model assigned to a_i. Now we can represent the probability P^* of a *formula* as a statistical average of probabilities over the states in AS:

$$P^*(A) = \sum_{a_i \in AS(a)} (P_{AS}(a_i) \times P_{a_i}([\![A]\!]_i))$$

We can represent the probability of A's being a theorem – $P^*(\vdash A)$ – as an average of $P_{a_i}([\![A]\!]_i^{th})$s, and similarly with regard to rules.

I am, of course, not presenting this solution as a general solution to the problem of logical omniscience in the different ways it appears in a probabilist context. Rather, I think as do others working in probabilist theories, that various modelling tasks require different tweaks to the basic model. This is a central feature of the way in which scientific modelling operates in practice.

8 Half-baked remarks: The apriority of logic

Anti-exceptionalism about logic is the doctrine that logics are to be thought of as being on a par with scientific theories. On this view, logic is up for revision and logical knowledge (or logical beliefs) are not a priori (Hjortland, forthcoming). What I say in this paper is compatible with some form of anti-exceptionalism – it presents a formal theory that can be taken to be an anti-exceptionalist methodology. The view of this paper, however, is not committed to an anti-apriorism in the epistemology of logic, nor of course is it committed to apriorism.

C.I. Lewis held that all choice between logical systems is pragmatic (C. Lewis, 1932). That is, we must appeal to theoretical virtues such as simplicity, intuitiveness, and strength to justify the acceptance of one logical system over another. This sort of justification is a priori in the sense that it does not appeal to empirical data about the things that logic is about (i.e. everything) but it is also defeasible. We may come across new evidence that another logical system, perhaps that we have not considered, better exemplifies these virtues. In general, a priori justification need not be conclusive.

What is perhaps more interesting is the problem of defining the notion of apriority in the current theory. In (Mares, 2011) I claim that one cannot define the notion of a priori outside a more general epistemological framework. A notion of absolute apriority can be defined easily in any probabilisitic theory. The proposition that is just the total set of worlds is absolutely a priori in an agent's belief state and a sentence that expresses the complete set of worlds in every belief state is absolutely a priori is a very strong sense.

There is, however, room in the theory for weaker senses of 'a priori'. For example, I can adopt a notion of *relative apriority*. A formula A is a priori relative to a logic L if and only if $N_L \subseteq [\![A]\!]$. On this definition, all theorems of L are a priori relative to L. If we strenghten this notion to say that $[\![A]\!]$ needs to be a superset of the theorems of L on all models, then only theorems of L would be a priori relative to L, but the weaker version is more interesting. It is an agent-relative notion of relative apriority.

Clearly, other definitions of 'a priori' are possible in this framework, but I will leave this topic until after I have the opportunity to give it further thought.

9 Concluding remarks

In this paper I have set out a formal epistemology for logic choice. The theory in this paper locates the choice between alternative logics within a widely accepted and well-understood epistemological framework – probabilism. This seems right. Reasoning about logic and debates concerning logic seem unremarkable in the sense that they are commonplace and seem to be extensions of our standard belief-forming practices. Whether probabilism or some other form of belief revision is correct is not the real issue here. Rather, the fact that very little logic needs to be presupposed in order to construct a theory of belief revision that can treat logic choice. This bodes well for the creation of theories that will represent as rational reasoning about other apparently fundamental beliefs and values.

Acknowledgements. I am grateful to Greg Restall and the other members of the programme committee for inviting me to address the CLMPS and for organising this session on philosophical logic. I also thank the members of the audience for their helpful comments, especially Raymundo Morado, Francesco Paoli, Arnon Avron, and Giovanna Corsi.

Bibliography

Betz, G. (2012). On degrees of justification. *Erkenntnis, 77*, 237–272.
Betz, G. (2013). Degrees of justification, Bayes' rule, and rationality. In F. Zenker (Ed.), (pp. 135–146). Dordrecht: Springer.
Boghossian, P. (2001). How are objective epistemic reasons possible? *Philosophical Studies, 106*, 1–40.
Fine, K. (1975). Vagueness, truth and logic. *Synthese, 30*, 265–300.
Gärdenfors, P. (1988). *Knowledge in flux: Modelling the dynamics of epistemic states*. Cambridge, MA: MIT Press.
Goheen, J., & Mothershead, J. (Eds.). (1970). *Collected papers of Clarence Irving Lewis*. Stanford, CA: Stanford University Press.
Grove, A. (1988). Two modellings for theory change. *Journal of Philosophical Logic, 17*, 157–170.
Hjortland, O. (forthcoming). Anti-exceptionalism about logic. *Philosophical Studies*.
Lewis, C. (1921). The structure of logic and its relation to other systems. *The Journal of Philosophy, 18*, 505–516. Read at the conference of the American Philosophical Association, December 1920. Reprinted in Goheen, & Mothershead, 1970, pp 371-382. Page references are to the reprinted version.
Lewis, C. (1932). Alternative systems of logic. *The Monist, 17*, 481–507. Reprinted in Goheen, & Mothershead, 1970, pp 400-419. Page references are to the reprinted version.
Lewis, D. (1979). Scorekeeping in a language game. *Journal of Philosophical Logic, 8*, 339–359.
Mares, E. (2002). A paraconsistent theory of belief revision. *Erkenntnis, 56*, 229–246.
Mares, E. (2011). *A priori*. Durham: Acumen-Routledge.
Mares, E. (2014). Belief revision, probabilism, and logic choice. *Review of Symbolic Logic, 7*, 647–670.
Paoli, F., & Restall, G. (2005). The geometry of non-distributive logics. *The Journal of Symbolic Logic, 70*, 1108–1126.
Plumwood, V. (1993). The politics of reason: Towards a feminist logic. *Australasian Journal of Philosophy, 71*, 436–462.
Restall, G. (2000). *Introduction to subsructural logics*. London: Routlege.
Scott, D. (1971). On engendering an illusion of understanding. *Journal of Philosophy, 68*, 787–807.
Segerberg, K. (1995). Belief revision from the point of view of doxastic logic. *Bulletin of the IGPL, 3*, 535–553.
Shogenji, T. (2013). Reductio, coherence, and the myth of epistemic circularity. In F. Zenker (Ed.), (pp. 165–184). Dordrecht: Springer.
Spohn, W. (2012). *The laws of belief: Ranking theory and its philosophical applications*. Oxford: Oxford University Press.
Urquhart, A. (1978). A topological representation theorem for lattices. *Algebra Universalis, 8*, 45–58.

van Benthem, J. (2011). *Logical dynamics of information and interaction*. Cambridge: Cambridge University Press.
Zenker, F. (Ed.). (2013). *Bayesian argumentation: The practical side of probability*. Dordrecht: Springer.

Author biography. Edwin Mares is a professor of philosophy at Victoria University of Wellington. He has written more than sixty articles and chapters, largely on philosophical logic, and he written the books, Relevant Logic (Cambridge, 2004), A Priori (Routledge, 2011), and (with Stuart Brock) Realism and Antirealism (Routledge, 2007).

6 An introduction to logical nihilism

GILLIAN RUSSELL *

Abstract. The spectre of logical nihilism—the view that there is no logic—has arisen recently in the literature on logical pluralism. Pluralists say that there is more than one correct logic, monists that there is only one. But could there be none at all? This paper presents an argument for the view that there is no logic, offering counterexamples even to apparently secure principles like conjunction elimination and identity. It develops a nihilist first order model theory, and which then reveals one problem with the argument for logical nihilism—though not, it is argued, one which should comfort the anti-nihilist

Keywords: logical pluralism, nihilism, generality in logic, model theory, context-sensitivity, counterexamples, logical consequence, conjunction elimination, identity.

Logical nihilism is the view that there is no logic. It's the limit on a spectrum which contains logical monism—the traditional view that there is exactly one logic—and logical pluralism—the view that there is more than one, popularised by (Beall & Restall, 2006) as well as (Varzi, 2002), (Field, 2009), and (Shapiro, 2014).[1] There are

---

*University of North Carolina, Chapel Hill

[1] As pluralists usually note, there are many ways to understand "logic" and on some senses it is quite trivial that there are many logics. The dispute gets interesting when we reserve "logic" for specifications of the entailment relation on a language and count specifications which agree on the extension of that relation as agreeing on the logic. The pluralist is then committed to the existence of many entailment relations, and (my kind of) nihilist to that relation having no relata. There is not much recent mainstream literature on nihilism, but Aaron Cotnoir has kindly showed me a draft of his unpublished paper on this topic. Mortensen, 1989 argues for two relatives of logical nihilism, one based on the idea that nothing is necessary, the other on the idea that nothing is true in all mathematical models. Mortensen, on my reading, is a nihilist about logical truth, though I am unsure whether he would generalise this to logical consequence. Nihilism is

a few ways to make the idea more precise, but on my favoured interpretation, logical nihilism says that there are no *laws* of logic, i.e. no pairs of premise-sets and conclusions such that the premises logically entail the conclusion, or equivalently, that the extension of the *logical entailment* relation is the empty set. The nihilist says that no mater what Γ and φ are,

$$\Gamma \nvDash \varphi.$$

This might seem extreme. Many non-classical logicians reject particular laws of logic, such as the law of excluded middle or explosion, but they are usually at pains to hang on to the laws that remain; overly weak logics threaten to be neither interesting nor useful. The nihilist gives up on *all* of these laws, including modus ponens and the principle of non-contradiction ($\vDash \neg(\varphi \wedge \neg\varphi)$) as well as less controversial principles, such as conjunction elimination ($\varphi \wedge \psi \vDash \varphi$) and identity ($\varphi \vDash \varphi$).

In this short paper I am going to argue that the view is much more reasonable than it might at first seem, present one argument for nihilism—including counterexamples to supposedly "safe" laws like conjunction elimination and identity—and show how to give a nihilist first-order model theory. This will reveal one limitation of the argument for nihilism, but it is not one, I will argue, which should comfort the anti-nihilist.

1 Nihilism, generality and self-defeat

Philosophers sometimes object to nihilism on the grounds that it is that it is self-defeating. The idea is that any argument for nihilism must *use* logic, and so if the conclusion is true, the argument for it must have used something to which it was not entitled; the truth of the conclusion would undermine the argument's force.

In response I wish to point out that even if there are no logical *laws*, it can still be the case that particular instances of familiar forms are unproblematic. For example, this instance of modus ponens

Snow is white → grass is green.
Snow is white.

Grass is green

might be perfectly acceptable to the nihilist—in the sense that the truth of the premises guarantees that of the conclusion—even if she believes that modus ponens is not a law of logic, thanks to some recherché counterexamples involving the truth predicate.

---

also touched on in (Bueno & Shalkowski, 2009) and discussed in terms of models in a recent paper by Estrada-González, 2012.

To claim that modus ponens is *logically* valid is to make a claim of great generality, whereas to claim that the argument above preserves truth is merely to claim that some *instance* of modus ponens is good. The nihilist can accept many instances of modus ponens, so long as they also think that there are some instances which are mistaken.

An analogy with intuitionism or dialetheism can be useful here: intuitionists deny that the law of excluded middle is a law of logic, but even they can reasonably appeal to it when not concerned with statements about infinite collections. (Iemhoff, R., 2013) And dialetheists reject explosion ($\varphi, \neg\varphi \vDash \psi$) as a law of logic—they might cite counterexamples that substitute Liar sentences for φ—and yet they may still think that it is ok to use explosion when we are doing simple number theory, or working out what to have for dinner. In a similar spirit, a nihilist can allow that all the moves in an argument are just fine—they never take one from truth to falsehood—even while they deny that they are instances of completely general, counterexample-free laws of logic. In fact, the generality of logic can make nihilism more appealing. Who is to say that further research might not give us good reason to think that there are sentences which, when substituted for φ and ψ in some erstwhile law, give us true premises but a non-true conclusion?[2]

2 An argument for logical nihilism

It is one thing to show that a view is not self-defeating, and another to give a positive argument for it. Our argument *for* nihilism begins by noting that if we artificially limit the substitution class of sentences, the result can be an artificially strong logic—one which weakens as soon as we lift the restrictions. For example, suppose we were only allowed to substitute *true* sentences for φ and ψ in the fallacious argument form $\varphi \to \psi, \psi \vDash \varphi$ (affirming the consequent). Whatever we substituted for the conclusion φ would be true and the fallacious principle would appear valid.

To get a more accurate picture we need to lift any artificial restrictions on what we substitute for sentence variables. In this case, we need to be able to substitute sentences which have the truth-value *false*, and this will allow us to substitute a true sentence for ψ and a false one for φ, obtaining a counterexample.

But have we have lifted the restrictions far enough? Consideration of sentences containing empty names, or predicates with indeterminate extension, or even future con-

---

[2]One might be tempted to say that it is the fact that we have *proved*, say, the law of excluded middle that rules out the possibility of later counterexamples, no matter how general the theorem is. But the usual model theoretic proofs of, say, the law of excluded middle are not held to be proofs by intuitionists; (moreover a certain kind of intuitionist thinks there are Kripke-style models which are counterexamples to the law of excluded middle—these are simply not considered to be counterexamples by the classical logicians.) Anyway, proofs of logical laws do not generally have the status required to settle the matter.

tingents, leads some philosophers to think that some sentences are *neither* true nor false, and so perhaps we should allow substitution of sentences which are *neither*. We might adjust our logic by adopting the strong Kleene valuations for the connectives and now it seems clear that if we substitute a *neither* sentence for φ in $\varphi \lor \neg\varphi$ we get an instance of the law of excluded middle which is not true—and hence that principle is not a logical truth after all.[3] Indeed, this approach has the same consequences for all laws in which there is nothing to the left of the turnstile including $\vDash \varphi \to \varphi$ and $\vDash \neg(\varphi \land \neg\varphi)$.

We still have the laws with something on the left, including modus ponens and disjunctive syllogism. But the dialetheist thinks that we only have some of these because we are still artificially restricting our substitution class. She thinks that once the language gets rich enough—say, admitting of truth and other semantic predicates—it becomes possible for sentences to be both true and false. (Priest, 1987/2006) Allowing such substitutions gives us counterexamples to both disjunctive syllogism and modus ponens.[4]

Combining the Strong Kleene and LP valuations of the connectives, the result of allowing sentences to be *both* and *neither* is the logic of First Degree Entailment (FDE). FDE is closer to logical nihilism than either LP or K_3, but it is not nihilism yet. Among the laws FDE retains are:

$$
\begin{array}{rcll}
\varphi \land \psi & \vDash & \psi & (\land E) \\
\varphi, \psi & \vDash & \varphi \land \psi & (\land I) \\
\varphi & \vDash & \varphi \lor \psi & (\lor I) \\
\varphi & \vDash & \varphi & (ID)
\end{array}
$$

Can the substitution class can be widened yet further to allow counterexamples to even these? I plan to show that it can and I will focus my attention on $(\land E)$ and (ID), on the grounds that these are sometimes take to be the safest and most secure laws of logic. The thought will be that if we can given counterexamples to even these principles, then we can give counterexamples to any logical laws.[5]

---

[3] I assume throughout this paper that the designated values are the ones that include truth e.g. *true* and *both*.

[4] For both (DS) and (MP) we can let φ be *both* and ψ be *false*. Then all the premises come out true (though some may be false as well) and the conclusions are false.

[5] Conjunction elimination is so sacrosanct—or perhaps just so uninteresting—that even the most radical of logicians sometimes see it as secure: "I think it just false that all principles of inference fail in some situation. For example, any situation in which a conjunction holds, the conjuncts hold, simply in virtue of the meaning of ∧." Priest, 2006, p. 202-3 Perhaps that it because it is analytic, or true in virtue of the meaning of 'and'. But claims of analyticity in the formal sciences have a bad track record. We're often less confident of our claims about meaning (and so claims about truth in virtue of meaning) claims than we are of the basic principles of logic. "Attempts are made to fence off purely logical claims as in some sense *analytic*, in a sense that would make them uncontroversial, whereas metaphysical claims are correspondingly synthetic, and inherently liable to controversy. The history of logic tells against any such

($\wedge E$) is sometimes taken to be definitional of $\wedge$. (Carnap, R., 1937, p. xv) (Gentzen, G., 1964) And even if it isn't, one might think that the reoccurrence of φ on the left and the right of the turnstile, along with the standard truth-conditions for conjunction, are sufficient for the correctness of ($\wedge E$). If $\varphi \wedge \psi$ is true, then the truth-conditions for $\wedge$ say that φ is true, giving us the truth of the right-hand side.

There are two ways a nihilist might try to resist here. The first is to dispute the definition and/or truth-clause for $\wedge$. There are logics where conjunction is given an alternative (non-equivalent) definition, and one might even think that one of these gives a better account of the truth-conditions of the English word 'and'. (Restall, 2002) Two problems with with this response are, first, that it is vulnerable to the charge of 'changing the subject'; the question was about ($\wedge E$), not a different rule written with similar looking symbols. (Quine, 1986) The second problem is that that the new logic will have laws for $\wedge$ too—just different ones—and so this strategy would remove obstacles to nihilism only to set up new ones.

A more promising approach is to allow the substitution of some odd kinds of expression, for example, to get a counterexample to conjunction elimination we might consider sentences whose truth-value depended upon *whether or not they were in the scope of a conjunction.* Imagine, for instance, that we had a predicate 'con-white' which took the same extension as 'white' while embedded in a conjunction, but the null-set when not so embedded. Then the following argument would have true premises and a false conclusion:

Snow is con-white $\wedge$ grass is green.

Snow is con-white.

Using a similar strategy to get a counterexample to (ID), we might consider sentences whose truth-value depends upon whether they function as a premise or as a conclusion in an argument. Suppose we had a predicate *prem-white*, whose extension matches the extension of *white* when the sentence appears in the premises to an argument, but is the null set otherwise. Then the following argument would have a true premise but a false conclusion:

Snow is prem-white.

Snow is prem-white.

We are quite used to considering context and how it effects the truth-value of sentences and their logic, though usually the aspects of context we are interested in are agent-identity and the time or place of the speech act—things which do not change across

---

contrast. All major logical principles have been rejected on metaphysical grounds."Williamson, 2013, p. 146

the argument. (Kaplan, 1989) The proposal now is simply to look at expressions whose extension varies with *sentential context*, which may.[6] In the next section we will look at a way to introduce predicates like this into the model theory for a standard first-order logic.

3 Nihilist model theory

We will introduce the new models by making adjustments to the interpretation function of a classical, first-order Tarski model. We begin with a simple first-order language, whose non-logical expressions are individual constants ($a, b, c,$ ...etc.) and predicates of arity $n > 0$ ($P, Q, R,$...etc). The logical expressions are the truth-functors $\neg, \wedge, \vee$ and $\rightarrow$, and the quantifier $\forall$. We will also need individual variables, $x, y, z,$ etc. Models are pairs of a domain and an interpretation function, $\langle D, I \rangle$, with D a non-empty set and I a function which assigns extensions to all the non-logical expressions (elements of the domain to individual constants and elements of D^n to non-logical predicates of arity $n > 0$.)

In classical models the truth-values of atoms are determined independently of their position in any larger sentence. That is the natural result of the fact that interpretation functions in classical models do not take sentence position, or anything that varies with it, as an argument. We can change that. But first, some terminology: let a predicate's *atom* be the shortest formula of which it is a part. An atom is in the *direct scope* of a truth-functor iff the smallest *proper* super-formula containing the atom has that functor as its main connective.[7] For example, in the formula $Pa \wedge Qb$, P's atom is Pa, which is in the direct scope of the conjunction. In the formula Pa, P's atom is Pa and it is not in the direct scope of any truth-functor, and in $\neg(Pa \rightarrow (Qb \wedge Rc))$, P's atom is Pa and this is in the direct scope of the conditional, but not in the direct scope of $\neg$ or $\wedge$.

For our first counterexample we need only three sentence position values: L, R, and S. L is the position of a predicate whose atom is in the direct scope of some binary truth-functor and whose atom appears to the left of that truth-functor. R is the position of a predicate whose atom is in the direct scope of some binary truth-functor and whose atom appears to the right of that truth-functor. And S is for a predicate whose atom is not in the direct scope of any binary truth-functor.[8] Now let $I(\alpha, X)$ (the interpretation

---

[6]The idea of such sensitivity to sentential context can already be found in (Frege, 1985/1892); Frege thinks that the extension of a sentence varies depending on whether or not it is embedded in a propositional attitude ascription.

[7]The smallest superformula containing the atom is the atom itself, but that is not a proper superformula.

[8]For the purposes of these definitions we assume that our truth-functors are $\leq$ 2-place, and that binary functors are written between their arguments. We would probably want numerical position-values if we were dealing with languages which contained $>$ 2-place truth-functors.

An introduction to logical nihilism

function of the model) be a function which takes a non-logical expression, α, and a sentence position, X, and yields an appropriate extension.

Now I could be a constant function of sentence position for some predicates. For example, where the domain is the natural numbers, this is a possible interpretation for a predicate P:

$$I(P,L) = \{0,1,2\}$$
$$I(P,R) = \{0,1,2\}$$
$$I(P,S) = \{0,1,2\}$$

Here the position of P's sentence atom makes no difference to P's extension. But we can also have new kinds of predicates, such as *LeftP*, *RightP*, and *SoloP*, whose extension varies with sentence position:

$I(LeftP,L) = \{0,1,2\}$ $I(RightP,L) = \emptyset$ $I(SoloP,L) = \emptyset$
$I(LeftP,R) = \emptyset$ $I(RightP,R) = \{0,1,2\}$ $I(SoloP,R) = \emptyset$
$I(LeftP,S) = \emptyset$ $I(RightP,S) = \emptyset$ $I(SoloP,S) = \{0,1,2\}$

We extend I to a valuation function V with the following definition:

1. $V(\Pi t_1,...t_n) = 1$ iff $I(\langle[t_1],....[t_n]\rangle) \in I(\Pi,X)$, where Π is an n-place predicate and $t_1,...,t_n$ are terms.[9]
2. $V(\neg \varphi) = 1$ iff $V(\varphi) = 0$
3. $V(\varphi \wedge \psi) = 1$ iff $V(\varphi) = 1$ and $V(\psi) = 1$
4. $V(\varphi \rightarrow \psi) = 1$ iff $V(\varphi) = 0$ or $V(\psi) = 1$
5. $V(\varphi \vee \psi) = 1$ iff $V(\varphi) = 1$ or $V(\psi) = 1$ or both
6. $V(\forall \xi \varphi) = 1$ iff for all $u \in D$, $V_{g^u_\xi}(\varphi) = 1$[10]

With the exception of clause 1. which mentions the new style of interpretation function, these are the familiar classical truth clauses; no mention is made of the sentence position variable. But since the extension of a predicate, and hence the truth-value of an atomic sentence, can depend on sentence position, the truth-value of a complex sentence can still be affected by sentence position. It's conceivable, for example, that $Fa \wedge Gb$ could be true, while $Gb \wedge Fa$ is false.[11]

---

[9] Atomic formulas do not officially have parentheses, but I will sometimes add them for readability.

[10] g is a variable assignment and g^u_ξ one just like g except that the element $u \in D$ is assigned to the variable ξ. I will say more about variable assignments in the next section, but the details can be ignored until then.

[11] This would give us a counterexample to the commutativity of conjunction: $Left(a) \wedge Right(b) \nvDash Right(b) \wedge Left(a)$. It would model the invalidity of 'This atom is on the left and this atom is on the right. *Therefore:* This atom is on the right and this atom is on the left'.

For the purposes of illustration, let's assign an interpretation to the individual constants a, b and c: suppose that whether X=R, L or S, $I(a,X) = 0$, $I(b,X) = 1$, and $I(a,X) = 2$.

Now consider $V(Pa \wedge RightP(a))$. $V(Pa) = 1$ since $[a] = I(a,L) = 0$ and $0 \in I(P,L)$, so the left-hand side of the conjunction is true. On our specified interpretation $I(RightP, R) = \{0, 1, 2\}$ and $I(a, R) = 0$. Since $0 \in I(Right, R)$, the right-hand side of the conjunction is true as well. Applying the truth-clause for $\wedge$ we see that the entire conjunction is true, i.e. $V(Pa \wedge RightP(a)) = 1$

Next consider $V(RightP(a))$. On our specified interpretation $I(RightP, S) = \emptyset$, and $I(a, S) = 0$. We note that $0 \notin \emptyset$ and use the truth-clause for atomic sentences to conclude that $RightP(a)$ is false, i.e. $V(RightP(a)) = 0$. The interpretation I, and more generally the model of which it is a part, is a counterexample to ($\wedge E$). So ($\wedge E$) fails in at least one case, and hence is not a logical law.

The very same interpretation also provides counterexamples to ($\wedge I$) and the commutativity of conjunction, but in order to give a counterexample to (ID) we need to adapt our interpretations slightly, and allow not just position in a supersentence to determine truth-value, but also position in the argument. Suppose, for example, that we used the sentence positions P and C (for 'premises' and 'conclusion') as position arguments for I, and that our langauge contained predicates $PremP$ and $ConcP$, whose extensions would vary with that position in the argument. This would allow us to define an interpretation, and a fortiori a model, on which $PremP(a) \vDash PremP(a)$ had a true premise but false conclusion, and give us a counterexample to (ID).

4 Sentences containing only logical expressions

The above models provide counterexamples to familiar logical principles by expanding the available interpretations for the non-logical expressions in a language. If our language is the simple one specified above, every sentence in it contains non-logical predicates and so is amenable to truth-value change by this method. Still, there are extensions of the language which contain sentences that lack a non-logical predicate or constant. Suppose, first, that we add a logical identity predicate, =. We make no change to the structure of the models, but extend the valuation function as follows:

$$V(t_1 = t_2) = 1 \text{ iff } [t_1] = [t_2]$$

The new symbol permits us to form the sentence $\forall x(x=x)$. This, one might think, will be a logical truth as usual. It doesn't *contain* any non-logical constants and so messing with the interpretation function will not affect its truth-value. Moreover, second, we might add the 0-place truth-functor $\top$, and the valuation clause:

$$V(\top) = 1$$

⊤ is a sentence all on its own, and the valuation function ensures that it is true in all models, and so a logical truth. Perhaps there are logical truths after all, if our language is rich enough? Moreover, T allows us to say more: there are valid arguments with premises on the left of the turnstile;

$$\top \wedge \top \vDash \top$$

My responses to these two objections to strict nihilism will be different. To the first, I follow (Williamson, 2013) in holding that our variable assignments should assign the same kinds of objects to individual variables (and predicate-variables, if we are working in a second order logic) as our interpretation function assigns to individual constants (predicates). For this reason, when we add quantifiers and the corresponding variable assignments used in their interpretation, variable assignments too should take position in the sentence as an argument, in addition to the variable itself. Of course this makes space for the variable on the left of the identity sign to be assigned a different element from the variable on the right, and hence for the assignment to make $x = x$ false. $\forall x(x = x)$ would then fail too.

But this response does not cover ⊤ or $\top \wedge \top \vDash \top$. Here I think the objector is correct: this method of generating counterexamples to the usual logical principles does not cover ⊤, and so the argument has not taken us all the way to nihilism. I hold, however, that this is little comfort. The reason nihilism is troubling, if it is, is that it threatens the usefulness of logic. Nihilists will not be using their logic to do exciting metatheory, specify deductive theories, develop ambitious logicist projects, or provide a systematic approach to modal metaphysics. But if this is the problem with nihilism then it is similarly a problem for a related view that I will call logical *minimalism*: logical minimalism is the view that there are *hardly any* laws of logic. The view that the only laws of logic are constructed from truth-functional logical connectives *alone* (with no variables or non-logical expressions at all) is certainly such a view. Though it is not total nihilism, it might as well be.

Acknowledgements. This paper has benefited from discussion at the DEXII at UC Davis, the 4th California Metaphysics Workshop at USC, the OSU/Maribor/Rijeka Philosophy Conference in Dubrovnik 2015, CLMPS 2015, the 9th meeting of the German Society for Analytic Philosophy, the Society for Exact Philosophy meeting at McMaster University and a colloquium talk in the philosophy department at Columbia University. I'm especially grateful for helpful comments from Axel Arturo Barceló, Chris Blake-Turner, Aaron Cotnoir, Liam Kofi Bright, Teresa Kouri, Thomas Hofweber, Kirk Ludwig, Chris Menzel, Catarina Dutilh Novaes, Christopher Peacocke, Graham Priest, Stephen Read, Greg Restall, Stewart Shapiro, Achille Varzi,

Timothy Williamson and Richard Zach. This research was partly done under the PA-PIIT project IA401015.

Bibliography

Beall, J., & Restall, G. (Eds.). (2006). *Logical pluralism*. Oxford: Oxford University Press.
Bueno, O., & Shalkowski, S. A. (2009). Modalism and logical pluralism. *Mind*, *118*(470), 295–321.
Carnap, R. (1937). *The logical syntax of language*. London: Keagan Paul.
Estrada-González, L. (2012). Models of possibilism and trivialism. *Logic and Logical Philosophy*, *21*(2), 175–205.
Field, H. (2009). Pluralism in logic. *The Review of Symbolic Logic*, *2*(2), 342–359.
Frege, G. (1985/1892). On Sense and Nominatum. In A. P. Martinich (Ed.), *The philosophy of language* (4th). Oxford: Oxford University Press.
Gentzen, G. (1964). Investigations into logical deduction. *American Philosophical Quarterly*, *1*(4), 288–306.
Iemhoff, R. (2013). Intuitionism in the philosophy of mathematics. In Zalta, E. N. (Ed.), *The Stanford Encyclopedia of Philosophy* (Fall 2013). Metaphysics Research Lab, Stanford University.
Kaplan, D. (1989). Demonstratives: An essay on the semantics, logic, metaphysics, epistemology of demonstratives and other indexicals. In J. Almog, J. Perry, & H. Wettstein (Eds.), *Themes from Kaplan*. Oxford: Oxford University Press.
Mortensen, C. (1989). Anything is possible. *Erkenntnis*, *30*(3), 319–337.
Priest, G. (1987/2006). *In contradiction* (2nd). Oxford: Oxford University Press.
Priest, G. (2006). *Doubt truth to be a liar*. Oxford: Oxford University Press.
Quine, W. V. O. (1986). *Philosophy of logic*. Cambridge, Mass.: Harvard University Press.
Restall, G. (2002). *An introduction to substructural logics*. London: Routledge.
Shapiro, S. (2014). *Varieties of logic*. Oxford: Oxford University Press.
Varzi, A. C. (2002). On logical relativity. *Philosophical Issues*, *12*(1), 197–219.
Williamson, T. (2013). *Modal logic as metaphysics*. Oxford: Oxford University Press.

Author biography. Gillian Russell is a Professor in the philosophy department at UNC Chapel Hill. She is the author of *Truth in Virtue of Meaning: A Defence of the Analytic/Synthetic Distinction* and the editor, with Delia Graff Fara of the *Routledge Companion to the Philosophy of Language* and with Greg Restall, of *New Waves in Philosophical Logic*. She works in the philosophy of logic and language, and in epistemology.

7 Intensional logic before Leibniz

PAUL THOM [*]

Abstract. Following on work by Klaus Glashoff, Leibniz's intensional reading of categorical propositions is interpreted within a formal framework in which intensions are related by inseparability and incompatibility, and some intensions are designated as quiddities. This same framework is used to demonstrate some of the main claims made in the modal logics of Avicenna and Robert Kilwardby. Those two logics are differentiated from each other by the different truth-conditions they assign to modal propositions, and by the different assumptions they make about quiddities. It is shown that on Leibniz's intensional reading, categorical propositions are equivalent to one of the main types of modal proposition distinguished by the medievals.

Keywords: Intensional logic, Leibniz, Avicenna, Kilwardby, syllogistic.

1 Introduction

Leibniz knew that a predicate's intension can be specified through the predicate's 'superconcepts', i.e. through the concepts that are inseparable from it as 'animal' is inseparable from 'man'.

Leibniz takes as his model of universal affirmative propositions a statement predicating a genus of its species. The genus is one factor in the definition of the species, the differentia being the other factor:

---

[*]The University of Sydney

> Thus as the notion of the genus is the part, the notion of the species is the whole, for it is composed from the genus and the differentia. (Glashoff, 2002, p.163. My translation.)

He thought that such propositions should be given an intensional reading:

> The common manner of statement concerns individuals, whereas Aristotle's refers rather to ideas or universals. For when I say *Every man is an animal* I mean that all the men are included amongst all the animals; but at the same time I mean that the idea of animal is included in the idea of man. 'Animal' comprises more individuals than 'man' does, but 'man' comprises more ideas or more attributes: one has more instances, the other more degrees of reality; one has the greater extension, the other the greater intension. (Leibniz, 1875–1890, p.495. Lenzen translation.)

According to this reading the truth of a categorical proposition depends on a relationship between the intension of the subject and that of the predicate. Klaus Glashoff illustrates this relationship by reference to the a-proposition, which Leibniz analyses as stating that everything 'positively contained' in the predicate is positively contained in the subject, and everything 'negatively contained' in the predicate is negatively contained in the subject. (Glashoff, 2010, p.267.)

Leibniz worked out a mathematical representation of these relationships. As Glashoff puts it:

> Leibniz's basic idea was to assign to each 'simple' concept a prime number and to each 'composed' concept the product of prime numbers. This idea leads directly to his definition of the U.A.-*propositio* – namely, that Axy is true just in case y divides x. (Glashoff, 2002, p.172.)

The truth-condition of the particular negative is the negation of that for the universal affirmative. But Leibniz had difficulties in formulating the truth-conditions for the universal negative and particular affirmative propositions. He first tried supposing that the particular affirmative *Some A is B* is true iff either *All A is B* or *All B is A* is true. In his arithmetical representation, what this comes to is that the number associated with A divides the number associated with B or vice versa. (Glashoff, 2002, p.173.) But he rejected this thought on the ground that a species might fall under two different genera neither of which was subalternate to the other.

Glashoff documents Leibniz's attempts to surmount this difficulty. In his final account of the truth-conditions for propositions with subject x and predicate y, Leibniz associated with each term an ordered pair of characteristic numbers, one marked as positive, the other negative. For example, $+s - \sigma$ might be associated with x, and $+p - \pi$ with y. As Glashoff explains it, the idea is that s is the product of prime numbers of elementary properties *contained* in x, that is, of all elementary properties y such that *All x is y* holds true. On the other hand, σ is the product of all primes belonging to those elementary properties z such that *No x is z* holds. Similarly for p and π. s and σ are stipulated to be 'apt', i.e. they cannot have a common divisor; they must be relatively prime. Similarly for p and π. Glashoff points out that these pairs correspond to the rational numbers. (Glashoff, 2002, p.177.) Using apt pairs of numbers, Leibniz gives the following truth-conditions for universal affirmative and particular affirmative propositions:

> Universal Affirmative. $\mathbf{a}((s,\sigma),(p,\pi))$ if and only if: p divides s and π divides σ.
> Particular Affirmative. $\mathbf{i}((s,\sigma),(p,\pi))$ if and only if: s and π as well as σ and p are relatively prime. (Glashoff, 2002, p.178.)

The i-proposition means that nothing positively contained in one term is negatively contained in the other (Glashoff, 2010, p.267.) Glashoff explains the conditions under which the truth-condition for the i-proposition follows from that for the a-proposition (Glashoff, 2002, p.183.)

My interest in this paper is not with the mathematics but with the underlying philosophical analysis in Leibniz's theory: his appeal to 'ideas' or 'concepts' as the intensions of predicates, his quantification over them (as in his phrase 'everything positively contained in the predicate'), and his claim that relations between these 'ideas' provide the truth-conditions for categorical predications read intensionally (e.g. the claim that the universal affirmative is true iff everything positively contained in the predicate is positively contained in the subject and everything negatively contained in the predicate is negatively contained in the subject).

2   A formal framework for a logic of intensions

With this is mind we need to devise a formal framework in which individual intensions can be named and quantified over. The framework needs to contain predicate constants which when attached to names of intensions are capable of producing truths. These predicate constants need to include some which will be useful for generating formal representations of (at least some) categorial propositions.

As a formal framework I shall suppose an extension of a predicate logic with identity, containing predicates $A, B, C, \ldots$. We augment this logic's object language in three ways. First, we add names corresponding to the predicates—α corresponding to A, β to B, γ to $C, \ldots$; these names are intended to name the intensions of their associated predicates; thus, α names the intension of the predicate A, and so on. Second, we add a one-place predicate Q which, when supplemented by the name of an intension, forms a proposition. The intended reading of Q is '_ is a quiddity'. In the Aristotelian philosophical tradition a quiddity is whatever is an appropriate answer to the philosophical question 'What is it? (*quid est*). Third, we add two-place predicates $\Leftarrow, \Downarrow$, from which propositions are formed by the addition of an ordered pair of names of intensions. The predicates are intended to express relations of inseparability and incompatibility between intensions. These two-place predicates are subject to the following postulates:

P1. $\kappa \Leftarrow \kappa$
P2. $(\kappa \Leftarrow \lambda \wedge \lambda \Leftarrow \mu) \supset \kappa \Leftarrow \mu$
P3. $\kappa \Downarrow \lambda \supset \lambda \Downarrow \kappa$
P4. $(\kappa \Downarrow \lambda \wedge \lambda \Leftarrow \mu) \supset \kappa \Downarrow \mu$.
P5. $\kappa \Leftarrow \lambda \supset \neg \kappa \Downarrow \lambda$

P5 effectively restricts consideration to intensions that are self-consistent. The other postulates reflect an intuitive understanding of inseparability and incompatibility among intensions.

Quiddities may be thought to be related in ways that are more restricted than the relations that govern intensions in general. For example, P6 may be thought to hold for quiddities, though it is not true in general that all pairs of intensions have to be related by either inseparability or incompatibility.

P6. $(Q\kappa \wedge Q\lambda) \supset (\kappa \Downarrow \lambda \vee \kappa \Leftarrow \lambda \vee \lambda \Leftarrow \kappa)$.

Even more specific than the notion of a quiddity is the notion of a *nature*:

D1. $N\kappa =_{def} Q\kappa \wedge \forall \lambda [(Q\lambda \wedge \kappa \Leftarrow \lambda) \supset \lambda \Leftarrow \kappa]$

A nature is a most specific quiddity.

Using D1 and P6, we can prove the theorem:

T1. $(N\kappa \wedge Q\lambda) \supset (\kappa \Downarrow \lambda \vee \lambda \Leftarrow \kappa)$

Proof. Suppose κ is a nature and λ a quiddity. Since by D1 natures are quiddities, we can apply P6 and infer that κ stands to λ in a relation of either inseparability or incompatibility. But since κ is a nature, by D1 if $\kappa \Leftarrow \lambda$ then $\lambda \Leftarrow \kappa$. So either κ and λ are incompatible or λ is inseparable from κ.

In order to test the satisfiability of a set **S** of propositions in the object language, we make the following additions to the usual requirements for a model. There is a domain **D** of intensions, and a subdomain **E** (**E** ⊂ **D**) of quiddities. To each name of an intension in **S** is assigned a member of **D**. To each of the predicates $A, B, C, \ldots$ in **S** is assigned a subset of **D**. To each name of an intension κ such that the proposition '$Q\kappa$' is in **S**, is assigned a member of **E**, these being the only members of **E**. To $\Leftarrow$ is assigned a set r_1, to $\Downarrow$ a set r_2, where r_1 and r_2 are sets of ordered pairs of members of **D**.

Let us now look at two medieval logicians who, in different ways, appealed to intensional considerations in their treatment of modal logic. The individual analyses of modalised categorical propositions proposed by our medieval authors will be given a formal representation in our extended predicate logic. Both of the medieval logicians I will discuss proposed a distinctive way of theorising modalised categorical propositions, i.e. propositions of the following eight forms:

La: Every _ is necessarily ...
Le: No _ is possibly ...
Li: Some _ is necessarily ...

Lo: Some _ is not possibly ...
Ma: Every _ is possibly ...
Me: No _ is necessarily ...
Mi: Some _ is possibly ...
Mo: Some _ is not necessarily ....

3 Avicenna (d. 1037)

In his late work *Pointers and Reminders* the great Persian logician Avicenna presents the following threefold division of predicates:

> Predicates may be essential, implicate accidental, and separable accidental. Let us begin by defining (*ta'rif*) the essential. [Some] predicates are ... constitutive of their subjects. ... by "constitutive" ... [I mean] the predicate which the subject needs for the realisation of its quiddity, and which is intrinsic to its quiddity, a part of it. This is like being a figure for a triangle, or corporeality for man.... conceiving body as body [and] ... triangle as triangle. (Street, forthcoming; Dunyā, 1971, p.151.)

Though he refers to predicates 'body' and 'triangle' here, it is clear that Avicenna is talking about these predicates in respect of their *intensions*. This is clear from his phrases 'body as body and 'triangle as triangle'. His idea is that one intension may be essential to another, or it may be an implicate of the other without being essential to it, or it may be separable from it.

He elaborates on the latter two cases:

> As for the non-constitutive implicate (and it is singled out by the term 'implicate' (*lazim*), even though the constitutive is also an implicate), it is that which is associated with (*yashabu*) the quiddity while not being a part of it, like [the sum of the angles of] a triangle being equal to the sum of two right angles. (Street, forthcoming; Dunyā, 1971, p.158.)

By an implicate of a given intension Avicenna appears to mean an intension that is inseparable from it; implicates of a given intension are contrasted with intensions that are not inseparable from it. He also uses a notion of a quiddity. I shall take his talk of what is intrinsic to a quiddity, or a part of the quiddity, to mean what is inseparable from the quiddity.

Avicenna divides propositions expressing a necessary connection between predicate and subject into three classes: those where the necessity is absolute, those where the necessity is relative to the essence of the subject, and those that are relative to a description under which the subject is presented:

> Necessity may be absolute (*'ala l-iṭlaq*), as in *God exists*. Or it may be connected (*mu'allaqa*) to a condition (*šarṭ*). The condition is either

perpetual [relative] to the existence of the substance [of the subject] (*ḏat*), as in *Man is necessarily a rational body* ... or the duration (*dawam*) of the subject's being described with what is set down with it, as in *All mobile things are changing*. (Street, forthcoming; Dunyā, 1971, p.265.)

He gives an example of a substantial necessity-proposition:

> For example, some bodies by necessity are moving, that is, as long as the substance of those some [bodies] exists. (Street, forthcoming; Dunyā, 1971, p.292.)

He has in mind the heavenly bodies, part of whose essence is to be in motion.

Plausibly, we can capture Avicenna's idea of propositions expressing descriptional necessities in our extended predicate calculus by adopting the following formal representations for universal necessity- and possibility-propositions with subject A and predicate B, 'All A is necessarily B' (for example, 'Everything walking is necessarily moving'), 'No A is possibly B', 'All A is possibly B', 'No A is necessarily B':

La: $\beta \Leftarrow \alpha$
Le: $\beta \Downarrow \alpha$
Ma: $\forall \kappa [\alpha \Leftarrow \kappa \supset \neg \beta \Downarrow \kappa]$
Me: $\forall \kappa [\alpha \Leftarrow \kappa \supset \neg \beta \Leftarrow \kappa]$.

L stands for necessity, M for possibility. α stands for the intension of 'A', β for the intension of 'B'. Thus, the La-proposition is true iff the intension of the predicate is inseparable from the intension of the subject; the Me-proposition is true iff the intension of the predicate is separable from every intension from which the intension of the subject is inseparable. And so on for the others.

Particular L- and M-propositions can be accounted for as the contradictories of universal propositions of the dual modality.

Given these formal representations of descriptional necessity- and possibility-propositions, we can show that the Le-proposition is convertible, as is the Li-proposition, and we can demonstrate the validity or invalidity of syllogistic inferences constructed from descriptional propositions. I take as an example the syllogism Barbara LML ('Every A is possibly B, every B is necessarily C, so every A is necessarily C'), which is invalid.

T2. (i) *Le-conversion* **is valid, (ii)** *Li-conversion* **is valid, (iii)** *Barbara LML* **is invalid – where all propositions are descriptional**

What is required to be proved is that
(i) $\beta \Downarrow \alpha$ is deducible from $\alpha \Downarrow \beta$;
(ii) $\exists \kappa [\beta \Leftarrow \kappa \wedge \alpha \Leftarrow \kappa]$ is deducible from $\exists \kappa [\alpha \Leftarrow \kappa \wedge \beta \Leftarrow \kappa]$;
(iii) $\{\forall \kappa [\alpha \Leftarrow \kappa \supset \neg \beta \Downarrow \kappa], \lambda \Leftarrow \beta, \neg \lambda \Leftarrow \alpha\}$ is satisfiable.

Proof. (i) P3.

(ii) This is trivial, because what is required to be proved is that $\exists \kappa [\beta \Leftarrow \kappa \wedge \alpha \Leftarrow \kappa]$ is deducible from $\exists \kappa [\alpha \Leftarrow \kappa \wedge \beta \Leftarrow \kappa]$.

(iii) Let $\mathbf{D} = \{d_1, d_2, d_3\}$, $\mathbf{E} = \emptyset$. Let r_1 be $\{\langle d_1, d_1 \rangle, \langle d_2, d_2 \rangle, \langle d_2, d_3 \rangle, \langle d_3, d_2 \rangle, \langle d_3, d_3 \rangle\}$. Let $r_2 = \emptyset$.

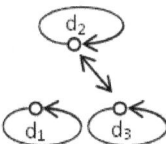

Figure 1. Counter-example to Barbara LML with all propositions descriptional

In Figure 1 unfilled circles represent intensions, arrows represent relations of inseparability. Since the only intension from which d_1 is inseparable is d_1 itself, and d_2 is compatible with d_1 (since $r_2 = \emptyset$), d_2 is compatible with every intension from which d_1 is inseparable, so $\forall \kappa [\alpha \Leftarrow \kappa \supset \neg \beta \Downarrow \kappa]$ is satisfied. Since $\langle d_3, d_2 \rangle \in r_1$, $\gamma \Leftarrow \beta$ is satisfied. Since $\langle d_3, d_1 \rangle \notin r_1$, $\neg \gamma \Leftarrow \alpha$ is satisfied. All the postulates are satisfied: there are no incompatibilities, so P3, P4 and P5 are trivially satisfied; every intension is inseparable from itself; transitivity of inseparability is not violated; there are no quiddities, so P6 is trivially satisfied. This model is illustrated in Figure 1.

The model postulates three intensions, none of which is a quiddity. Each intension is inseparable from itself. The second and third are mutually inseparable. Every intension is compatible with itself and with the others. We could let the first intension be 'walking', the second and third 'humming'. Since 'humming' is compatible with every intension from which 'walking' is inseparable (viz. from 'walking' itself), 'Everything walking is possibly humming' is true. Since 'humming' is inseparable from 'humming', 'Everything humming is necessarily humming' is true. But 'Everything walking is necessarily humming' is not true, since 'humming' is not inseparable from 'walking'.

Now we come to Avicenna's substantial predications. What does Avicenna mean by the predicate's being 'perpetual [relative] to the existence of the substance of the subject'? I shall take this to mean (i) that the intension of the subject-term is a quiddity, and (ii) that the intension of the predicate-term is inseparable from every nature from which the intension of the subject-term is inseparable. Given this, we represent the La-proposition 'Every A is necessarily B' by the formula $Q\alpha \wedge \forall \kappa [N\kappa \supset (\alpha \Leftarrow \kappa \supset \beta \Leftarrow \kappa)]$. Bearing in mind that La- and Mo-propositions are contradictories, as are Le- and Mi-propositions, and that the La- and Le-propositions must entail the Li- and Lo-propositions, we can adopt the following representations for universal necessity- and possibility-propositions:

La: $Q\alpha \wedge \forall \kappa [N\kappa \supset (\alpha \Leftarrow \kappa \supset \beta \Leftarrow \kappa)]$
Le: $Q\alpha \wedge Q\beta \wedge \forall \kappa [N\kappa \supset (\alpha \Leftarrow \kappa \supset \beta \Downarrow \kappa)]$
Ma: $Q\alpha \wedge \forall \kappa [N\kappa \supset (\alpha \Leftarrow \kappa \supset \neg \beta \Downarrow \kappa)]$

Me: $Q\alpha \supset \forall\kappa[N\kappa \supset (\alpha \Leftarrow \kappa \supset \neg\beta \Leftarrow \kappa)]$.

Given these formal representations, we can prove:

T3. (i) *Le-conversion* **is valid, (ii)** *Li-conversion* **is invalid, (iii)** *Barbara LML* **is valid – where all propositions are substantial.**

What needs to be proved is that
(i) $Q\beta \wedge Q\alpha \wedge \forall\kappa[N\kappa \supset (\beta \Leftarrow \kappa \supset \alpha \Downarrow \kappa)]$ is deducible from
$Q\alpha \wedge Q\beta \wedge \forall\kappa[N\kappa \supset (\alpha \Leftarrow \kappa \supset \beta \Downarrow \kappa)]$;
(ii) $\{Q\alpha \wedge \exists\kappa[N\kappa \wedge \alpha \Leftarrow \kappa \wedge \beta \Leftarrow \kappa], \neg Q\beta \vee \neg\exists\kappa[N\kappa \wedge \beta \Leftarrow \kappa \wedge \alpha \Leftarrow \kappa]\}$ is satisfiable,
(iii) $Q\alpha \wedge \forall\kappa[N\kappa \supset (\alpha \Leftarrow \kappa \supset \gamma \Leftarrow \kappa)]$ is deducible from
$Q\alpha \wedge \forall\kappa[N\kappa \supset (\alpha \Leftarrow \kappa \supset \neg\beta \Downarrow \kappa)]$ and $Q\beta \wedge \forall\kappa[N\kappa \supset (\beta \Leftarrow \kappa \supset \gamma \Leftarrow \kappa)]$.

Proof. (i) Suppose $Q\alpha \wedge Q\beta$ and $\forall\kappa[N\kappa \supset (\alpha \Leftarrow \kappa \supset \beta \Downarrow \kappa)]$. Then $\forall\kappa[N\kappa \supset (\neg\beta \Downarrow \kappa \supset \neg\alpha \Leftarrow \kappa)]$. So, by P5, $\forall\kappa[N\kappa \supset (\beta \Leftarrow \kappa \supset \neg\alpha \Leftarrow \kappa)]$. So, by T1, $\forall\kappa[N\kappa \supset (\beta \Leftarrow \kappa \supset \alpha \Downarrow \kappa)]$. Hence, $Q\beta \wedge Q\alpha \wedge \forall\kappa[N\kappa \supset (\beta \Leftarrow \kappa \supset \alpha \Downarrow \kappa)]$.

(ii) Let **D**=$\{d_1,d_2\}$, **E**=$\{d_1\}$, where d_1 is assigned to α and d_2 is assigned to β. Since d_1 is the only member of **E**, **N**= $\{d_1\}$. Let r_1 be $\{\langle d_1,d_1\rangle, \langle d_2,d_1\rangle, \langle d_2,d_2\rangle\}$. Let $r_2= \varnothing$. Since $d_1\in$ **E** and $d_1\in$ **N** and both $\langle d_1,d_1\rangle$ and $\langle d_2,d_1\rangle$ are in r_1,
$Q\alpha \wedge \exists\kappa[N\kappa \wedge \alpha \Leftarrow \kappa \wedge \beta \Leftarrow \kappa]$ is satisfied. Since
$d_2\notin$ **E**, $\neg Q\beta \vee \neg\exists\kappa[N\kappa \wedge \beta \Leftarrow \kappa \wedge \alpha \Leftarrow \kappa]$ is satisfied. P1 is satisfied: r_1 contains both $\langle d_1,d_1\rangle$ and $\langle d_2,d_2\rangle$. So is P2: whenever $\langle d_i,d_j\rangle$ and $\langle d_j,d_k\rangle$ are in r_1 so is $\langle d_i,d_k\rangle$. P3-P5 are satisfied: r_2 is null. P6 is satisfied: d_1 is the only member of **E**, and $\langle d_1,d_1\rangle$ is in r_1.

A could be 'human', B 'capable of laughter'. The model then postulates that 'human' is an essential predicate from whose intension the intension of 'capable of laughter' is inseparable. 'Some human is necessarily capable of laughter' is true because the intension of 'human' is a quiddity and there is a nature (namely that quiddity) from which the intensions of both 'human' and 'capable of laughter' are inseparable. But 'Something capable of laughter is necessarily human' is not true, because the intension of 'capable of laughter' is not a quiddity.

(iii) Suppose $Q\alpha \wedge \forall\kappa[N\kappa \supset (\alpha \Leftarrow \kappa \supset \neg\beta \Downarrow \kappa)]$ and
$Q\beta \wedge \forall\kappa[N\kappa \supset (\beta \Leftarrow \kappa \supset \gamma \Leftarrow \kappa)]$. Then by T1, $Q\alpha \wedge \forall\kappa[N\kappa \supset (\alpha \Leftarrow \kappa \supset \beta \Leftarrow \kappa)]$. So by P2, $Q\alpha \wedge \forall\kappa[N\kappa \supset (\alpha \Leftarrow \kappa \supset \gamma \Leftarrow \kappa)]$.

T3 agrees with Avicenna's results. He accepts the conversion of the Le-proposition (Dunyā, 1971, p.344). But he rejects Li-conversion, arguing that 'Something laughing is necessarily human' is true, since 'Everything laughing is necessarily human' is true, but 'Something human is necessarily laughing' is not true (Dunyā, 1971, p.335). And he regards Barbara LML as valid where the propositions are substantial. In a context of considering first-figure syllogisms with a possibility minor premise, he says:

And if *All Bs are As necessarily* [sc. the major premise], then the truth is that the conclusion is necessary. (Street, forthcoming; Dunyā, 1971, p.393.)

T4. In the absence of P6, *Barbara LML* is invalid for substantial propositions.

What is required to be proved is that in the absence of P6
$\{Q\alpha \land \forall\kappa[N\kappa \supset (\alpha \Leftarrow \kappa \supset \neg\beta \Downarrow \kappa)], Q\beta \land \forall\kappa[N\kappa \supset (\beta \Leftarrow \kappa \supset \gamma \Leftarrow \kappa)], \neg Q\alpha \lor \neg\forall\kappa[N\kappa \supset (\alpha \Leftarrow \kappa \supset \gamma \Leftarrow \kappa)]\}$ is satisfiable.

Proof. Let $\mathbf{D}=\mathbf{E}=\{d_1,d_2,d_3\}$. Let r_1 be $\{\langle d_1,d_1\rangle, \langle d_2,d_2\rangle, \langle d_2,d_3\rangle, \langle d_3,d_2\rangle, \langle d_3,d_3\rangle\}$. Let r_2 be $\varnothing$.

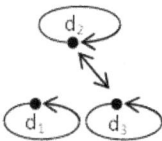

Figure 2. Counter-example to Barbara LML for substantial propositions, in the absence of P6

The model is illustrated in Figure 2, where filled circles represent quiddities, and arrows represent relations of inseparability.

Since $d_1 \in \mathbf{E}$, and every intension from which it is inseparable (namely d_1 itself) is inseparable from it, by D1 $d_1 \in \mathbf{N}$. Since $d_2 \in \mathbf{E}$ and $d_3 \in \mathbf{E}$ and every intension from which d_2 is inseparable (namely d_2 and d_3) is inseparable from it, $d_2 \in \mathbf{N}$. Similarly, $d_3 \in \mathbf{N}$. So $\mathbf{N}=\mathbf{E}$. $Q\alpha \land \forall\kappa[N\kappa \supset (\alpha \Leftarrow \kappa \supset \neg\beta \Downarrow \kappa)]$ is satisfied, because $d_1 \in \mathbf{E}$ and d_2 is compatible with every nature from which d_1 is inseparable (since r_2 is null). $Q\beta \land \forall\kappa[N\kappa \supset (\beta \Leftarrow \kappa \supset \gamma \Leftarrow \kappa)]$ is satisfied, because $d_2 \in \mathbf{E}$ and the only members of $\mathbf{N}$ from which d_2 is inseparable are d_2 and d_3, and d_3 is inseparable from both d_2 and d_3. $\neg Q\alpha \lor \exists\kappa[N\kappa \land \alpha \Leftarrow \kappa \land \neg\gamma \Leftarrow \kappa]$ is satisfied, because there is a nature from which d_1 is inseparable but from which d_3 is separable (viz. d_1).

Now for the postulates. P1 and P2 are satisfied, since the reflexivity and transitivity of inseparability are not violated. P3–P5 are satisfied trivially, since there are no incompatibilities. But P6 is not satisfied: there is a pair of quiddities which are not related by either inseparability or incompatibility, namely d_1 and d_2 (also d_1 and d_3).

Since Avicenna accepts Barbara LML, our analysis provides support for the view that he accepted P6.

Historians of Arabic logic have suggested that we should understand Avicenna's La-proposition to have an *ampliated* subject (e.g., Street, 2002, pp.129-160, 134). That is to say, the intension of the stated subject-term *A* should be understood to be 'compatible with *A*'. It turns out that our formalisation has the effect of ampliating the subject of the La-proposition in this fashion. Let us call a proposition with an ampliated subject-term an L'a -proposition; and let us represent such propositions formally as follows.

L'a: $Q\alpha \wedge \forall \kappa[N\kappa \supset (\neg\alpha \Downarrow \kappa \supset \beta \Leftarrow \kappa)]$.

We can then prove that the L'a-proposition is equivalent to the La-proposition. That the L'a-proposition entails the La-proposition is evident from P5. It remains to prove that the La-proposition implies the L'a.

T5. $Q\alpha \wedge \forall\kappa[N\kappa \supset (\neg\alpha \Downarrow \kappa \supset \beta \Leftarrow \kappa)]$ is deducible from $Q\alpha \wedge \forall\kappa[N\kappa \supset (\alpha \Leftarrow \kappa \supset \beta \Leftarrow \kappa)]$.

Proof.

1. $Q\alpha \wedge \forall\kappa[N\kappa \supset (\alpha \Leftarrow \kappa \supset \beta \Leftarrow \kappa)]$ [Assumption
2. $Q\alpha$ [1.
3. $\forall\kappa[N\kappa \supset (\alpha \Leftarrow \kappa \supset \beta \Leftarrow \kappa)]$ [1.
4. $\forall\kappa[N\kappa \supset (\neg\alpha \Downarrow \kappa \supset \beta \Leftarrow \kappa)]$ [2., 3., T1, D1
5. $Q\alpha \wedge \forall\kappa[N\kappa \supset (\neg\alpha \Downarrow \kappa \supset \beta \Leftarrow \kappa)]$ [2., 4.

4 Robert Kilwardby (d. 1279)

As a second example of a medieval logician for whom modal syllogistic is based on intensional considerations, I briefly mention the English Dominican Robert Kilwardby. In his commentary on Aristotle's *Prior Analytics* (Thom & Scott, 2016) Kilwardby distinguishes two types of necessity-proposition. One type requires merely that the predicate (or, what it signifies) be inseparable from the subject; the other makes a stronger claim, namely that there is a *per se* relationship between predicate and subject:

> ...propositions like this, which have the name of an accident as subject are not necessity-propositions *per se* but only incidentally. For a *per se* necessity-proposition requires the subject *per se* to be something under the predicate. But when it is said 'Every grammarian of necessity is a man', the subject is not *per se* something under the predicate. But it is granted to be necessary because 'grammarian' is not separated from that which is something under 'man'. (Thom & Scott, 2016, p.130.)

When Kilwardby speaks of what is 'under' the subject, he means not individual objects, but intensions. This is evident from the fact that he adopts what has been called the heterodox reading of the *dici de omni* principle (Malink, 2013, p.64). This is to say that he takes the statement 'All A is B' to be true only if for every common term C: if all C is A then all C is B (Thom & Scott, 2016, p.xxvii) Given this, something's being 'under' the subject-term involves a relation of inseparability between intensions.

We see from the quoted passage that according to Kilwardby the subject of a *per se* necessity-proposition cannot be an accidental term like 'grammarian'. He contrasts

Intensional logic before Leibniz

accidental with substantial terms:

> For every term is either substantial or accidental .... And I call those terms *substantial* which signify in the manner of an underlying subject, those *accidental* that do so in the manner of an accident belonging to something else. (Thom & Scott, 2016, p.418.)

A term's substantiality or accidentality is a matter of its signification, i.e. of its intension. A substantial term, unlike an accidental term, has a quiddity as its intension. Thus, we can represent what he calls 'incidental' universal affirmative necessity-propositions simply as statements to the effect that the intension of the predicate is inseparable from the intension of the subject (like Avicenna's descriptional necessities).

Per se necessity-propositions, on the other hand, reduce to what Aristotle calls *per se* predications:

> For necessary propositions reduce to some mode of *per se* inherence, following Aristotle's statement in *Posterior Analytics* I that 'Only *per se* inherences are necessary. (Thom & Scott, 2016, p.160.)

The reference is to Aristotle's *Posterior Analytics* I.6 74b12. According to the doctrine expounded in the early chapters of the *Posterior Analytics*, not only the subject but also the predicate of a *per se* predication belongs in the underlying subject's essence, being either a genus or a differentia of that subject. So Kilwardby is committed to requiring that both terms of an La-propositions have quiddities as their intensions. Given this, the La-proposition can be represented as $Q\alpha \wedge Q\beta \wedge \forall \kappa [N\kappa \supset (\alpha \Leftarrow \kappa \supset \beta \Leftarrow \kappa)]$.

Among *per se* necessity-propositions, Kilwardby makes a semantic distinction between universal affirmatives and universal negatives:

> A universal affirmative necessity affirms the predicate only of those things that are actually under the subject, not of those for which it's contingent to be under the subject. .... But it is otherwise with a negative necessity-proposition .... the predicate in that proposition is also actually denied, under a modality of necessity, of all things that are under the subject, and of all things for which it's contingent to be under the subject. (Thom & Scott, 2016, pp.514, 520.)

By 'being contingently under the subject' Kilwardby means 'being possibly under the subject'. He says that 'the possible and the contingent are convertible in all things' (Thom & Scott, 2016, p.100), meaning that there is a generic sense of 'contingent' which is equivalent to 'possible':

> ...the possible is not stated in the division of the contingent as a specific member of the contingent, but in order to designate a mode of the contingent, namely a mode of taking it in general as a genus. (Thom & Scott, 2016, p.146.)

I take him to be referring to those intensions that are compatible with the subject's intension. Thus the Le-proposition can be represented as $\forall \kappa [N\kappa \supset (\neg \alpha \Downarrow \kappa \supset \beta \Downarrow \kappa)]$.

In order to devise suitable formal representations of the Ma- and Me-propositions, we have to take into account Kilwardby's claim that in the case of *per se* necessity-propositions an Li-proposition 'Some A is necessarily B' implies the disjunction of the La- propositions 'Every A is necessarily B' and 'Every B is necessarily A':

> For among necessities, to state that the particular is true is the same as stating the universal, on account of the necessary relationship of the terms. (Thom & Scott, 2016, p.160.)

In order for this claim to be correct, the Li-proposition would have to entail that both of its terms have quiddities as their intension. For, if the Li-proposition had that entailment, then Kilwardby's claim about Li-proposition's entailing one or other of the corresponding La-propositions would be correct, provided we can assume P6. Suppose that the Li-proposition is true. Then $Q\alpha \wedge Q\beta$. But then, by P6, $\alpha \Downarrow \beta \vee \alpha \Leftarrow \beta \vee \beta \Leftarrow \alpha$. The first disjunct can be excluded, because the Li-proposition must at least entail $\exists \kappa [\alpha \Leftarrow \kappa \wedge \beta \Leftarrow \kappa]$; and if $\alpha \Downarrow \beta$ and $\beta \Leftarrow \kappa$ then by P4 $\alpha \Downarrow \kappa$, which by P5 entails $\neg \alpha \Leftarrow \kappa$. So we have $\alpha \Leftarrow \beta \vee \beta \Leftarrow \alpha$, which by P2 implies $\forall \kappa [N\kappa \supset (\alpha \Leftarrow \kappa \supset \beta \Leftarrow \kappa)] \vee \forall \kappa [N\kappa \supset (\beta \Leftarrow \kappa \supset \alpha \Leftarrow \kappa)]$, i.e. (since $Q\alpha \wedge Q\beta$) one or other of the La-propositions corresponding to our initial Li-proposition is true.

Notice, however, that we can arrive at the same conclusion even without P6. We could, for example, make the weaker assumption P7.

P7. $(N\kappa \wedge Q\lambda \wedge Q\mu \wedge \lambda \Leftarrow \kappa \wedge \mu \Leftarrow \kappa) \supset (\lambda \Leftarrow \mu \vee \mu \Leftarrow \lambda)$.

P6 states that one of a pair of quiddities must be inseparable from the other if the two are compatible. P7 states that one must be inseparable from the other if both are inseparable from a common nature. P1–P5 provide no reason to prevent two quiddities being compatible even if there is no nature from which both are inseparable. Thus P7 is weaker than P6.

Suppose, then, that the Li-proposition is true. Then $Q\alpha \wedge Q\beta$. But then, by P7, $(N\kappa \wedge \alpha \Leftarrow \kappa \wedge \beta \Leftarrow \kappa) \supset (\alpha \Leftarrow \beta \vee \beta \Leftarrow \alpha)$. The Li-proposition must entail $\exists \kappa [N\kappa \wedge \alpha \Leftarrow \kappa \wedge \beta \Leftarrow \kappa]$. So, $\alpha \Leftarrow \beta \vee \beta \Leftarrow \alpha$, which by P2 implies $\forall \kappa [N\kappa \supset (\alpha \Leftarrow \kappa \supset \beta \Leftarrow \kappa)] \vee \forall \kappa [N\kappa \supset (\beta \Leftarrow \kappa \supset \alpha \Leftarrow \kappa)]$, i.e. (since $Q\alpha \wedge Q\beta$) one or other of the La-propositions corresponding to our initial Li-proposition is true.

Accordingly, I propose a minimal interpretation of Kilwardby in which P6 is dropped in favour of P7, and the Li-proposition is represented as $Q\alpha \wedge Q\beta \wedge \exists \kappa [N\kappa \wedge \alpha \Leftarrow \kappa \wedge \beta \Leftarrow \kappa]$. The Me-propositions must then be represented as $\neg Q\alpha \vee \neg Q\beta \vee \neg \exists \kappa [N\kappa \wedge \alpha \Leftarrow \kappa \wedge \beta \Leftarrow \kappa]$.

The Lo-propositions needs to follow from the Le-proposition. This will be the case if we represent it as $\exists\kappa[N\kappa \wedge \neg\alpha \Downarrow \kappa \wedge \beta \Downarrow \kappa]$ provided that we can assume $\exists\kappa N\kappa$. Accordingly I propose that Kilwardby requires an extra postulate:

P8. $\exists\kappa N\kappa$.

Given this representation of the Lo-proposition, the Ma-proposition must be represented as $\forall\kappa[N\kappa \supset (\neg\alpha \Downarrow \kappa \supset \neg\beta \Downarrow \kappa)]$.

> La: $Q\alpha \wedge Q\beta \wedge \forall\kappa[N\kappa \supset (\alpha \Leftarrow \kappa \supset \beta \Leftarrow \kappa)]$
> Le: $\forall\kappa[N\kappa \supset (\neg\alpha \Downarrow \kappa \supset \beta \Downarrow \kappa)]$
> Ma: $\forall\kappa[N\kappa \supset (\neg\alpha \Downarrow \kappa \supset \neg\beta \Downarrow \kappa)]$
> Me: $\neg Q\alpha \vee \neg Q\beta \vee \neg\exists\kappa[N\kappa \wedge \alpha \Leftarrow \kappa \wedge \beta \Leftarrow \kappa]$.

Given these formalisations, and assuming that all necessity-propositions are *per se*, we can prove the following theorem:

T6. (ii) *Le-conversion* is valid, (iii) *Li-conversion* is valid, (vi) *Barbara LML* is **invalid**.

What is required to be proved is:
(i) $\forall\kappa[N\kappa \supset (\neg\beta \Downarrow \kappa \supset \alpha \Downarrow \kappa)]$ is deducible from $\forall\kappa[N\kappa \supset (\neg\alpha \Downarrow \kappa \supset \beta \Downarrow \kappa)]$;
(ii) $Q\beta \wedge Q\alpha \wedge \exists\kappa[N\kappa \wedge \beta \Leftarrow \kappa \wedge \alpha \Leftarrow \kappa]$ is deducible from $Q\alpha \wedge Q\beta \wedge \exists\kappa[N\kappa \wedge \alpha \Leftarrow \kappa \wedge \beta \Leftarrow \kappa]$;
(iii) $\{\forall\kappa[N\kappa \supset (\neg\alpha \Downarrow \kappa \supset \neg\beta \Downarrow \kappa)], Q\beta \wedge Q\gamma \wedge \forall\kappa[N\kappa \supset (\beta \Leftarrow \kappa \supset \gamma \Leftarrow \kappa)],$
$\neg Q\alpha \vee \neg Q\gamma \vee \neg\forall\kappa[N\kappa \supset (\alpha \Leftarrow \kappa \supset \gamma \Leftarrow \kappa)]\}$ is satisfiable.

Proof.

> (i) Trivial.
> (ii) Trivial.
> (iii) We can re-use the model that was used in **T4**. See Figure 2. Let **D**=**E**= $\{d_1, d_2, d_3\}$. Let r_1 be $\{\langle d_1, d_1\rangle, \langle d_2, d_2\rangle, \langle d_2, d_3\rangle, \langle d_3, d_2\rangle, \langle d_3, d_3\rangle\}$. Let $r_2 = \varnothing$. By D1, **N**= $\{d_1, d_2, d_3\}$.

This model satisfies $\forall\kappa[N\kappa \supset (\neg\alpha \Downarrow \kappa \supset \neg\beta \Downarrow \kappa)]$, because every nature is compatible with d_1 and with d_2, since r_2 is null. The model satisfies $Q\beta \wedge Q\gamma \wedge \forall\kappa[N\kappa \supset (\beta \Leftarrow \kappa \supset \gamma \Leftarrow \kappa)]$, because $d_2 \in \mathbf{E}$ and $d_3 \in \mathbf{E}$ and d_3 is inseparable from every nature from which d_2 is inseparable (namely, d_2 and d_3). And the model satisfies $\neg Q\alpha \vee \neg Q\gamma \vee \neg\forall\kappa[N\kappa \supset (\alpha \Leftarrow \kappa \supset \gamma \Leftarrow \kappa)]$, because there is a nature (d_1) from which d_1 is inseparable and d_3 is separable (since $\langle d_3, d_1\rangle$ is not a member of r_1).

T6 agrees with Kilwardby's results. He accepts Le-conversion and Li-conversion:

> First [Aristotle] describes the conversion of necessity-propositions, saying that they convert just like assertorics. For the universal negative converts in its terms without qualification; but the universal affirmative and the particular affirmative convert to a particular affirmative. (Thom & Scott, 2016, p.122.)

(He thinks Aristotle is right.) And here he is on the invalidity of Barbara LML:

> Now, the reason why the necessity, or even the assertoric, doesn't follow is this. A universal affirmative necessity affirms the predicate only of those things that are actually under the subject, not of those for which it's contingent to be under the subject. For the proposition 'Every man of necessity is an animal' doesn't say that whatever can be a man is an animal, but whatever is a man is an animal. And accordingly when a proposition of this type is stated as major premise and the minor is a contingency, an actual affirmation of those things which are taken contingently under the middle does not follow. And accordingly neither an affirmative necessity nor an affirmative assertoric follows, because in both cases the predicate is actually affirmed of the subject. (Thom & Scott, 2016, p.520.)

5 Leibniz

Returning now to Leibniz, we prove:

T7. The Leibniz-Glashoff analysis of (non-modal) a- and e-predications is equivalent to our analysis of Avicenna's descriptional necessity-propositions and Kilwardby's incidental necessity-propositions.

Proof. The Leibniz-Glashoff analysis of 'Every A is B' is 'Everything positively contained in the predicate is positively contained in the subject, and everything negatively contained in the predicate is negatively contained in the subject'. The analysis of 'No A is B' is 'Something positively contained in one term is negatively contained in the other'. Taking positive containment as the converse of inseparability, and negative containment as incompatibility, the Leibniz-Glashoff truth-condition for *All A is B* becomes:

LA. $\forall \kappa [\kappa \Leftarrow \beta \supset \kappa \Leftarrow \alpha] \wedge \forall \kappa [\kappa \Downarrow \beta \supset \kappa \Downarrow \alpha]$.
By P1, LA implies $\beta \Leftarrow \alpha$.
By P2, $\beta \Leftarrow \alpha$ implies $\forall \kappa [\kappa \Leftarrow \beta \supset \kappa \Leftarrow \alpha]$; and by P4, $\beta \Leftarrow \alpha$ implies $\forall \kappa [\kappa \Downarrow \beta \supset \kappa \Downarrow \alpha]$.

The Leibniz-Glashoff truth-condition for *No A is B* is

LE. $\exists \kappa [\kappa \Leftarrow \alpha \wedge \kappa \Downarrow \beta] \vee \exists \kappa [\kappa \Leftarrow \beta \wedge \kappa \Downarrow \alpha]$
By P3 and P4, LE implies $\alpha \Downarrow \beta$.
By P1, $\alpha \Downarrow \beta$ implies LE.

In this light we can say that the analysis Leibniz gives of categorical propositions (according to Glashoff) is equivalent to the meaning Avicenna assigned to descriptional necessity-propositions and Kilwardby to incidental necessity-propositions. But the more sophisticated analyses developed by the medievals—the substantial necessity-propositions of Avicenna and the *per se* necessity-propositions of Kilwardby—find no parallel in Leibniz. At the same time, Leibniz's mathematical representation of categorical propositions, to the best of my knowledge, has no precedent in the medieval logicians.[1]

---

[1] I thank Tony Street for his generous help with the Avicenna material.

Bibliography

Dunyā, S. (Ed.). (1971). *Avicenna. al-Išārāt wal-Tanbīhāt (Pointers and Reminders)* (2nd ed.). Cairo: Dār al-Maʿārif.

Glashoff, K. (2002). On Leibniz's characteristic numbers. *Studia Leibnitiana, 34*(2), 161–184.

Glashoff, K. (2010). An intensional Leibniz semantics for Aristotelian logic. *The Review of Symbolic Logic, 3*(02), 262–272.

Leibniz, G. W. (1875–1890). *Die Philosophischen Schriften von Gottfried Wilhelm Leibniz Vol. 5*. Gerhart, C.I. (Ed.) Berlin, Germany: Weidman.

Malink, M. (2013). *Aristotle's modal syllogistic*. Cambridge, MA: Harvard University Press.

Street, T. (Ed.). (forthcoming). *Avicenna's logic explained. A translation of Naṣīr al-Dīn al-Ṭūsī's Ḥall Mushkilāt al-Ishārāt*. Oakland, CA: University of California Press.

Street, T. (2002). An outline of Avicenna's syllogistic. *Archiv für Geschichte der Philosophie, 84*(2), 129–160.

Thom, P., & Scott, J. (Eds.). (2016). *Robert Kilwardby, Notule libri Priorum, Part 1*. Oxford, UK: Oxford University Press.

Author biographies. Paul Thom is a Fellow of the Australian Academy of the Humanities. He has published several books on the history of medieval logic in both the Latin and Arabic traditions, as well as books on the philosophy of the performing arts. His most recent publication is a two-volume critical edition (with John Scott) and translation of the *Prior Analytics* commentary of Robert Kilwardby (d.1279).

Part B

General Philosophy of Science

8 On some French probabilists of the twentieth century: Fréchet, Borel, Lévy

MARIA CARLA GALAVOTTI [*]

Abstract. In the first half of the Twentieth century a number of authors active in distant parts of Europe and in different areas of scientific research shared a probabilistic approach to science and knowledge in general, albeit embracing different interpretations of probability. My "Probabilistic Epistemology: a European Tradition"[1] focussed on the work of Polish logician Janina Hosiasson, British mathematician Frank Plumpton Ramsey and geophysicist Harold Jeffreys, Italian statistician Bruno de Finetti, and German philosopher of science Hans Reichenbach, arguing that one can speak of a European tradition in probabilistic epistemology.

To clarify the matter, by *probabilistic epistemology* I mean the view that probability is an essential ingredient of knowledge, and that induction is a fundamental component of the scientific method. Such a view is grounded in the conviction that certainty of knowledge and completeness of information are unachievable, and as Patrick Suppes clearly stated, "it is the responsibility of a thoroughly-worked-out empiricism to include an appropriate concept of uncertainty at the most fundamental level of theoretical and methodological analysis. Probabilistic methods provide a natural way of doing so" (Suppes, 1984, p. 99).

The purpose of this paper is to expand on my earlier work by adding to the picture the French mathematicians Maurice Fréchet, Émile Borel and Paul Lévy, all of whom advocated a probabilistic approach to epistemology, bringing new evidence that probabilistic epistemology was widespread throughout Europe in the first half of the

---

[*]Department of Philosophy and Communication, University of Bologna, Italy
Via Zamboni 38 40126 Bologna, Italy
Email: mariacarla.galavotti@unibo.it

[1] See Galavotti (2014).

Twentieth century. Apart from that, the philosophy of probability embraced by such outstanding mathematicians seems worthy of attention in itself. Given the scant literature on the topic, confined to its technical aspects, the present analysis will broaden the picture of the debate on the foundations of probability.

Keywords: probability, epistemology, frequentism, subjectivism.

1 European probabilism in the first half of the twentieth century

Before addressing the philosophy of probability of the French authors, I will briefly sketch the main traits of the position embraced by Richard von Mises, Hans Reichenbach, and Bruno de Finetti to give some substance to criticism moved against these authors by the French probabilists examined in the second part of the paper.

1.1 Richard von Mises' frequentism

Richard von Mises (1883-1953) gave great impulse to the debate on the foundations of probability in the first decades of the last century. Evidence of this is the fact that right at the beginning of the opening lecture (*conference d'introduction*) delivered at the "Colloque consacré a la théorie des probabilités" held in 1937 in Geneva, Maurice Fréchet credits von Mises with "having awakened interest in questions previously addressed in a fragmentary way" (Fréchet, 1938b, p. 19).

Von Mises is deemed the main representative of the frequency theory of probability. According to this view, probability is a characteristic of phenomena that can be empirically analysed by means of observed frequencies. It is defined as the limit of the relative frequency of a given attribute observed in the initial part (sample) of an indefinitely long sequence of repeatable events. A key tenet of this interpretation is that probability values are in general unknown, but can be approached by means of frequencies, as the number of observed elements increases.

The core of von Mises' theory is the notion of *collective* defined as follows: "A collective is a mass phenomenon or a repetitive event, or, simply, a long sequence of observations for which there are sufficient reasons to believe that the relative frequency of the observed attribute would tend to a fixed limit if the observations were indefinitely continued. This limit will be called the probability of the attribute considered within the given collective" (von Mises, 1951/1957, p. 15). In order to qualify as a collective, a sequence has to: (1) be indefinitely long, (2) exhibit frequencies that tend to a limit, and (3) be random. Randomness rests on the method of "place selection", which consists in extracting sub-sequences from the original sequence by considering only the place that each member occupies in the sequence, while ignoring their distinctive properties. Each place selection is defined by a rule that states for any element of the sequence whether it ought to be included in the sub-sequence or not. For instance, the sub-sequence obtained by picking all members whose place number in the sequence is a prime number would satisfy the place selection method. Von Mises defines randomness as *insensitivity to place selection*, which obtains when "the limiting values of the relative frequencies in a collective must be independent of all possible place selections" (von Mises, 1951/1957, p. 25).

Having defined collectives along these lines, von Mises proceeds to formulate the principles of probability theory in terms of collectives by means of the operations of selection, mixing, partition and combination. Adoption of this conceptual machinery is intended to secure probability a foundation that is both empirical and objective. The author maintains that probability applies only to collectives, as suggested by the title of one section of the first chapter of *Probability, Statistics and Truth*: "First the collective - then the probability" (von Mises, 1951/1957, p. 18). A major drawback of von Mises' position is that it makes no sense to apply probability to single events. He openly admitted this, deeming single-case probability simply meaningless.

A remarkable feature of von Mises' perspective is that it embodies a probabilistic epistemology open to indeterminism. The idea is that the developments brought into physics by quantum mechanics have imposed the need to ground the whole edifice of science on a statistical conception of nature, granting indeterminism the same plausibility traditionally attached to determinism. Deeply convinced that probability and statistics offer the most powerful heuristic tool for investigating reality, von Mises heralds a probabilistic approach to the building of scientific knowledge.[2] Such a conviction inspires the closing passage of *Probability, Statistics and Truth*, where the author writes that "starting from a logically clear concept of *probability*, based on experience, using arguments which are usually called *statistical*, we can discover *truth* in wide domains of human interest" (von Mises, 1951/1957, p. 220).

Much of the debate on von Mises' work has focussed on the notion of randomness. After criticism advanced by a number of authors including Alonzo Church, Abraham Wald, Jean Ville, Arthur Copeland and others, von Mises' unrestricted notion of randomness was abandoned in favour of a weaker notion entailing a restricted domain of place selections. Another problematic aspect of von Mises' version of frequentism is the impossibility of applying probability to single events, for it is undeniable that in many areas, including everyday life and the social sciences, the need to speak of single-case probabilities is widely felt. The same holds for Quantum Mechanics, where one speaks of single atoms, particles, and so on; it was precisely in order to solve this problem that in the 50s of the last century Karl Popper put forward the so-called propensity interpretation of probability.[3]

---

[2] See von Mises (1951/1968) for more on von Mises' probabilistic epistemology.

[3] See Galavotti (2005) for more on the propensity theory, as well as on the authors dealt with in the first part of this paper.

1.2 Hans Reichenbach's frequentism

Hans Reichenbach (1891-1953) worked out a version of frequentism that strays in various ways from von Mises'. In particular, Reichenbach sets himself the task of developing a flexible theory suitable for wide applicability, in which it makes sense to talk of single-case probabilities. In addition, Reichenbach adopts a notion of randomness limited to a restricted domain of selections, together with a notion of "practical limit" relative to "sequences that, in dimensions accessible to human observation, converge sufficiently and remain within the interval of convergence"(Reichenbach, 1949/1971, p. 347). He also develops a theory of induction, together with an argument for its justification.

Reichenbach shares von Mises' conviction that it is probability, not truth, that should be put at the core of a sound reconstruction of scientific knowledge, because "the ideal of an absolute truth is an unrealizable phantom" (Reichenbach, 1937, p. 90). Stress on action, prediction, and practice confers a pragmatist flavour to Reichenbach's view of knowledge: he explicitly acknowledges his debt towards Charles Sanders Peirce and William James in connection with the theory of meaning, which revolves around the tenet that "there is as much meaning in a proposition as can be utilized for action" (Reichenbach, 1938, p. 80).

Reichenbach names *Rule of induction* the canon by which probability is obtained as the limit of observed frequency, and emphasizes that any probability attribution is a *posit*, namely "a statement with which we deal as true, although the truth value is unknown" (Reichenbach, 1949/1971, p. 373). Posits differ depending on whether they occur within primitive or advanced knowledge. Reichenbach calls *primitive* the state of knowledge in which no prior knowledge of probabilities is available, and *advanced* the state of knowledge in which prior probabilities are available. In the first case, the Rule of induction yields prior probabilities, and posits are called *blind*; in the second, the probability calculus can be used to combine prior probabilities, and the posits so obtained are called *appraised*. The interplay between blind and appraised posits generates *the method of concatenated inductions*, which is intrinsically *self-corrective*, because the Rule of induction guarantees convergence of probability estimates made by its means.

The self-corrective character of the method of concatenated inductions lies at the core of Reichenbach's pragmatic justification of induction: inductive inference, and more precisely the Rule of induction, is justified because it provides the best possible guide to the future. As Reichenbach wrote: "it is a method of which we know that if it is possible to make statements about the future we shall find them by means of this method" (Reichenbach, 1949/1971, p. 475).

A major point of divergence with von Mises is the fact that Reichenbach does not consider single-case probabilities meaningless, and he attempts to accommodate them within the frequentist outlook. The idea is that posits regarding single events receive a weight from the probabilities attached to the reference class to which the event in question has been assigned. The crucial issue here is the choice of the reference class, which must obey a requirement of *homogeneity*. A reference class is homogeneous if it includes all the properties that are taken to be relevant to the event under study. This is obviously a very strong requirement, hardly ever met in practice because - apart from cases falling directly under the scope of scientific theories - one can never be sure that all relevant information has been taken into account. As a matter of fact, this requirement clashes with Reichenbach's intent to develop a version of frequentism suited to a wide range of applications both in science and everyday life.[4]

1.3 Bruno de Finetti's subjectivism

Starting from the late 1920s, Bruno de Finetti (1906-1985) developed a radical form of probabilism that can be described as a blend of pragmatism and the kind of empiricism that is today called anti-realism. It moves from a rejection of the notions of truth, determinism and "immutable and necessary" laws, to reaffirm a conception of science as a product of human activity, deeply imbued with probability. De Finetti identifies the aim of science with prediction, and regards (subjective) probability as the best possible tool for making good forecasts.

According to the subjective interpretation, probability is a quantitative expression of the degree of belief in the occurrence of an event, entertained by a person in a state of uncertainty. It is taken as a primitive notion having a psychological foundation, which requires an operative definition in order to be measured and used in practice. A well-known method for measuring degrees of belief is the betting scheme, according to which probability expresses the conditions under which someone is ready to bet on the occurrence of an event. This method, which is endowed with a long tradition dating back to the Seventeenth century, is deemed by Frank Plumpton Ramsey - the other "father" of the subjective interpretation - "fundamentally sound" although "insufficiently general" and "necessarily inexact",[5] because it suffers from problems such as the diminishing marginal utility of money, and the personal (greater or less) disposition to gambling. In view of this, both Ramsey and de Finetti stressed that subjective probability can be given an operational definition also by means of other methods: Ramsey adopted a system of preferences, and de Finetti turned to penalty methods like Brier's rule.[6] The cornerstone of the subjective theory is the notion of coherence, for as Ramsey and de Finetti showed, coherent degrees of belief obey the rules of additive probabilities. Put otherwise: the laws of probability can be derived from the assumption of coherence.

All coherent probability functions are admissible for upholders of the subjective theory. This means that once coherence is guaranteed disagreement is admitted. In other

---

[4]For more on Reichenbach's position see Galavotti (2011a).
[5]See Ramsey's "Truth and Probability" in Ramsey (1931), reprinted in Ramsey (1990).
[6]See de Finetti (1970/1975).

words, for subjectivists probability evaluations are not univocally determined by evidence, and the estimation of probability also depends on subjective elements such as experience and personal abilities, in addition to empirical evidence. With increasing evidence, though, the opinions of different people will converge. The result, known as "de Finetti's representation theorem" (although the author always refused to call it a "theorem"), proves that convergence between subjective probability and observed frequencies is assured by the adoption of exchangeability in connection with Bayes' rule.

Having said that, it must be added that for de Finetti it does not make sense to regard the Bayesian method - or the inductive method, which for him amounts to the same thing - as self-corrective. Bayesian reasoning entails updating probability evaluations in the light of new evidence, but updating should not be construed as approximation to true probabilities. De Finetti entrusted this conviction to the statement "Probability does not exist", printed in capital letters in the Preface of the English edition of *Theory of Probability*. Such a statement, often mentioned and just as often misunderstood, has been taken to imply that all coherent probability evaluations are on a par. This is not so: albeit rejecting as metaphysical the idea that probability is an objective property of phenomena, de Finetti took very seriously the problem of the objectivity of probability evaluations, namely the problem of devising methods that enable successful predictions to be made. In this spirit, he maintained that both "(1) the objective component, consisting of the evidence of known data and facts; and (2) the subjective component, consisting of the opinion concerning unknown facts based on known evidence" (de Finetti, 1974, p. 7) are essential ingredients of probability evaluations.

2 The French milieu

This section of the paper tackles the philosophy of probability of the French mathematicians Fréchet, Borel and Lévy, all of whom gave important contributions to various branches of mathematics and the theory of probability.[7] In addition, they were actively involved in the lively debate on the foundations of probability carried on in a number of publications and congresses in the first half of the twentieth century. Particularly important in that connection was the *International Congress of Mathematicians* that took place in Geneva in 1937. The meeting hosted a *Colloque consacré a la théorie des probabilités* that brought together a number of renowned mathematicians, statisticians and scientists, including among others Bruno de Finetti, William Feller, Maurice Fréchet, Richard von Mises, Werner Heisenberg, Eberhard Hopf, Jerzy Neyman, George Polya, J.F. Steffensen, Francesco Cantelli and Abraham Wald. Another important gathering was the *XVIII Congrès international de philosophie des sciences* held in Paris in 1949, which included one session entitled *Calculus*

---

[7] See von Plato (1994) for the contribution given by Borel, Fréchet and Lévy to the theory of probability.

of probabilities, where some of the most outstanding probabilists of the time, including Maurice Fréchet, Émile Borel, Bruno de Finetti, Jerzy Neyman, Jean Ville and Paul Lévy, delivered papers devoted to technical as well as philosophical aspects of the notion of probability. The proceedings of both of these conferences were published in the periodical *Actualités scientifiques et industrielles*. Also worth mentioning is a monographic issue of the journal *Dialectica* published in 1949 under the title *The probable knowledge*, collecting, among others, papers by Émile Borel, Bruno de Finetti, George Polya, Corrado Gini, Paul Lévy, Padrot Nolfi, M.S. Bartlett, and Subrahmanyan Chandrasekhar.

2.1 Maurice Fréchet's "modernized axiomatic theory"

Professor of general mathematics and the calculus of probabilities at the Sorbonne University in Paris, Maurice Fréchet (1978-1973) is considered the founder of the theory of abstract spaces, and gave outstanding contributions to topology and functional analysis. He organized a series of lectures at the École Normale Superieure and at the Institut Poincaré, where in 1935 Bruno de Finetti was invited to deliver the lecture course later published under the title "La prévision: ses lois logiques, ses sources subjectives". Although he did not share de Finetti's subjective approach, Fréchet thought highly of him, and the two entertained a correspondence in the course of which, among other things, Fréchet called de Finetti's attention to the work of Ramsey, pointing out the similarity between their views on probability.[8] Moreover, it was Fréchet who in 1939 suggested de Finetti adopt the term "exchangeability" instead of "equivalence", which he had used until then.[9]

Fréchet is deeply convinced that the notion of probability should be addressed from the standpoint of its applications. To accomplish that task the mathematics of probability cannot do the whole job, and must be backed up by philosophy. The idea is that probability should provide a guide to action and decision, and hence must be applicable to the problems encountered in all sorts of practical situations. Moreover, probability should be amenable to "verification" by observing the success of predictions made by its means. Fréchet endorses Augustin Cournot's tenet that the application of probability to real phenomena requires going beyond mathematics, and quoting a passage pointed out to him by Paul Lévy states that "in order to go from the idea of an abstract relationship to that of a law that can be useful in the realm of phenomena, *mathematical reasoning* [...] *is obviously insufficient. One needs to appeal to other notions, to other principles of knowledge; in a word: one needs philosophical criticism*" (Fréchet, 1938a, p. 43). According to Fréchet, a similar attitude was shared by Henri Poincaré

---

[8] See Box 6 of *Bruno de Finetti Collection; Archives of Scientific Philosophy*, Hillman Library of the University of Pittsburgh.

[9] This is mentioned by de Finetti in his "farewell lecture" at the University of Rome, see de Finetti (1976, p. 283).

and Francesco Cantelli. Emphasis is on the need to build a bridge between the abstract, mathematical theory of probability and its applications, or between a schema representing reality and the corresponding elements of reality.

Fréchet does not regard the problem as peculiar to probability theory, but shared by all the empirical sciences. In this spirit, he thinks that the evaluation of probability is similar to the measurement of a physical magnitude, and such an analogy plays a crucial role in his conception of probability, labelled a *modernized axiomatic theory*. His theory revolves around the tenet that probability is to be construed "as a physical magnitude attached to an event and a category of trials, of which the frequencies of the event in a great number of trials are approximated measures" (Fréchet, 1951, p. 5). When dealing with probability one proceeds "exactly as in the experimental sciences where measures are *generally* approximate values of physical magnitudes" (Fréchet, 1939-1940, p. 12), the difference being that in probability, as opposed to other disciplines, the precision of measurement increases as the number of observations grows. This process results from a combination of empirical and axiomatic considerations, in which a fundamental role is played by what Fréchet calls *inductive synthesis*. In order to accomplish an inductive synthesis one proceeds as follows: "The axiomatic theory aims at - and allows - certain unknown probabilities *p* to be derived from other known probabilities *p'*. The interpretation of probability allows each *p'* to be calculated approximately on the basis of the observation of certain frequencies *f'* and to put each *p* approximately equal to a frequency *f*. In the end, one will succeed in calculating some frequencies, whose direct observation could be impossible or difficult, starting from other frequencies *f'* which are easily observed" (Fréchet, 1946, p. 150).

The peculiarity of the procedure Fréchet describes is that the application of the axioms should be preceded by the act of verifying whether in practical situations the axioms apply to the particular class of events under study. In that sense, the measurement of probability requires a synthesis between the axiomatic theory and practical situations. According to Fréchet, "In this synthesis intuition and contact with reality are the main directions to follow and therefore rigour is not supreme. Applied to probability, this leads us to conclude from the practical statistical processes that any frequency is to be considered an approximate measure of one *physical constant* attached to an event E and to a category C of trials" (Fréchet, 1939-1940, p. 11). The inductive synthesis so described is the cornerstone of the interpretation of probability heralded by Fréchet, and is meant to represent the element of novelty implicit in the author's expression "modernised axiomatic theory". As repeatedly emphasized in his writings, Fréchet's interpretation has a *practical* import, and falls outside the axiomatic theory to which it is linked by means of inductive synthesis.

Obviously, the crucial step in such a synthesis is the choice of the proper "category of trials". In this regard, Fréchet faces a reference class problem similar to that besetting Reichenbach's attempt to apply the frequency theory to single case evaluations. Fréchet admits that such a choice depends on subjective elements. However, wanting to retain an objective notion of probability, he emphasizes that the subjective elements neither concern nor affect the probability itself, but rather the way a certain problem is formulated in a given context. It is at that stage that the choice of the reference class is made, depending on the information available and the purpose being pursued. Fréchet observes that such a choice "cannot go beyond our knowledge, but it could use only part of it; an act of volition [...] *decides* the choice of the category of trials" (Fréchet, 1938a, p. 50). It is the purpose for which probability is being evaluated that guides the choice of the reference class, after deciding which of all the known features of a given phenomenon are deemed relevant and picked as characterizing the category of trials. Fréchet emphasizes that the choice of the reference class has nothing to do with probability, which "comes into play after the event and the category of trials have been chosen, at which point it is entirely determined and its value is independent of the person who made the choice" (Fréchet, 1938a, p. 50).

Fréchet's answer to the question whether probability is objective or subjective is that "what is subjective is not the *value* of probability, it is the *problem* that is being posed, namely the choice that is made of the category of trials in connection to which one calculates the probability of an event" (Fréchet, 1938a, p. 50). To the subjective interpretation of probability Fréchet opposes the view that "all of my information comes into play with the choice of the category of trials, namely in connection with the specification of the problem, not with its solution" (Fréchet, 1946, p. 146). The probability of an event must be kept separate from someone's degree of belief in its occurrence in the same way as "a distinction is to be made between the correct solution to a problem and the solution given by some student, which can vary from one student to another" (Fréchet, 1946, p. 143).

In that regard, Fréchet claims agreement with Harold Jeffreys' tenet that probability values are univocally determined by evidence and that one must distinguish between "objective probability and subjective degrees of belief, which are but personal estimates of the same" ((Fréchet, 1946, p. 146), footnote 2).[10] Fréchet emphasizes that in order to approach the value of probability of an event, subjective estimates "must be corrected", and this can only be done by recourse to observed frequencies. Therefore "those probabilists who by assimilating probabilities with degrees of belief thought that they could avoid appealing to frequencies, will be forced to turn to them when asked to justify the success of their interpretation" (Fréchet, 1946, p. 145). This, and other assertions, such as the claim that subjectivists "are just happy with grading a state of mind" (Fréchet, 1951, p. 17) suggest that Fréchet fell into the misunderstanding pointed out in Section 2.3, by which upholders of the subjective interpretation fix probability values without taking frequencies into account. As already observed, this is the result of a misrepresentation of the subjective approach. An additional criticism against subjectivism concerns the notion of coherence, to which Fréchet objects that most people do not behave coherently. Instead of "rational" coherent degrees of belief, most of the time they retain "irrational probabilities" that do not satisfy coherence. Therefore, to Fréchet's eyes the subjective interpretation of Savage and de

---

[10] See Galavotti (2003) for more on Jeffreys' views on probability and probabilistic epistemology.

Finetti refers to a *fictitious* agent, not to the way in which real agents behave in practice (see Fréchet, 1954).

Albeit sharing his objective approach to probability, Fréchet is also critical of Reichenbach. The crucial difference between their perspectives is that Fréchet does not impose a homogeneity requirement on the choice of the reference class, leaving it to the context. Fréchet stresses that any problem can be tackled from different angles, and therefore there is no one single reference class to be taken as suitable in a given situation. Borrowing an example from Borel, he discusses the probability of death of an individual, observing that it has a different meaning for his doctor and for his insurance company, so that the first will choose the population of "people of the same age, same weight, same blood pressure, same lung capacity, etc." while the insurance company will consider the population of insured having the same age. "The values of the corresponding probabilities can differ without ceasing to be compatible. [...] the probabilities evaluated by the doctor and the insurance company can be assimilated to two distinct physical magnitudes. Their approximate measure will be obtained by means of statistics properly made but obtained from different populations" (Fréchet, 1938a, pp. 50–51).

Fréchet also raises a more general objection to the frequency interpretation of von Mises and Reichenbach, observing that "those who advocate the 'proceed to the limit' definition think that they have in this way made the theory nearer practice. In fact, they have made it more remote." (Fréchet, 1939-1940, p. 17). To substantiate this claim, Fréchet borrows an example from Bruno de Finetti: take an unlimited sequence of trials where the frequency of Tails observed in n trials tends to $1/2$ when n increases, but in which one has obtained Tails 10,000 times. This could be a collective, in which the probability of Tails is $1/2$, however such a probability would not be $1/2$ for whatever experimenter (see Fréchet, 1939-1940, pp. 17–18). Although frequencies are the essential ingredient for estimating probabilities, for Fréchet estimates are not to be obtained by proceeding to the limit as the number of observed cases increases. Frequency is not part of the definition of probability, it is rather the tool that allows probability values to be approximated in practice. Fréchet repeatedly emphasizes that his position is a *practical interpretation* according to which frequency is an "empirical measure of probability" (Fréchet, 1951, p. 12). Unlike von Mises' interpretation which includes "a measure of probability in the axiomatic theory" (Fréchet, 1951, p. 12) and embodies a constructive definition of probability, Fréchet's viewpoint is meant as a descriptive account falling outside the axiomatic theory.

To conclude, Fréchet's major concern is that probability must provide a guide to action and decision. To accomplish that task it must be applicable to problems encountered in science as well as in practical situations occurring in everyday life. Such a concern imbues both Fréchet's concept of probability and the criticism he moves against frequentism and subjectivism.

2.2 Émile Borel's moderate subjectivism

A leading mathematician, Émile Borel (1871-1956) gave substantial contributions to analysis and group theory, did seminal work in probability theory, and developed an original philosophy of probability. Between 1925 and 1939 he published a series of monographs under the collective title *Traité du calcul des probabilités et ses applications*, ending with the essay *Valeur pratique et philosophie de la probabilité*, where his views on the nature of probability are spelled out with great clarity.

Borel takes a probabilistic attitude towards science and knowledge in general: "probability lies at the core of scientific knowledge" since "the value of all scientific results can be assessed only by means of a probability coefficient" (Borel, 1939/1952, p. 10). The theory of probability is of paramount importance because "not only does it possess the same practical and philosophical value of all other scientific theories, but it is the basis of all our knowledge" (Borel, 1939/1952, p. 11). Borel was not only a successful scientist but also a politician and a social reformer - among other things he was for twelve years a member of the Chamber of Deputies and in 1925 served as minister of the Navy.

Convinced that probability is also of great value for life and that "the practical value of probability can surpass that of the rest of human knowledge" (Borel, 1939/1952, p. 42), Borel was actively engaged in pedagogical reforms aimed at educating people on probability from their youth, because in this way "one will reduce the persistence of many prejudices" (Borel, 1939/1952, p. 10). He struggled to counteract the widespread resistance to thinking in probabilistic terms, and in an effort to convince people that "there are only statistical truths" (Borel, 1907/2014, p. 1083) argued that "the mathematical answer to be given to many practical questions is a coefficient of probability. Such an answer will not seem satisfactory to many minds, who expect certainty from mathematics. This is a very bad inclination; it is utterly regrettable that the education of the public is, in this respect, so little advanced; this may be due to the fact that the mathematics of probability remains a subject of near universal ignorance, even though every day it intrudes a bit more into everyone's life (various insurance policies, mutual aid societies, retirement pensions, etc.). A coefficient of probability constitutes a perfectly clear answer, corresponding to an absolutely tangible reality. Some minds will maintain that they 'prefer' certainty; they might as well 'prefer' that 2 plus 2 were 5" (Borel, 1907/2014, p. 1087). This long passage has been reported here as evidence of Borel's deep concern for the theoretical and practical value of probability in science and life too.

To start with, Borel calls attention to the fact that the theory of probability differs from other sciences, and from other branches of mathematics as well. For one thing, other magnitudes - including numbers belonging to arithmetic - are measured by means of a measurement unit that can vary, and according to the chosen unit they can be said to be large or small. This is not so for probability where the unit of measurement is unique, and by convention probability 1 equals certainty. Therefore "we have an absolute scale for measuring the degree of smallness of probabilities" (Borel, 1939/

1952, p. 6). In addition, the theory of probability differs from the other sciences because of the nature of its object, given that "by its very nature" probability "cannot claim to give us certainties" (Borel, 1939/1952, p. 4). While from an abstract or axiomatic point of view, once some principles have been postulated, probability values are well defined, when one moves from theory to practice the import of such values is uncertainty "precisely because the very object of deductions and calculations is simply a probability, not a number, a length, a time, as in Arithmetic, Geometry, Astronomy" (Borel, 1939/1952, p. 5).

For Borel, probability is endowed with a different value depending on the body of information available within a given context. Typically, probability has a more objective meaning in science, where its assessment is grounded on a strong body of information shared by the community of scientists. By contrast, probabilities attached to individual judgments can have "different values for different individuals" (Borel, 1924/1964, p. 50). In a review of Keynes' *Treatise on Probability* originally published in 1924 Borel criticizes Keynes for concentrating only on the probability of judgments, overlooking the application of probability to science. Incidentally, he attributes such an attitude to English rather than continental authors who are deemed more attentive to the progress of science, especially physics. As Borel emphasizes: "the probability that an atom of radium will explode tomorrow is, for the physicist, a constant of the same kind as the density of copper or the atomic weight of gold. Albeit these constants are always at the mercy of the progress of physical-chemical theory, they are constants in the present state of science" (Borel, 1924/1964, p. 50).

Borel does not share Keynes' tenet that there are probabilities which cannot be evaluated numerically. In that connection he rather agrees with subjectivists, with whom he also shares the conviction that "the method of betting permits us in the majority of cases a numerical evaluation of probabilities that has exactly the same characteristics as the evaluation of prices by the method of exchange" (Borel, 1924/1964, p. 57). The betting method provides the operative tool by which one can build a bridge between probability and action, as it "can be applied to all verifiable judgments; it allows for a numerical evaluation of probabilities with a precision quite comparable to that with which one evaluates prices" (Borel, 1924/1964, p. 57).

Borel endorses Reichenbach's adhesion to the pragmatist principle that the meaning of a proposition lies with its practical consequences. In Borel's words: "a proposition has practical interest for men only insofar as it can influence their actions" (Borel, 1939/1952, p. 89). At the same time, he moves a number of objections against the frequency theory. In particular, he objects to Reichenbach's solution to the single case problem that the homogeneity requirement would lead us to consider "classes that contain so few elements that the concept of frequency no longer applies" (Borel, 1939/1952, p. 87). The more detailed the description of a single case is, the more evident the differences with other cases of the same kind, so that "one will find that probability is defined in a way that is less and less precise the better the case at hand is known" (Borel, 1939/1952, p. 88). Albeit agreeing with Reichenbach that single case probability evaluations are of vital importance for the sake of practical applications,

Borel deems that frequencies are useful, but other elements also come into play. He makes the example of a doctor asked to predict the probability of survival of a patient that had contracted a certain disease. Surely the doctor will consider the frequency of deaths among people with the same illness in a given period of time, but he will likewise take into account additional information considered relevant in the light of his own experience. Borel's conclusion is that single case probability attributions result from the concurrence of empirical information, especially frequencies, plus personal experience and common sense.

Given that subjective ingredients are part of probability evaluations, the possibility that based on the same body of empirical information two people can come up with different assignments is admitted, precisely as it was by subjectivists. When it comes to measuring the probability of single events, Borel holds that "the probability of a single case is defined subjectively by the conditions under which an agent is ready to bet on the occurrence or non-occurrence of an event" (Borel, 1939/1952, p. 105). It is noteworthy in that connection that he addresses the reader to de Finetti's "La prévision, ses lois logiques, ses sources subjectives".[11]

Having so accounted for the probability of single events Borel feels the need to address the issue of defining an objective notion of probability. Objective probabilities, he claims, "can be defined as probabilities whose value is the same for a certain number of individuals who are well informed on the conditions of the aleatory event" (Borel, 1939/1952, p. 105). It should not pass unnoticed that Borel's concept of objective probability strongly resembles that put forward by Frank Ramsey, who defines the probabilities occurring in physics as being objective in the sense "that everyone agrees about them, as opposed e.g. to odds on horses" (Ramsey, 1990, p. 106). The idea is that the assessment of probability occurring in physics is constrained by theories that have gained the assent of the scientific community after a good deal of evidence in their favour has been collected. Late in his life, de Finetti also admitted that in some areas of science, like physics, probability assignments stand on "more solid grounds" than those belonging to everyday life, but he did not develop that idea (de Finetti, 1995/2008, p. 63). Apart from that remark, de Finetti did not pay much attention to the notion of objective probability, convinced that subjective probability can do the whole job. As a matter of fact, Bruno de Finetti praises Borel for holding that probability must be referred to the single case and can be measured sufficiently well by means of the betting method. At the same time, de Finetti criticizes Borel's eclectic attitude according to which probability can take an objective as well as subjective value (see de Finetti, 1939).

It is noteworthy that after defining objective probability as recalled above Borel adds that "should an event, like the throw of a die, be repeatable a great number of times under the same conditions, the theory of repeated trials tells us that the limiting value of the frequency equals the probability; this will give us a verification, *not a definition*"

---

[11] See de Finetti (1937/1964).

(italics added, Borel, 1939/1952, p. 105). This claim suggests a parallel with the distinction drawn by de Finetti between the *definition* and the *evaluation* of probability, stating that while probability is by definition the expression of subjective degree of belief, its evaluation is a complex procedure which results from "the conjunction of both objective and subjective elements at our disposal" (de Finetti, 1973, p. 366). Like de Finetti, Borel stresses that what makes the difference among probability evaluations is the amount and kind of evidence backing them, but he refuses to conflate the meaning of probability with the objective elements - be it frequencies and/or symmetries - that are part of that evidence.

To sum up, Borel can be deemed a subjectivist whose conception of probability has much in common with de Finetti's, but his position is more moderate due to his admission that probability assessments made in the context of sciences like physics have an objective character. However, Borel does not attach a realistic meaning to objective probability, taking instead an attitude closer to the pragmatist idea of objectivity as agreement among members of the scientific community.

2.3 Paul Lévy's "rationalistic theory"

Paul Lévy (1886-1971), professor of analysis at the École Polytechnique in Paris, gave important contributions to various branches of mathematics, including functional analysis and probability theory. In *Calcul des probabilités* 1925 Lévy addresses the foundations of probability, putting forward the viewpoint he calls "rationalistic theory" (*théorie rationaliste*), later summarized in a short article entitled "Les fondements du calcul des probabilités", appearing in 1949 in the already mentioned issue of the journal *Dialectica* dealing with "probable knowledge". Like Fréchet and Borel, Lévy is concerned with the *practical meaning* of probability, taking for granted that its mathematical aspects are beyond dispute. Clarifying the relationships between the mathematical principles (axioms) and their practical applications is precisely the task Lévy attaches to the debate on the foundations of probability, which must be of interest to both philosophers and mathematicians.

Lévy takes a sympathetic attitude towards the notion that probability has a subjective component, and rejects von Mises' frequency theory, called *empirical*. In his words: "probability, essentially subjective, traces a clear-cut distinction between what one knows and what one does not know but can be modified by some information that was previously unavailable" (Lévy, 1925, p. 14). The proper tool for such updating is Bayes' rule, to which Lévy devotes one section of the first chapter of *Calcul des probabilités*.

The meaning of probability varies according to the problem addressed: one thing is to talk about the probability of obtaining a double six when two dice are thrown, and another thing is to talk about the probability of contracting a certain disease in a given situation. While the second problem "is of interest to statisticians and doctors [...] mathematicians and philosophers should primarily concern themselves with games of chance, where the concept of probability can be seen in its purity" (Lévy, 1949, p. 57). The meaning of probability in the realm of chance games is "the expectation of a frequency": we expect that after a considerably long series of throws each side of a die will come up with a frequency close to ⅙, unless we have reason to believe otherwise, namely assuming that the die is unbiased. What we expect is not that the die comes up *exactly* ⅙th of the times it has been thrown, but that this result obtains *approximately*. More precisely, we expect that the deviation between the observed and the expected frequency is not too large. If not, we must revise our expectations. Conversely, if the deviation between the observed frequency and our expectation is small, our initial expectation is confirmed, together with the assumption backing it. Fréchet calls attention to the analogy between this way of proceeding and what is commonly done in geometry where all reasoning is based on perfect solids which are not to be found in reality, but good approximations are considered satisfactory. This inspires Lévy's claim that "only subjective probability is liable to offer a schematization of probability which is as useful to the probabilist as the consideration of perfect solids is to the geometrician. One can discuss the intuitive character of subjective probability [...] But I think that everyone must acknowledge the interest of subjective probability" (Lévy, 1949, pp. 58–59).

Against the pretence of frequentists of grounding the evaluation of probability on frequencies alone Lévy calls attention to the importance of intuition and individual judgment, on which rests the task of applying the laws of probability to practice. To the upholders of the frequency theory like von Mises, whom he labels "empiricists", Lévy objects that we only experience individual cases, and when a series of repeated trials is available "if in the result of such experiences we discover a characteristic which does not look random, but [...] predictable, this must be so by virtue of individual experience. It must be discovered. The empiricist denies it. Therefore I can only reckon empiricism as the refusal of a progress that rationalism has made without effort" (Lévy, 1949, p. 62). The fundamental mistake made by those who embrace the frequency theory is that of putting at the core of the definition of probability the relationship between probability and frequency, which is thereof assumed *a priori*, when it should be a matter for demonstration. In Lévy's words: "the empiricist cannot hope to demonstrate Bernoulli's theorem, because he starts with assuming the property that should be proved" (Lévy, 1949, p. 63).

To admit that subjective elements enter into the evaluation of probability is for Lévy simply a matter of taking a rationalistic attitude towards the issue of the interpretation of probability, which boils down to the problem of linking theory and practice. In Lévy's words, "There is no doubt that the rationalist has his own difficulties when it comes to dealing with empirical results. But he has never concealed such difficulties; to solve them he will employ all his good sense, he will not merely use a ready-made formula" (Lévy, 1949, p. 64). The rationalist's decisive advantage over the empiricist is that of having understood that the properties of a series must be explicated in terms of the properties of its elements. Lévy considers it crucial progress, and deems the empiricist's refusal to acknowledge it a "fundamental mistake" (Lévy, 1949, p. 64).

3 Closing remarks

As claimed at the outset, the work of the French mathematicians Fréchet, Borel and Lévy is best seen in the frame of a larger picture that includes a number of authors active in different fields and in different places, such as Frank Ramsey and Harold Jeffreys in Great Britain, Bruno de Finetti in Italy, Richard von Mises and Hans Reichenbach in Germany, and Janina Hosiasson in Poland, all of whom heralded a genuinely probabilistic view of knowledge. The same conviction was actually shared by many others, including authors like John Maynard Keynes and Ernest Nagel, thereby extending the picture considerably.

A first consideration suggested by the work of such authors is that the debate on the foundations of scientific knowledge is broader than the received view centred on the Vienna Circle that has long been dominant.

A further consideration that can be drawn from their writings is that the upholders of probabilistic epistemology share by and large a body of tenets ingrained in the pragmatist tradition, such as the stress on prediction as the main task of science, the idea that success is the canon for the justification of induction, the centrality of action in connection with the theory of meaning and the value of probability, and the attention paid to the practical applications of probability. This is evidence of the influence exercised by pragmatism on European epistemology. Reichenbach, Ramsey, Hosiasson and de Finetti all explicitly acknowledge the influence on their thought of pragmatist thinkers such as Peirce, James and the Italian Giovanni Vailati.[12] By contrast, the French probabilists do not refer to such authors, but their writings abound with references to Henri Poincaré. There is no doubt that Poincaré was a formidable source of inspiration for French mathematicians,[13] and exercised an influence that went far beyond technical aspects, deeply affecting their philosophy of probability whose pragmatist flavour is likely to descend from his work.

---

[12] See Galavotti (2011b) and Galavotti (to appear) for more on the influence of pragmatism on the debate on the foundations of probability in the last century.

[13] See for instance von Plato (1994), where Borel is regarded as Poincaré's successor not only academically, but also "in an intellectual sense" (p. 36).

Bibliography

Borel, É. (1952). Valeur pratique et philosophie des probabilités. In *Traité du calcul des probabilités et ses applications (1925-1939)* (Vol. 4(3)). Paris: Gauthier-Villars. (Original work published 1939)

Borel, É. (1964). Apropos of a treatise on trobability. In H. Kyburg, & H. Smokler (Eds.), *Studies in subjective probability* (pp. 45–60). New York-London: Wiley. (Original work published 1924)

Borel, É. (2014). An economic paradox: The sophism of the heap of wheat and statistical truths. *Erkenntnis, 79 (2014)*, 1081–1088. (Original work published 1907).

de Finetti, B. (1939). Punti di vista: Émile Borel. *Supplemento Statistico ai Nuovi Problemi di Politica, Storia, ed Economia*, 5, 61–71.

de Finetti, B. (1964). Foresight, its logical laws, its subjective sources. In H. Kyburg, & H. Smokler (Eds.), *Studies in subjective probability* (pp. 53–118). New York-London: Wiley. (Original work published 1937)

de Finetti, B. (1973). Bayesianism: Its unifying role for both the foundations and the applications of statistics. In *Bulletin of the International Statistical Institute, Proceedings of the 39th Session* (pp. 349–368).

de Finetti, B. (1974). The value of studying subjective evaluations of probability. In C.-A. S. Staël von Holstein (Ed.), *The concept of probability in psychological experiments* (pp. 1–14). Dordrecht-Boston: Reidel.

de Finetti, B. (1975). *Theory of probability*. New York: Wiley. (Original work published 1970)

de Finetti, B. (1976). Probability: Beware of falsifications! *Scientia*, 70, 282–303.

de Finetti, B. (2008). *Philosophical lectures on probability* (A. Mura, Ed.). Dordrecht: Springer. (Original work published 1995)

Fréchet, M. (1938a). Exposé et discussion de quelques recherches récentes sur les fondements du calcul des probabilités. *Colloque consacré a la théorie des probabilités, Deuxième partie. Actualités scientifiques et industrielles*, 735, 23–55.

Fréchet, M. (1938b). Les principaux courants dans l'évolution récente des recherches sur le calcul des probabilités. *Colloque consacré a la théorie des probabilités, Première partie. Actualités scientifiques et industrielles*, 734, 19–23.

Fréchet, M. (1939-1940). The diverse definitions of probability. *Erkenntnis*, 8, 9–23.

Fréchet, M. (1946). Les définitions courantes de la probabilité. *Revue Philosophique*, 71, 129–169.

Fréchet, M. (1951). Rapport général sur les travaux du colloque de calcul des probabilités. *Proceedings of the XVIII Congrès international de philosophie des sciences (1949), Actualités scientifiques et industrielles*, 1146, 3–21.

Fréchet, M. (1954). Un problème psychologique sur les probabilités subjectives irrationelles. *Journal de Psychologie Normale et Pathologique*, 52, 431–438.

Galavotti, M. C. (2003). Harold Jeffreys' probabilistic epistemology: Between logicism and subjectivism. *British Journal for the Philosophy of Science*, 54, 43–57.

Galavotti, M. C. (2005). *Philosophical introduction to probability*. Stanford: CSLI.

Galavotti, M. C. (2011a). On Hans Reichenbach's inductivism. *Synthèse, 181*, 95–111. doi:10.1007/s11229-009-9589-6

Galavotti, M. C. (2011b). Probability and pragmatism. In D. Dieks, W. J. Gonzalez, S. Hartmann, T. Uebel, & M. Weber (Eds.), *Explanation, prediction, and confirmation* (pp. 499–510). Dordrecht: Springer.

Galavotti, M. C. (2014). Probabilistic epistemology: A European tradition. In M. C. Galavotti, E. Nemeth, & F. Stadler (Eds.), *Philosophy of science in Europe; European philosophy of science* (pp. 77–88). Dordrecht: Springer.

Galavotti, M. C. (to appear). The ghost of pragmatism. Some historical remarks on the debate on the foundations of probability. In S. Pihlström, F. Stadler, & N. Weidtmann (Eds.), *Logical empiricism and pragmatism*. Dordrecht: Springer.

Lévy, P. (1925). *Calcul des probabilités*. Paris: Gauthier-Villars.

Lévy, P. (1949). Les fondements du calcul des probabilités. *Dialectica, 3*, 55–64.

Ramsey, F. P. (1931). *The foundations of mathematics and other logical essays* (R. B. Braithwaite, Ed.). London: Routledge and Kegan Paul.

Ramsey, F. P. (1990). *Philosophical papers* (H. Mellor, Ed.). Cambridge: Cambridge University Press.

Reichenbach, H. (1937). La philosophie scientifique: une esquisse de ses traits principaux. In *Travaux du IX Congrès International de Philosophie* (pp. 86–91). Paris: Hermann.

Reichenbach, H. (1938). *Experience and prediction*. Chicago-London: University of Chicago Press.

Reichenbach, H. (1971). *The theory of probability*. Reprint of the English edition, 1949. Original published in German 1935. Los Angeles: University of California Press. (Original work published 1949)

Suppes, P. (1984). *Probabilistic metaphysics*. Oxford: Blackwell.

von Mises, R. (1957). *Probability, statistics and truth*. Originally published in German in 1928. The 3rd German edition, 1951, is the definitive version and the basis of this English edition (a revision of the English translation of 1939). New York: Allen and Unwin. (Original work published 1951)

von Mises, R. (1968). *Positivism*. Reprint of 1951 edition. Original published in German 1939. Harvard: Harvard University Press. (Original work published 1951)

von Plato, J. (1994). *Creating modern probability*. Cambridge: Cambridge University Press.

Author biography. Maria Carla Galavotti is professor of Philosophy of Science, University of Bologna, Research Associate of the Centre for Philosophy of the Natural and Social Sciences, LSE, life member of the Center for the Philosophy of Science, University of Pittsburgh, and Clare Hall College, Cambridge, and member of the Leopoldina National Academy of Sciences of Germany. Her research focuses on the foundations and the history of probability, the nature of scientific explanation, prediction, and causality. She authored *Philosophical Introduction to Probability* (CSLI, 2005), and edited *Stochastic Causality* (eds. M.C. Galavotti, P. Suppes, D. Costantini, CSLI, 2001); *Cambridge and Vienna. Frank P. Ramsey and the Vienna Circle* (Springer, 2006), *New Directions in the Philosophy of Science* (eds. M.C. Galavotti, D. Dieks, W.J. Gonzalez, S. Hartmann, T. Uebel, M. Weber, Springer, 2014). Her collection of Ramsey's manuscripts entitled *Notes on Philosophy, Probability and Mathematics* (Bibliopolis, 1991) has become a classic reference book.

9 Patrick Suppes: From logic to probabilistic metaphysics

ANNE FAGOT-LARGEAULT *

for Alexandra

Keywords: probability, rationality, education, free will, measurement.

---

*Collège de France & Académie des sciences, Paris

Introduction

As a brief introduction, let me remind the reader of two facts. First a quotation from Patrick Suppes, quite meaningful in the context of our Congress of logic, methodology and philosophy of science : "I count **probability** as perhaps the single most important concept in the philosophy of science" (Suppes, 2002, p.14); hence the title of this presentation. Second, as our President Elliott Sober said in his inaugural welcome talk, P. Suppes was President of DLMPS from 1975 to 1979; and he had been very active and contributive in the preparation of the first DLMPS congress, that took place in 1960 on the Stanford campus. He really was one of us. The present paper attempts to summarize : (1) Suppes' background and career, (2) his style as a philosopher of science, (3) his probabilistic metaphysics.

1 Suppes' background and career

Patrick C. Suppes was born in Tulsa, Oklahoma, on March 17, 1922. His mother died when he was four years old. He was raised by his stepmother, who encouraged his intellectual curiosity, while his father would see him follow his own trail in the oil business. He graduated from Tulsa High School in 1939. Later on he liked to say that his public school education had been very influential, due to his being admitted in a six years experiment of accelerated education for able students.

Suppes began college in 1929 at the University of Oklahoma, found it boring, went to Chicago, and back in Tulsa. At the time his interests were mostly in physics. In 1942 he was recruited to serve in the Army Reserves, returned to the University of Chicago and graduated in a "special meteorology program". Then during WW2 he served as a meteorologist in the US Army from 1943 to 1946, first on the Solomon Islands, then in Guam. In a 'Self-profile' that he wrote in 1979[1] he mentions having read Aristotle on the Solomon Islands. Note that the importance in Suppes' philosophy of the notion of habit is clearly reminiscent of Aristotle.

After the war Patrick Suppes earned a PhD (1947-50) in Philosophy from the University of Columbia, New York. His director of studies was Ernest Nagel. The topic of the dissertation was action at a distance, according to Descartes, Newton, Boscovich, and Kant. Such a topic points to an interest for the history of philosophical problems, in so far as it helps looking for better solutions or approaches. Such was typically Suppes' attitude towards the history of philosophy: "We can learn from the past, but we can also improve upon it" (Suppes, 1996, p.110).

Suppes' career was simple, straight and extraordinarily productive. From the fall of 1950 on, he spent 64 years at Stanford University, living on campus, building his re-

---

[1] 'A Self-profile', p. 3-56, in: Raju J. Bogdan, 1979, *Patrick Suppes*. Hereafter quoted as 'Self-profile'.

search facilities, accompanying the university's evolution from the initial 'farm' to a vast and busy industrial city. As a professor he would teach in the Philosophy Department, and had links with the Departments of Mathematics, Statistics, Psychology, Education. As a researcher he created the *Stanford Institute for Mathematical Studies in the Social Sciences* (IMSSS, initially 'Ventura Hall'), of which he was director from 1959 to 1992. Interested, long before the personal computer existed, in the possibility of using computers to teach mathematics or language and facilitate classroom learning, he launched with IBM a system of *computer-assisted instruction* (CAI) in East-Palo-Alto that low-achieving students in an elementary school could use experimentally from 1966 on. He also founded a *Stanford Education Program for Gifted Youth*, which he managed from 1992 to 2010.

In his lifetime P. Suppes was gratified with many honors and distinctions. He was elected a member of the National Academy of Education (1965), of the National Academy of Sciences (1978), of the IIP (International Institute of Philosophy), etc. Those however were not his raison de vivre. When he was offered an opportunity to become President of Stanford University he refused, arguing that he liked "research better than bureaucracy" ('Selfprofile', p. 52).

He was strongly dedicated to international relations between philosophers, was often invited abroad, travelled easily, co-edited a number of collective volumes. As an example, he was in Paris in November 1979 at the invitation of Jules Vuillemin, and gave a series of four lectures at the College de France. Those were translated into French and published (Suppes, 1981a). He was at the College de France again in 2005 at the invitation of the author of the present paper. He then gave a well-attended public lecture on *"Neuropsychological Foundations of Philosophy"*, and participated in two seminars, having exciting exchanges with neuroscientists and neurosurgeons[2].

Formal talks were not his only, nor his favorite mode of communication. Planning to be in Paris in March, 2002, he welcomed the perspective of having an informal talk and discussion with philosophy students at the College de France. He suggested that we could take "Rationality and Freedom" as a theme. He sent us a series of papers he had written, some still unpublished. As our students had a hard time reading some of those papers, which were loaded with technicalities, Suppes sent an email:

> Very pleased to have the papers looked at prior to our meeting. Then we can have a real discussion rather than a lecture, which I would like very much. My use of ergodic theory in some of these papers is a bit heavy going for philosophers [...]. But don't be put off by the technical framework. The basic message is easy to communicate in discussion. And I do have something to say about free will. Perhaps the single best one-sentence formulation is this: Free will is a scientific problem, not a philosophical one. And the answer is positive, with no need to worry

---

[2] See: *Lettre du College de France*, February 2006, 16:8.

about compatibility with determinism or indeterminism. Look forward to seeing you. Pat[3]

A letter followed, with detailed instructions on what to read exactly, and how the discussion could be carried. A copy of the letter may remind the reader of his handwriting (Figure 1.).

2 Suppes' style as a philosopher of science

To the question as to what kind of philosopher of science was Patrick Suppes, there may be two answers. Suppes was a kind of *Janus bifrons*. Having regard to his conducting theoretical *vs.* empirical research, he was two different philosophers of science. Whether the one completed the other, or whether he evolved from one to the other, is not that clear. Let us opt for the second hypothesis.

"In my younger and more formalistic days..." (see: Bogdan, 1979, p.208). The young Suppes, Nagel's student, was an analytic philosopher, interested in the logic and language of science, talking about primitive notions, axioms, theorems, proofs, formal methods... He dreamed of doing for physics what Whitehead and Russell (1910–1913) did for mathematics:

> If time and energy permitted, I would like best to write a kind of Bourbaki of physics showing how set-theoretical methods can be used to organize all parts of theoretical physics and bring to all branches of theoretical physics a uniform language and conceptual approach. (Suppes, 1969, p.191).

The more mature Suppes proved a real talent and inclination for laboratory investigations. "I found that I had a natural taste for elaborate analysis of experimental data." ('Self-profile', p. 28). He would then appear as a pragmatic philosopher, empiricist, associationist, neobehaviorist (however, anti-reductionist)—occasionally trusting his intuitions—mostly interested in experimental psychology, and admirer of William James, whom he easily quoted (James, 1890/1950). Here is how he sees himself:

> I think of myself primarily as a philosopher of science, but to a degree that I think is unusual among professional philosophers I have had over the period of my career strong scientific interests. Much of this scientific activity could not in fact be justified as being of any direct philosophical interest. But I think the influence of this scientific work on my philosophy has been of immeasurable value. I sometimes like to describe this

---

[3]Copy of an email sent: Sun, 24 Feb 2002, by <psuppes@mail-csli.stanford.edu> (personal communication)

STANFORD UNIVERSITY, STANFORD, CALIFORNIA 94305-4115

PATRICK SUPPES
LUCIE STERN PROFESSOR OF PHILOSOPHY, EMERITUS
VENTURA HALL

26 Feb 2002

Dear Anne,

I suggest we divide our discussion into two parts. One part centered around the unpublished (but about to be published) manuscript "Rationality and Freedom." Relevant to this topic are especially the two earlier papers on "The nature and measurement of freedom" and "Arimizing freedom," as well as "Freedom and Uncertainty."

For Part II, I suggest the focus be on free will. Here the three most important papers are:
(1) "The transcendental character of freedom"
(2) "Voluntary Motion, biological computation and free will,"
(3) "Principles that Transcend experience: Kant's antinomies."

The remaining papers can be read when helpful.

Look forward to our discussion.

Best regards,
Pat

Figure 1.

influence in a self-praising way by claiming that I am the only genuinely empirical philosopher I know ('Self-profile', p. 45).

Continuity is claimed by Suppes, from the logic course taught at Stanford in the 1950es (Suppes, 1957/1967) to the monumental summa (Suppes, 2002) reflecting 40 years of research and published when he reached eighty years:

> Course notes materials were developed ... that conceptually followed the final chapter on the set-theoretical foundations of the axiomatic method of my *Introduction to Logic*, first published in 1957. (Suppes, 2002, p.xii).

The objective remained of achieving the unity of science through the progressive "reduction of one part of science to another" (Suppes, 2002, p.467). The task, however, appears difficult: "one of the puritanical themes of this book is that scientific reduction is hard work. To finish it off with details and explicit representation theorems is even harder. [...] The problem was too difficult for me."(Suppes, 2002, p.467-8). In other words, Suppes acknowledges that he never succeeded reducing completely a piece of science to another.

While the objective of working towards the **unity of science** through achieving the program of reduction is neither abandoned, nor disavowed, another program opens up, that may be named Suppes' Aristotelian program. It aims at searching and analysing information about the world, even though the pieces of knowledge that are collected may resist the constraints of logic and mathematics. At this point, philosophy of science admits of a mixture of formal and informal methods:

> It is difficult to predict the general future of axiomatic methods in the empirical sciences. [...] The use of such methods permits us to bring to the philosophy of science the standards of rigor and clarity that are an accepted part of the mathematical sciences. A conservative prediction is that they will continue to be applied in foundational work throughout this century, even surrounded by a context of informal philosophical or scientific analysis. Such a mixture of the formal and informal [...] is both desirable and necessary, in the sense that many significant ideas in the philosophy of science are not expressed in a way that makes them suitable to formulate in terms of systematic axioms. Experimental and statistical practices in all parts of science are the source of the largest numbers of such examples. (Suppes, 2002, p.49).

Admittedly, as a philosopher of science, Suppes himself is a **mixture**: the diversity of his interests goes far beyond logical analysis. In the (already cited) *Self-profile* that he gave of himself in 1979, he enumerates seven research areas, where he thinks he has contributed something of value:

> I have grouped the discussion of my research under seven headings: foundations of physics; theory of measurement; decision theory; foundations of probability, and causality; foundations of psychology; philosophy of

language; education and computers; and philosophy of science ('Self-profile', p. 9).

Let us take the example of the theory of measurement. Suppes contributed to build a set of methods of measurement for psychology (see: Suppes and Zinnes, 1963). He was especially happy with the initial theoretical paper he published with Dana Scott in 1958 (Scott and Suppes, 1958, Journal of Symbolic Logic, 23: 113-128): "A formal definition of a theory of measurement as a particular kind of class of relational systems is given and the general problem of axiomatizing a particular theory of measurement is studied." Measurement scales followed. He is now working on another subject, but he maintains that philosophy students should learn about measurement: "I continue to proselytise for the theory of measurement as an excellent source of precise but elementary methodology to introduce students to systematic philosophy of science" ('Self-profile', p. 16).

"My own empirical bent in philosophy is nowhere more clearly reflected than in my attitude toward the philosophy of language", Suppes wrote in 1979 ('Self-profile', p 40). Let us see how he attempted to elucidate the process of **learning**: learning how to speak and read, how to compute or conceptualize, as well as how to mount horses or bicycles.

> In 1956 my oldest child, Patricia, entered kindergarten and my interests in applications were once again stimulated, in this case to thinking about the initial learning of mathematical concepts by children. [...] In recent years the interest in mathematical concept formation has melded into my work on computer-assisted instruction ('Self-profile', p. 29-30).

While developing computer-aided learning experimental programs, he also looked into the history of philosophy:

> The concept of meaning has a much longer history in philosophy and logic than it does in psychology or in linguistics. It is possible to begin the philosophical story with Aristotle, but Frege and Tarski will do... ('Self-profile', p. 33).

He sees the meaning of a word, or phrase, as being context sensitive, and resting on a sort of sedimentation of numbers of associations in a multiplicity of linguistic environments. He certainly does not admit of the meaning of a word to be conceived as a stable unit, possibly a supernatural entity, a 'true' idea serving as an absolute reference for the variety of usages of the word.

> I have come to be skeptical of the long philosophical tradition of looking for various kinds of bedrocks of certainty, whether in epistemology, logic, or physics. Just as the natural notion of a person is not grounded in any hard and definite realization, and certainly not a physical one because of the continual fluctuation of the molecules that compose the body of a person, so it is with the meaning of expressions. ('Self-profile', p. 38).

Parenthetically, Suppes expected the use of computers in education to help the students think by themselves, and to free them from the tentative coercion of prestigious professors delivering an authoritative message from the pulpit.

> My vision for the teaching of philosophy is that we should use the new technology of computers to return to the standard of dialogue and intimate discourse that has such a long and honored tradition in philosophy. Using the technology appropriately for prior preparation, students should come to seminars ready to talk and argue. ('Self-profile', p. 44).

He certainly did not underestimate the difficulty of the programming. But he was optimistic about the future of the computer-teacher. Not all of his predictions have been confirmed.

> There are many good teachers and many bad ones. We understand some of the features that distinguish these two groups but we scarcely have a detailed constructive theory as to how to model the best and how to eliminate those features characteristic of the worst. [...] The computers that are used for instruction 20 years from now will, almost without question, no longer be silent but fully talking - and talking with a great deal of instructional sophistication. The silent computers that dominate the present scene will definitely be a thing of the past by the end of the century. (Suppes, 1981b, p. xxviii).

Rational decision and action is another topic that Suppes tackled with the tools of both mathematics and psychology. As a formal tool, he liked best the Bayesian model, the implicit assumption of which is that you should always choose so as to maximize your expected utility. He also praised the Aristotelian model, where it is assumed that you should always act "for good reasons". He insisted on not dissociating rational decision from rational action, and explained that a wise decision maker should take into account both whatever information he may gather on the actual state of affairs, and a clear notion of what he wants promoted:

> Both beliefs and values are essential ingredients of the expected utility model. A person with beliefs but no values does not know what to choose, and a person with values or feelings but no beliefs can easily choose foolishly. (Suppes, 1984, p.207).

But the standard models of the decision making process analyze decision making as a mathematical computation. That is a simplifying fiction. Explicit expected utility computations are rare. William James knew better than our economists. "The actual computations we do are fragmentary, occasional, contextual, driven by associations internal and external."[4] In other words:

---

[4] P. Suppes, 'Rationality and Freedom', manuscript, Paris, Collège de France, March 2, 2002. The manuscript is mentioned in Suppes' letter (Figure 1).

> A theory of rationality that is posited on some exemplary style of rational deliberation, conscious, measured and complete, is utterly mistaken as a psychological account of how any of us go about making decisions about practical problems or solving theoretical ones. (Suppes, 2005, II, 3).

But wouldn't a psychological account of our decisions end up with the certified report that what we call rationality reduces to psychological determinism? Not so, Suppes answers: "the Brownian movement has killed and buried determinism"[5]. Kant was mistaken. His 3rd antinomy[6] suggested that determinism is certain, free will is conjectural. As a matter of fact, free will is certain, while the question of determinism/indeterminism remains fuzzy:

> The Kantian thesis has been properly stood on its head. Free will, as exemplified in voluntary motion, is the hard empirical fact. Determinism (or, if you prefer, indeterminism) is the transcendental metaphysical assumption out of reach of detailed confirmation [...] I have taken a biological line of argument that makes free will a natural concept exhibited in the behavior of other species. This is clearly a scientific line of attack. (Suppes, 1994, pp.462, 466).

(Remember: "Free will is a scientific problem, not a philosophical one"—end of part 1). At that point, Suppes' message is that while Kant's program is clearly dead, an empirical approach of the notions of 'truth' or 'belief' can be seen as a modern version of Aristotle's program.

3 Suppes' probabilistic metaphysics

"Randomness is in nature, and not simply in our ignorance of true causes" (Suppes, 1984, p.23). When did Suppes start giving randomness a real, ontological dimension? At least as soon as 1974, when he gave a series of invited lectures in Sweden:

> In 1974, I gave the Hägerström lectures in Uppsala, Sweden, entitled 'Probabilistic Metaphysics'. In those lectures I took as my starting point Kant's criticism of the old theology; my purpose was to criticize various basic tenets of what I termed the new theology ('Self-profile', p. 47).

Following a suggestion from Amartya Sen, Suppes reworked his lectures. The book came out ten years later. In that book the attack on the "new theology" is straight and forceful. According to the author, the new theology, initiated by Immanuel Kant, and tacitly assumed by virtually all philosophers of science after Kant, asserts that:

---

[5] "Le mouvement brownien a sonné le glas du déterminisme", in: Suppes, 1981a, p.57.

[6] Kant, Critique of pure reason (transcendental dialectic, 3rd antinomy).

1. the future is determined by the past,
2. every event has a sufficient determinant cause,
3. knowledge must be grounded in certainty,
4. scientific knowledge can in principle be made complete,
5. scientific knowledge and method can in principle be unified (Suppes, 1984, p.2).

All five tenets will be found disputable. "We should be able to construct a general metaphysics or epistemology on other grounds" (Suppes, 1984).

For the construction of the new metaphysics, two elements are preserved from the old metaphysics. From Aristotle's *Metaphysics*, the analysis of 'being' is retained. Let us note that Suppes gives a remarkable account of the aristotelian notions of matter, form, and substance:

> Matter qua matter is purely potential and without attributes (Metaphysics, 1029a19). A substance has both form and matter. The nature of a substance is complex. It is neither simply the form nor the matter (Physics, 191a10; Metaphysics, 1043a15). There is no principle of individuation for matter qua matter...The principle of individuation for substances does not require sameness of matter for sameness of substance. For example, an animal is both intaking and excreting substance, but we still speak of the identity of that animal through time. (Suppes, 1984, p.5; see also Suppes' paper in *Synthese*, (Suppes, 1974)).

From Immanuel Kant the project of a critical philosophy is retained as a method. Unfortunately Kant had rejected the use of probability. Therefore his possible contribution to the contents of the new philosophy of science is nil.

Building a general philosophy of science is a huge enterprise. In the Introduction of his *Probabilistic Metaphysics*, Suppes gives a list of eight basic "metaphysical propositions". Four of these will be minutely argued, each in a whole chapter. These are:

1. "The fundamental laws of natural phenomena are essentially probabilistic rather than deterministic in character." (Suppes, 1984, chap. 2, p.34).

2. "In general, causality is probabilistic, not deterministic in character, and consequently no inconsistency exists between randomness in nature and the existence of valid causal laws." (Suppes, 1984, chap. 3, p.70).

3. "Certainty of knowledge—either in the sense of psychological immediacy, in the sense of logical truth, or in the sense of complete precision of measurement—is unachievable." (Suppes, 1984, chap. 4, p.99).

4. "There is no bounded fixed scientific theory toward which we are in general converging." (Suppes, 1984, chap. 5, p.117).

Assuming the "ontological character of randomness" (Suppes, 1984, p.208) entails that: 1) unified science, as cherished in the old days, was a dream; 2) learning is merely building habits; 3) there are limits to the rationality of our 'rational' strategies of action. Let us briefly comment on those three points.

1. The idea of **unified science** was promoted in the early 20th century by Neurath and others (Neurath, Carnap, & Morris, 1938, vol.1, part 1) with the ambition that psychology and other social sciences might be reduced to biology, biology could be reduced to physics and chemistry, so all sciences would converge, using the same (mathematical) language and methods. But science is plural and will remain plural. Suppes sketches a program adapted to "the plurality of science" (Suppes, 1984, chap.6):

> the rallying cry of unity followed by three cheers for reductionism should now be replaced by a patient examination of the many ways in which different sciences differ in language, subject matter, and method... (Suppes, 1984, p.125)

Plurality in science reflects the diversity in the world. Suppes' anti-reductionism is ontologically based, and not only an affair of methodology:

> the important central features of meaning and correctness of a program are in a strong sense irrelevant to the particular physical realization in which the program is embodied (Suppes, 1984, p.130).

You cannot reduce the mental conception and design of a program to its realization in a brain or a computer. Suppes wants psychology to be a science of the mind, as serious and dignified as physics.

2. How do we **learn**? How do we know that Rome is not the capital of France? "three processes—talking, listening, and reading—are the main methods by which information is transmitted and intellectual skills are learned." (Hintikka, Suppes, & Moravcsik, 1973, Preface). Don't we have to learn logic? How do our intellectual skills develop?

> It is misleading to say that we are making a deduction to arrive at the conclusion that 'Rome is the capital of France' is false .. [...] The approach to computation about such things and processes, characteristic of our minds, was well recognized by Hume, the godfather of the central mechanism of association, and already foreshadowed by Aristotle. What I shall insist on here is the universal role of association as the main method of computation in the brain (and in the mind, if you will) in dealing with ordinary experience. (Suppes, 2005, II, 3).

What about our being able to correct errors of judgement or of calculation, like mistakenly applying a probabilistic causal law to individual cases? "Certainly it would be ludicrous to think that there is a logically valid inference from the mean data to the individual data. But [...] ordinarily much of what I know about individuals is based upon generic causal relations." (Suppes, 1984, p.61).

3. **Rational behavior** cannot be entirely disconnected from its factual context (internal: moods, hormones; or external: weather, needs of our children, unexpected

situation, etc.). "Intentions and actions are afloat on a sea of random happenings." (Suppes, 1984, p.208). Most of the time we don't explicitly rationalize before doing. We decide without much thinking, because we trust our habits. For complex tasks we may use recipes or "justified procedures", as do the "rational cooks and carpenters". In important circumstances we will effectively compute, using for example the Bayesian model - note that the Bayesian theory provides a rational for changing beliefs, given the acquisition of new information; but an analysis of the process of information selection is missing, leaving the subject free to pay or not pay attention to possibly relevant information. Finally, trained subjects may have rational intuitions. "Intuitive judgment is a skill like jogging, playing tennis, or finding mathematical proofs. Even simpler and more universal examples are walking and talking." Suppes ambitioned to develop "a proper educational psychology for teaching those skills." (Suppes, 1984, p.218). Let him conclude:

> What we teach our students or ourselves about practical decision making cannot be wholly reduced to algorithms or even to implicit axioms, but we can, on the other hand, improve on Aristotle just because we can apply axioms and procedures as appropriate. The use of modern quantitative methods of decision making is necessarily limited but powerful when properly applied. The role of judgment and practical wisdom in applying such methods will continue to be of central importance. The tension between calculation, qualitatively justified procedures, and judgment will not disappear. Nor will its philosophical analysis. (Suppes, 1984, p.221).

Bibliography

Aristotle, Loeb Edition. (1968). *Metaphysics,* translated by H. Tredennick. *Physics,* translated by P.H. Wickstead & F.M. Cornford. Cambridge, Mass., USA: Harvard University Press.
Bogdan, R. J. (Ed.). (1979). *Patrick Suppes*. Dordrecht, Holland: Reidel.
Hintikka, J., Suppes, P., & Moravcsik, J. (Eds.). (1973). *Approaches to natural language: Proceedings of the 1970 Stanford Workshop on Grammar and Semantics*. Dordrecht, Holland: Reidel.
James, W. (1950). *The principles of psychology*. New York, NY, USA: Dover. (Original work published 1890)
Kant, I. (1781-87). *Kritik der reinen Vernunft.* Kant I. (1949). *Critique of Pure Reason* (F. M. Müller, Trans.). London, UK: Macmillan and Co.
Neurath, O., Carnap, R., & Morris, C. W. (Eds.). (1938). *International handbook of unified science*. USA: University of Chicago Press.
Scott, D., & Suppes, P. (1958). Foundational aspects of theories of measurement. *The Journal of Symbolic Logic*, 23(2), 113–128.
Suppes, P. (1967). *Introduction to logic*. Princeton, NJ, USA: Van Nostrand Company. (Original work published 1957)
Suppes, P. (1969). *Studies in the methodology and foundations of science: selected papers from 1951 to 1969*. Dordrecht, Holland: Reidel.
Suppes, P. (1974). Aristotle's concept of matter and its relation to modern concepts of matter. *Synthese*, 28, 27–50.
Suppes, P. (1981a). *Logique du probable* (H. Rouanet, Trans.). Postface by Jules Vuillemin. Paris, France: Flammarion.
Suppes, P. (1981b). *University-level computer-assisted instruction at Stanford: 1968-1980*. CA, USA: Institute for mathematical studies in the social sciences, Stanford University. (also see: Collected works of Patrick Suppes, online).
Suppes, P. (1984). *Probabilistic metaphysics*. Oxford, UK: Blackwell.
Suppes, P. (1994). Voluntary motion, biological computation and free will. In P. A. French, T. E. Uehling, & H. K. Wettstein (Eds.), *Midwest studies in philosophy volume XIX: Philosophical naturalism* (pp. 452–467). Ind., USA: University of Notre Dame Press.
Suppes, P. (1996). The aims of education. In A. Neiman (Ed.), *Philosophy of education 1995* (pp. 110–126). Urbana, Ill., USA: Philosophy of Education Society, University of Urbana-Champaign.
Suppes, P. (2002). *Representation and invariance of scientific structures*. Stanford, CA, USA: CSLI Publications.
Suppes, P. (2005). Neuropsychological foundations of philosophy. Public lecture given at Collège de France, Paris on 7th of November 2005.
Suppes, P., & Zinnes, J. L. (1963). Basic measurement theory. In R. D. Luce, R. R. Bush, & E. Galanter (Eds.), *Handbook of mathematical psychology* (Vol. 1.1, pp. 1–76). New York, NY, USA: John Wiley and Sons (repr. 1967).

Whitehead, A. N., & Russell, B. (1910–1913). *Principia mathematica*. Cambridge, UK: Cambridge University Press.

Author biography. Anne Fagot-Largeault, PhD (Stanford 1971), MD (Paris 1978), Docteur ès lettres et sciences humaines (Doctorat d'Etat, Paris 1986). **Currently** emeritus professor of the College de France (philosophy of life science). Member of the French Academy of Sciences, of the International Academy of Philosophy of Science; associate member of the Royal Academy of Belgium (humanities section). Member of the College of experts of the French Biomedicine Agency for the research on human embryos and embryonic stem cells. Member (2007-2011) of the Executive Committee of the DLMPS. Has been a member of Data and Safety Monitoring Committees for international clinical trials in aids, cancer, cellular therapy.

Career: teacher of philosophy in high school (1961-1966); assistant professor at the university of Paris-12 (1971-87); full professor in philosophy of science at the universities of Paris-10 (1987-95) and Paris-1 (1995-2001); inaugural lesson at the College de France in 2001. Medical practice at the Henri Mondor hospital (Assistance publique de Paris, 1978-2003) mostly as a psychiatrist in the emergency room.

Research: centered on the philosophy of life sciences, developed along three lines: (1) diagnostic reasoning, inductive logic, statistical and probabilistic methods, heuristic procedures; (2) investigation of causal links, validation of causal hypotheses, causal explanation, evolution and the ontology of becoming; (3) ethics of medical practice and investigation, methodology of clinical & of epidemiological research, epistemology of bio-medical sciences, bio-medical anthropology.

Publications: see www.college-de-france.fr.

10 Logical empiricist reconstructions of theoretical knowledge

WILLIAM DEMOPOULOS [*]

Abstract. I argue that the logical empiricist account of theoretical knowledge exhibits a fundamental misconception about the character of the claims theories express. This is the idea that the application of truth to theoretical claims—whether they involve unobservable entities or the structure of space and time—rests on considerations that show them to possess an arbitrariness that does not attach to non-theoretical, empirical statements. This idea was completely explicit in the case of logical empiricism's conventionalist account of theoretical claims about the geometry of space and time, but it emerges as an unintended consequence of its partial interpretation account of theories about entities which transcend observation. The misconception stems from an incorrect assessment of the epistemic warrant theoretical claims enjoy and an incorrect assessment of the basis for our confidence in existence claims involving unobservable entities. The view I advocate allows that the value of a physical theory is often instrumental and independent of whether the theory is even approximately true. But this concession to instrumentalism is compatible with the idea that a theory's instrumental value can consist in facilitating the discovery of salient truths about reality, even a part of reality that is entirely hidden from observation. The argument to this conclusion rests on an analysis of the methodology of theory-mediated measurement and the role this methodology plays in securing fundamental existence claims of the kind we associate with Jean Perrin in the case of molecular reality and J. J. Thomson in connection

---

[*]The University of Western Ontario, London, Canada N6A 5B8

with the constitution of cathode rays. The larger study from which the present paper is drawn is an extended presentation of that argument, while the following discussion focuses on some of the main considerations that motivated its development.

Keywords: structuralist thesis, partial interpretation, logical empiricism, model-theoretic argument.

1 Introduction

The logical empiricists articulated what is arguably the first systematic development of a theory of theories. Their approach derived from the perspective afforded by modern logic and Hilbert's work on the axiomatic method. It was designed to address two key developments, the first concerned with abstract principles, the second with existence claims. In the case of abstract principles, Einstein's theories of relativity demonstrated the need to revise our conception of the epistemological status of the geometry of space and the chronometry of time; to understand the nature and epistemological significance of such a revision was pressing. As for existence claims, the early twentieth-century experimental success of atomism established the appeal to unobservable entities as a permanent feature of physical theory and raised the question of the nature of the support that attaches to claims which purport to be about entities that transcend observation.

The logical empiricist reconstruction of relativity was a corrective to the pretensions of the rationalist adherence to the synthetic a priori and the naiveté of an earlier empiricism; it was distinguished by its conventionalist interpretation of geometry and chronometry. The search for an account of theoretical claims which appeal to unobservable entities culminated in the partial interpretation view of theories, one of whose principal goals was to address the prima facie challenge to empiricism that such claims represent by clarifying their empirical status.

My focus is the partial interpretation reconstruction of theories and its elaboration by Carnap using the notion of the Ramsey-sentence of a theory. I will touch on work of David Lewis which is often seen as a contribution to this tradition, and I will articulate a central difficulty for the partial interpretation reconstruction that is suggested by one of Hilary Putnam's "model-theoretic arguments." The difficulty consists in the failure on the part of the partial interpretation and related reconstructions to provide an adequate account of our epistemic access to theoretical domains. To address this issue it is necessary to delve more deeply into the methodology by which such access has in fact been secured. My goal here is to motivate the importance of the problem of epistemic access. The larger study from which this paper is drawn is an extended discussion of this issue and the role theory-mediated measurement plays in its resolution.

2 The partial interpretation account of theories

The partial interpretation account of theories holds that the vocabulary of a theory consists of an observational and a theoretical component, where the application of the observation-theory distinction is based on whether a vocabulary item is held to apply to entities in the intended domain of the theory which are observable—in which case the item belongs to the *observation vocabulary*—or unobservable, in which case it belongs to the *theoretical vocabulary*. In the classical formulation of Carnap (1956, pp. 41–42 and pp. 46–48), this distinction partitions vocabulary items into just these two classes. (The possibility of a third class of terms—*mixed vocabulary* items that are understood to apply to both observable and unobservable entities—will be addressed later.)

Given the foregoing distinction in vocabulary, the sentences of the language of a theory are divided into three classes: one consisting of sentences which are generated from just the observation vocabulary, another of sentences generated from just the theoretical vocabulary, and a third consisting of sentences generated from the combined observation and theoretical vocabularies. These are, respectively, the *observation sentences*, *theoretical sentences*, and *correspondence rules* of the language. A *theory* is the conjunction of a selection of theoretical sentences and correspondence rules. There are no special restrictions on the logic of a theory, and it may be either first-order or higher-order; the notion of a theory may also be generalized in various ways that are irrelevant to the conceptual issues on which I intend to focus.

Historically, the partial interpretation account of theoretical knowledge derives from the idea that theoretical terms are introduced by sentences which, taken by themselves, are indistinguishable in their epistemic status from the statements of a pure mathematical theory. Theoretical statements share with the statements of a mathematical theory the property that their interpretation is responsible only to the logical category of their constituent nonlogical constants. This view of theoretical statements is a consequence of the fact that the partial interpretation account is a continuation and extension to the theories of physics of the axiomatic tradition that Hilbert initiated in pure mathematics. Especially influential was Hilbert's contention that the primitives of a mathematical theory are whatever satisfies its axioms. This contention—that the postulates of a theory "implicitly define" its primitive notions—swept away the subjective associations that characterized an older tradition's understanding of a mathematical theory's primitives, even in the case of geometry, where they were thought to have a familiar "intuitive" content. The partial interpretation account sought to extend Hilbert's analysis of mathematical theories to physics by providing an account of the empirical content of the theoretical statements of physics that is based on the connections between theoretical terms and observation terms that are expressed by correspondence rules.

One can see in this brief sketch the two characteristic theses of the partial interpretation view: the first, its claim that only the observation vocabulary is completely understood; and the second, the correlative claim that the interpretation of the theoretical vocabulary is limited by constraints which depend only on the logical category of the theoretical terms and whatever restrictions the true observation sentences impose on the domain of unobservable entities over which the theoretical sentences and correspondence rules are evaluated. I will refer to this second claim as the *structuralist thesis*. We have yet to explain how the partial interpretation view conceives the relation between interpretations, true interpretations, and truth.

3 Carnap on Ramsey sentences and the explicit definition of theoretical terms

Carnap's mature reconstruction of the language of science[1] builds on and extends the partial interpretation view of theories. The central notion of this account is the *Ramsey sentence* of a theory: the sentence formed by replacing theoretical terms by (new) variables of the appropriate logical category, then closing the resulting formula by adding an existential quantifier for each of the new variables. It is a very short step from the two characteristic theses of the partial interpretation account of theories to the notion that a partially interpreted theory's Ramsey sentence captures its "factual content": the Ramsey sentence is observationally equivalent to the theory in the sense that any argument from the partially interpreted theory to a sentence of the observation language can be recovered using the Ramsey sentence instead.[2] Notice that the Ramsey sentence's use of variables in place of uninterpreted theoretical terms simply makes explicit the commitment of the partial interpretation account to the structuralist thesis. As Ramsey expressed it:

> So far ... as *reasoning* is concerned, that the [transforms of the theoretical sentences and correspondence rules which constitute the matrix[3] of the Ramsey sentence of the theory] are not complete propositions makes no difference, provided we interpret all logical combinations as taking place within the scope of a [single existential] prefix. ...For we can reason about the characters in a story just as well as if they were really identified,

---

[1] This is the reconstruction expounded in (Carnap, 1963) and extended in various ways to be described below in (Carnap, 1961). The publication date of (Carnap, 1963) is a poor guide to the work's date of composition since the publication of the volume in which it appeared was delayed for many years.

[2] The Ramsey sentence of a theory, like one of its Craig transcriptions, eliminates theoretical vocabulary, but unlike a Craig transcription, it retains the connections between observable properties and relations which are mediated by their association with theoretical properties and relations. This is why the difficulties which Hempel (1965, pp. 214–216) shows to be a necessary feature of the elimination of theoretical vocabulary for reconstructions based on the notion of a Craig transcription are not difficulties for Ramsey-sentence-based reconstructions. For further discussion see Demopoulos (2013, Chapter 7.2) and Putnam (2012).

[3] By the matrix of a Ramsey sentence I mean the formula which results when the "new" existential quantifiers are deleted.

provided we don't take part of what we say as about one story, part about another. (Ramsey, 1929, p. 232)

Carnap's mature reconstruction refines the doctrine of partial interpretation in two principal respects. As we have already noted, Carnap explicates the factual content of a partially interpreted theory in terms of its Ramsey sentence. But Carnap took things a step further by combining his account of the factual content of a theory with an explication of theoretical analyticity—analyticity relative to a theory—in terms of what has come to be known as the *Carnap sentence* of a theory: the conditional whose antecedent is the theory's Ramsey sentence and whose consequent is the partially interpreted theory. Before Carnap the distinction between the factual and analytic (and hence, non-factual) components of a theory followed the distinction between postulates and definitions. But since this distinction is inherently arbitrary, its utility for a dichotomy that is supposed to reveal our factual commitments may be doubted.

The Carnap sentence is justifiably regarded as analytic because it is a kind of "implicit definition" of the theoretical vocabulary, one that is provably non-factual in the sense that the only observation sentences it logically implies are logical truths. And as John Winnie (Winnie, 1970) later showed, the Carnap sentence, like a proper definition, satisfies a special noncreativity condition that is similar to the noncreativity condition that is customary for proper *explicit* definitions.[4]

Carnap advanced the Ramsey sentence not just as a clarification of the partial interpretation view of theories, but as a correct representation of how scientists understand their theoretical claims. They intend, Carnap held, an "indeterminate" claim, one that may have many interpretations under which it comes out true. As scientists understand them, theoretical claims are indeterminate as to the interpretation of their theoretical vocabulary and *any* representative class or relation which makes true the Ramsey sentence of the theory to which the claim belongs is as acceptable as any other. To narrow down the interpretation any further than is demanded by the truth of the Ramsey sentence would, for Carnap, violate the intentions of the scientist who constructed the theoretical system.

In (Carnap, 1961), which is one of his last papers on theoretical terms, Carnap converts the *implicit* definition of theoretical terms by the Carnap-sentence into a sequence of *explicit* definitions of them. But these explicit definitions do not eliminate—and were not intended by Carnap to eliminate—the indeterminateness of his earlier account. Indeed, Carnap formulates his explicit definitions in what he calls a "logically indeterminate" language. The language L_ε which he employs is a standard first- or higher-order language enriched with Hilbert's ε-operator and the extensional axioms which govern its use. There are two such axioms governing the use of Hilbert's ε-operator. Given a formula Fx in one free variable, the first axiom tells us that if there is something satisfying Fx, then there is an "ε-representative" of F, denoted '$\varepsilon_x(Fx)$', that is selected

---

[4]For an exposition of these matters, see (Demopoulos, 2007) and (Gupta, 2009).

by the choice function which interprets the ε-operator. The second axiom tells us that if the formulas Fx and Gx are extensionally equivalent, their ε-representatives are the same. That it should be possible to apply Hilbert's ε-operator to the Ramsey-sentence reconstruction of theories is a consequence of Carnap's observation that the Carnap sentence of a theory can be derived from a sentence that is in the same form as the first of the axioms for the ε-operator.

For Carnap the principal virtue of this proposal is that it incorporates the convenience of having the use of a theoretical vocabulary while retaining the characteristic indeterminateness of that vocabulary that is the hallmark of the partial interpretation view and of his mature reconstruction in terms of Ramsey and Carnap sentences. Thus he writes that the theoretical postulates and correspondence rules "are intended *by the scientist who constructs the system* to specify the meaning of [a theoretical term] to just this extent: if there is an entity satisfying the postulates, then [the term] is to be understood as denoting one such entity. Therefore the definition [of a theoretical term by means of Hilbert's ε-operator] gives to the indeterminate [theoretical term] just the intended meaning with just the intended degree of indeterminacy" (Carnap, 1961, p. 163, emphasis added).

4 A proposal of David Lewis and two theorems of John Winnie

David Lewis's (1970) is sometimes credited with having refined Carnap's and Ramsey's reconstructions and to have improved on Carnap's approach to the explicit definition of theoretical terms by showing how it might be possible to avoid multiple interpretations of the theoretical vocabulary under which the theory comes out true. Lewis maintained that allowing for what he calls "multiple realizations" concedes too much to instrumentalism. Lewis does not say why multiple realizablity is a concession to instrumentalism, but let us for the moment grant the point and consider how he makes the case that there are many theories which, if realizable, are *uniquely* realized. Lewis is clear that he must provide an independent defense of this contention, since the possibility of there being just one realization appears to be excluded by two theorems of Winnie (1967). Modulo the conceptually unimportant technical restriction that not all theoretical properties and not all theoretical relations are universal, Winnie shows that on the partial interpretation view of theories, if a theory has one realization, there is always another; and if a theory is realizable at all, it is arithmetically realizable.

Lewis's response to Winnie rests on two features of his conception of the language in which theories are formulated. First of all, Lewis follows the partial interpretation account by dividing the vocabulary of a theory into two parts, which he calls the "O-vocabulary" and the "T-vocabulary" of the theory. However, Lewis's "O-T distinction" is not the distinction between observation and theoretical vocabulary of the partial interpretation account. Lewis's distinction concerns *old* vocabulary, vocabulary which is understood prior to the formulation of the vocabulary-introducing theory; and the contrast Lewis's distinction draws between old and T- or *new* vocabulary has nothing to do with observation or observability. In principle, Lewis's O-T distinction could be completely orthogonal to the observational-theoretical distinction of Carnap and the doctrine of partial interpretation. A second, related, difference involves Lewis's notion of an "O-mixed term." This is a notion that does real work for Lewis, but before explaining it, some further background regarding the partial interpretation view and its relation to Lewis's O-T distinction is necessary.

In his exposition of the partial interpretation reconstruction, Winnie includes in addition to the observation and theoretical predicates a separate category of *mixed predicates*, predicates that apply to both observable and unobservable entities. Lewis has the notion of a *mixed term*, and of special importance are those he calls "O-mixed" terms. These are terms which, like Winnie's mixed predicates, can apply to both observable and unobservable entities. But their characterization—unlike Winnie's characterization of observation and theoretical predicates—has nothing to do with observability. And while Winnie's mixed predicates are distinguished from observation predicates, Lewis's O-mixed terms count as O-terms, and as such, are assumed to be fully understood whether they apply to observable or unobservable entities; therefore the interpretation of O-terms—whether they are unmixed and refer only to observable entities, or are mixed and refer also to unobservable entities—must be preserved as we pass from one realization of a theory to another.

The situation is altogether different for Winnie and for the standard view. As we will soon see in greater detail, when Winnie establishes the existence of alternative realizations by means of a permutation map, it is only the entities in the observable part of the domain that cannot be permuted, and it is only the interpretation of the observation predicates—predicates which apply only to observable entities—that must be the same in any realization of a partially interpreted theory. No such requirement applies to the interpretation of theoretical predicates; nor does it apply to mixed predicates.

Since Winnie's permutation map is defined as the identity on observable entities, it is trivially true that observable relations are isomorphic to their images under his mapping. But although it is trivially true that Winnie's map is an isomorphism between observable relations, it is not part of Winnie's argument that *every* one-one map from the observable part of the domain onto itself can be extended to an isomorphism on the properties and relations which interpret the observation predicates of the theory. However the situation is different for one-one maps from the unobservable part of the domain onto itself and the relations which interpret the theory's theoretical predicates.

To establish that a partially interpreted theory has many models if it has one model, Winnie uses a construction based on a mapping from and onto the domain of a model of the theory that permutes at least one pair of unobservable entities so that it changes the image, under the mapping, of the interpretation of at least one theoretical predicate. By the structuralist thesis—and this is the observation Winnie's proof rests on—the theoretical predicates and relations can always be understood so that the relations which interpret them are *defined* by their images under an arbitrary one–one mapping from and onto the unobservable part of the domain. As for the properties and relations which interpret mixed predicates, their images are unchanged only by the action of the mapping from and onto the observable part of the domain. Such a construction defines the theoretical and mixed properties and relations of a new structure as the images of the properties and relations of a structure we know to be a model of the theory. Since the new properties and relations are *by construction* isomorphic to the properties and relations of a model of the theory, the structure which they define must also be a model of the theory.

To see why Winnie's argument does not affect Lewis's claims about uniqueness of realization, let us suppose that electrons and protons are unobservable entities, that electrons are (strictly) smaller than protons, and that 'smaller than' is a term of Lewis's O-mixed vocabulary; any realization must therefore preserve the interpretation of 'smaller than.' It follows that the construction of a realization which, like Winnie's, interchanges an electron and a proton, is ruled out. In fact, the only case in which Lewis's and Winnie's notions of realization coincide is the case in which Lewis's O-terms and T-terms have (respectively) only observable and only unobservable entities in their extensions, and where therefore there are no mixed terms. In such a case, even on Lewis's account, a theory must have more than one realization if it has any realization at all. We might put this by saying that Lewis's approach to theories and the definition of theoretical terms is at best only "accidentally" affected by Winnie's results.

To sum up our discussion of Winnie and Lewis, Winnie's mixed predicates form a special category distinct from observation predicates, and only the observable entities in the interpretation of mixed and observation predicates are unaffected by his permutation map. But for Lewis mixed terms can be classified as O-mixed terms, and in order to avoid Winnie's permutation argument, it suffices that there should be a suitably rich collection of O-mixed terms. In fact it may suffice that there should be *one* term that stands for a relation which is such that all the various types of theoretical entity are comparable in terms of this relation. In our example of electrons and protons, it is sufficient that the particles of one kind are (strictly) smaller along some dimension than those of the other—assuming, of course, that *smaller than* along this dimension is picked out by an O-mixed term. It is therefore not at all implausible that for Lewis a realizable theory can always be extended by the introduction of O-mixed terms and appropriate postulates involving them to a theory that is *uniquely* realizable.

Lewis's deployment of the Ramsey sentence and a modified form of the Carnap sentence shares a strong formal affinity with Carnap's reconstructions: Lewis's adoption of the Ramsey sentence suggests a commitment to the idea that a model of our understanding of new or T-terms—including the case in which the new terms are associated with the introduction of a new class of entities—is adequate if it captures the inferential connections of the statements containing the new terms with the sentences in the O-vocabulary. But although Lewis's old vocabulary may well include "observation" terms, the connections between T-terms and O-terms which Lewis's mixed statements express is not explicitly proposed as a connection with observation vocabulary, but merely as a connection with vocabulary that is understood. And for Lewis, by contrast with the partial interpretation or Ramsey-sentence reconstructions of the logical empiricists, so far as our understanding of *old* vocabulary is concerned, it might be based entirely on our grasp of the inferential connections into which its items enter—independently of whether or not these include inferential connections with observation sentences.[5] It is clear therefore that despite their formal affinity, the absence of an epistemological motivation underlying Lewis's reconstruction of theories makes his account very different from both the partial interpretation reconstruction and from Carnap's various refinements of it.

Carnap and the advocates of partial interpretation take it as a desideratum of an adequate reconstruction of theoretical knowledge that it should address the empirical basis of theoretical claims. On the partial interpretation reconstruction, this problem is addressed by the provision of an explanation of our understanding of theoretical claims in terms of the connection correspondence rules establish between theoretical and observation vocabulary. Carnap's Ramsey-sentence reconstruction dissolves the problem of how we come to understand the meanings of terms which apply to unobservable entities by *eliminating* theoretical terms in favor of variables. But this dissolution of the problem is merely an emendation—not a rejection—of the partial interpretation view, an emendation that preserves the structuralist thesis. Indeed, Carnap's transition to the Ramsey sentence reconstruction rests on the recognition that the partial interpretation view subscribes to this thesis. For if, in addition to whatever restrictions the true observation sentences impose on the domain of a model of a partially interpreted theory, the constraints on the interpretation of theoretical vocabulary appeal only to the logical categoryof the theoretical terms, then there can be no objection to their replacement by variables, and the problem of accounting for how theoretical terms are understood then simply disappears. As for the empirical basis for theoretical *claims*—as opposed to our understanding of theoretical *vocabulary*—and the explanation of their difference from the claims of pure mathematics, modulo the elimination of theoretical vocabulary, these issues are addressed by Carnap's Ramsey-sentence reconstruction much as they are by the partial interpretation reconstruction. Instead of appealing to the association of theoretical claims with observation sentences by the mediation of correspondence rules, the empirical basis of such claims is accounted for by the association of the Ramsey-sentence transforms of theoretical claims with observation sentences that is effected by the Ramsey-sentence transforms of correspondence rules.

The centrality of the problem of the empirical basis of theoretical claims to the logical empiricists' reconstruction of theories serves not only to distinguish their ap-

---

[5] See (Gupta, 2013, especially section 11) for a critical perspective on "inferential-role" models of meaning.

proach from Lewis's. It also shows why it is so misleading to characterize their account—whether partial interpretation, Ramsey-sentence, or one of Carnap's later reconstructions—to have incorporated a "syntactic view" of theories. For the logical empiricists, theories are indeed linguistic objects. But to call the logical empiricist account "syntactic" misses the fact that it is first and foremost a reconstruction of theoretical knowledge that purports to illuminate how observation bears on the empirical character and evidential support of theoretical claims. Evidently, neither goal can be successfully addressed without going beyond syntax—as indeed the principal logical empiricist reconstructions do when they assume that the observation language is interpreted.

This contrasts with the situation in mathematical logic where a theory is defined as a set of sentences in a language about which we assume *only* that its syntax and underlying logic are completely explicit. The logical tradition is also motivated by an epistemological problem which is no less fundamental to its notion of a theory than the epistemological motivation underlying logical empiricism is to its conception of a theory. The logical tradition sought to show that a *finitary* notion of proof suffices for the reconstruction of all mathematical reasoning—even reasoning within theories whose intended interpretation is over an infinite domain of arbitrarily large cardinality. But here the restriction to syntax in the characterization of a theory is entirely natural, since it is essential to the successful positive solution of the problem which motivates the logical tradition that a theory should be represented as a purely syntactic object. Although the logical empiricist approach to theories was profoundly influenced by the logical tradition, its goals were different: it sought to build a platform for the representation of the theoretical claims of physics that would be capable of illuminating their content and the basis on which they are understood and evaluated. In particular, it sought to show how observation must be a central component of an adequate empiricist solution to this problem. The nature of the questions the logical empiricist approach sought to address demanded—and were recognized by its proponents as demanding—a notion of theory that includes more than the purely syntactic conception of the logical tradition.

5 Hilary Putnam's model-theoretic argument

We have seen how, by contrast with Lewis, Carnap was prepared to accept multiple realizability as a point *in favor* of partial interpretation and Ramsey-sentence reconstructions. In his "Reply to Hempel" in the Schilpp volume devoted to his work Carnap even went so far as to endorse an arithmetical interpretation of the Ramsey sentence of a theory as the *correct* understanding of what, on his reconstruction, the sentence asserts. As Carnap put it,

> I agree with Hempel that the Ramsey sentence does indeed refer to theoretical entities .... However, it should be noted that these entities are not unobservable physical objects like atoms, electrons, etc., but rather (at

least in the form of the theoretical language which I have chosen in [(Carnap, 1956, Section VII)] purely logico-mathematical entities, e.g. natural numbers, classes of such, classes of classes, etc. (Carnap, 1963, p. 963).

Carnap might well be understood therefore to have also anticipated and embraced the content of the second of Winnie's two theorems as an acceptable consequence of his view of the factual content of theories.

In light of these considerations, let us for the moment set to one side the issues connected with arithmetical interpretations and multiple realizablity and turn our attention to a closer examination of the question, 'How do theories, whose characteristic feature is that their theoretical claims transcend observation, acquire their empirical status?' This brings us to Hilary Putnam's model-theoretic argument, by which I mean his first such argument, the one developed in (Putnam, 1977). I should emphasize that my primary interest is the actual argument, rather than Putnam's uses of it; these are all more various than the application I will isolate. As I understand its significance, the argument shows that the answer to our question given by the doctrine of partial interpretation and its close descendants is incompatible with the thesis that when theories which transcend observation are true, they express salient truths about unobservable entities. The fault with all these views stems from their failure to satisfactorily address the basis for our epistemic access to theoretical domains.

The model-theoretic argument consists of a simple technical argument and an observation. The technical argument establishes that any model of a theory's observational consequences can be extended to a model of the theory's theoretical sentences and correspondence rules, where the domain of this extension is the standard domain of observable and unobservable entities. The argument which establishes this conclusion also supports an observation, namely, that on the partial interpretation reconstruction of theories, as well as Carnap's various refinements of it, the conditions under which a partially interpreted theory can be shown to be *satisfiable* suffice to show that the theory is *true*.

The simple technical argument exploits a construction like Winnie's. The conclusion of the argument follows from a well-known folklore result that assures us that any model of a theory's observational consequences can be extended to a model of the theory.[6] This is true, in particular, when the model of the theory's observational consequences is defined over the standard domain of observable entities. But so far as

---

[6]The result is reported in (van Benthem, 1978) as

Lemma 3.2 For any L_o–structure M, if M is a model of the L_o–consequences of T, then there exists an $L_o \cup L_t$ -structure N such that N is a model of T and the reduction of N to L_o is an elementary extension of M.

Here L_o and L_t, are first-order languages with equality all of whose vocabulary is, respectively, observational and theoretical. ($L_o \cup L_t$ is the language generated by the observational and theoretical vocabulary of L_o and Lt.) The lemma assumes that the vocabulary of the theory consists only of theoretical and observational terms.

the folklore result is concerned, the "theoretical" or "unobservable" part of the extension might well be "abstract"—even number-theoretic. Nevertheless, if we are given such a "partially abstract" model, we can use a construction based on a one-one onto mapping which is the identity on the observable part of the domain of this model and is arbitrary from its theoretical part onto the theoretical domain of non-abstract unobservable entities. To do so we require the additional and contingent assumption that the cardinalities of the theoretical domains of the two models are the same. Relative to this assumption, Putnam's argument proceeds by defining the theoretical properties and relations over the standard domain of observable and unobservable entities as the images of the properties and relations of the abstract model under a mapping that is the identity map on the observable part of the domain of the partially abstract model, and is arbitrary from its theoretical part onto the theoretical part of the domain of non-abstract theoretical entities. But since the defined properties and relations meet all the conditions that the partial interpretation account is capable of imposing on theoretical properties and relations, the argument shows not merely that the theory is satisfiable, but that, on the partial interpretation account, it is *true*.

The arguments of Winnie and Putnam both exploit the same technical idea in their respective definitions of the theoretical relations which interpret the theoretical predicates of a partially interpreted theory. But their arguments support what are conceptually very different difficulties for the view. Putnam's argument is not directed at the existence of *multiple* realizations; nor does it concern the existence of an *arithmetical* model of a partially interpreted theory. Rather, Putnam's argument takes us from a cardinality assumption, and the existence of what might well be an arithmetical model of the kind explored by Winnie, to the conclusion that, on the partial interpretation view, the fact that a theory is *satisfiable* over the intended domain of observable and theoretical entities suffices to show that it is *true*.

The significance of the model-theoretic argument has been the subject of an extensive discussion. But whatever the resolution of the many controversies the argument has generated involving "metaphysical realism", "intended" reference, or the "indeterminacy" of reference, it seems clear that the notion that the truth of what is asserted about unobservable entities might depend only on their *number* runs counter to one of the simplest and least contentious convictions of "realism" and, indeed, of common sense. This is the conviction that if a theory is true, this is because its theoretical claims have captured a salient aspect of the reality they seek to describe, an aspect that goes beyond any mere question of cardinality.

The partial interpretation account of theories claims to reconstruct the empirical status of a theory's theoretical statements using only the theory's logico-mathematical framework and the apparatus of correspondence rules. But when we are restricted to just these resources, the truth of theoretical claims reduces to their satisfiability in any sufficiently large model of the true observation sentences. And this shows that the reconstruction has failed to correctly represent the nature of the epistemological status of a theory's theoretical claims. It has failed because the epistemic basis for such an assertion of satisfiability is entirely different from what is required by an assertion of truth. The idea that the claim that a theory is true should depend only on a cardinality constraint and a logical argument fails to adequately separate the epistemic basis for the truth of the theoretical assertions of an empirical theory from the epistemic basis for the mere satisfiability of the "abstract" assertions of a purely mathematical theory over a given domain.

The conclusion we have just reached should perhaps have been anticipated, given the origin of the partial interpretation view in Hilbert's conception of the foundations of geometry. In his correspondence with Frege, Hilbert defended the idea that satisfiability in a sufficiently large domain is a suitable surrogate for the "truth" of a mathematical theory. But whatever its plausibility for theories of pure mathematics, the methodological demands we impose on the theoretical claims of physics cannot be captured by so weak a requirement, not at least if we wish to preserve the methodological difference between physics and pure mathematics. An advocate of partial interpretation might respond to this objection by recalling that a physical theory will qualify as true not if it is merely satisfiable, but only if it is satisfiable in a model which is an extension of a domain that forms the basis for a model of the true observation statements. By contrast, the domains which bear witness to the "truth" of a mathematical theory need not have any connection with such a model. For an advocate of partial interpretation, the theories of physics are true because they are *empirically adequate* in the sense that they have observational consequences, all of which are true; but a theory of pure mathematics is not necessarily associated with *any* observation language and is not required to be empirically adequate.

However this response misses the point of the model-theoretic argument as we have presented it: provided the domain over which a partially interpreted theory involving unobservable entities is interpreted includes the domain of the model of the true observation sentences, it is a consequence of the partial interpretation view that the method of argument by which we are able to establish the "truth" of a purely mathematical claim over a given domain also suffices to establish the truth of a theoretical claim.

The model-theoretic argument puts us in a position to see why—pace Lewis—the multiple realizablity which afflicts partial interpretation and Ramsey-sentence reconstructions is largely tangential to the question of realism. For suppose we are given a realization of the sort Putnam's argument shows is possible. Given such a model, we have seen how, following Lewis (1970), we might rule out alternative realizations by supplementing the partially interpreted theory with a judicious selection of O-mixed terms and appropriate assumptions involving them. Notice however that this is compatible with the possibility that a theory is true only because it has a realization that models its observational consequences and is the right size. So in light of Putnam's argument, uniqueness of realization is insufficient to ensure the widely held conviction that if our physical theories are true, this is because they succeed in isolating salient truths about the entities with which they deal, independently of whether they are observable or unobservable. In any case, when, long after his 1970 paper on theoretical terms, Lewis came to address Putnam's argument, he did not appeal to O-mixed terms to resolve the problem he took the argument to pose but based his reply on a distinction among possible realizations. The core assumption of Lewis's response is that a theory is *true* only if it is true relative to a realization whose properties and relations are *natural*. Since there is nothing in Putnam's construction of his interpretation of theoretical predicates which requires that they should be natural properties and relations, Lewis argued that the construction fails to show that the theory's theoretical claims are, in the relevant sense, true.[7]

It has been suggested that we should adapt Lewis's reply to the model-theoretic argument and supplement the partial interpretation account by restricting the class of admissible realizations to those that involve natural relations, thereby distinguishing, in the way Lewis proposes, true theories from theories that are merely satisfiable in a domain that extends a model of the true observation statements. Lewis's distinction might be further exploited to characterize true *empirical* theories as those that are not merely satisfiable in some such realization or other, but are true because they are true in a realization whose relations are natural. But we should be cautious about accepting Lewis's suggestion as an adequate response to the model-theoretic argument or as a guide for emending the partial interpretation view.

To begin with, Lewis's reply to Putnam leaves unresolved the problem of how we are able to make significant claims about relations that are *not* "natural". Even if we have no interest in theorizing about such relations, an adequate response to the model-theoretic argument should nevertheless explain how it is *possible* to do so without the assertion of the truth of such a theory collapsing into an assertion of its satisfiability over a domain—even a domain that extends the model of the observational consequences of the "theory" of such a natural relation. Indeed, as Fraser MacBride has remarked, on the assumption that we achieve knowledge of natural relations only with the progress of science—and perhaps only after many distractions involving non-natural relations—anyone following Lewis's suggestion *must* have an interest in how we manage to make significant, but as it happens, misguided claims about nonnatural relations.

But secondly, and more importantly, addressing Putnam's argument by appealing to Lewis's proposal obscures the difficulty the model-theoretic argument raises for partial interpretation and Ramsey-sentence reconstructions. The problem with these approaches is not their failure to designate certain properties and relations as natural, but the fact that they are too weak to explain the difference between the epistemological status of theoretical and purely mathematical claims. But then the partial interpretation framework for addressing how theories are warranted must *also* fail to capture the

---

[7]Lewis's response is developed in (Lewis, 1983) and (Lewis, 1984). See also (Merrill, 1980), which Lewis cites as having influenced his reply. (Psillos, 1999, pp. 67–68) has argued that Lewis's requirement of uniqueness is an anticipation of his later appeal to natural relations in his reply to Putnam. But it is clear from our earlier discussion that for Lewis all the heavy lifting to establish unique realizablity is done by assumptions which are independent of the requirement that the properties and relations which interpret the O- and T-vocabularies are natural, or that they "carve nature at its joints."

methodology by which claims about unobservables are established, and this shows that a different approach to these two issues is required. An emendation of the view based on Lewis's reply to Putnam only succeeds in recording the fact that we *do* distinguish mathematical claims from the theoretical claims of physics, but it has nothing to contribute to our understanding of the methodology by which we make this distinction. Nor does it contribute to our understanding of how we successfully gain epistemic access to theoretical domains in order to warrant our claims about them.

6 Conclusion

There are two questions that an adequate account of theoretical knowledge must address: (i) 'What is the methodology by which we gain epistemic access to theoretical domains?' and (ii) 'How are the theoretical claims of physics about such domains distinguished from the claims of pure mathematics?' In so far as partial interpretation and Ramsey-sentence reconstructions are unable to adequately address the second of these two questions, it is evident that they have failed to marshal the resources necessary for addressing the first. The burden of the larger study from which the present discussion is drawn is to explain a key physical methodology that makes theoretical domains epistemically accessible. This methodology clarifies how the properties and relations—more generally, the parameters—which enter into theoretical claims are empirically well-founded by robust theory-mediated measurements. Theoretical domains are epistemically accessible because the parameters which qualify their constituents are empirically well-founded. This account of epistemic accessibility is capable of supporting existential hypotheses involving entities which transcend observation without, however, necessarily also supporting—or even requiring—the truth of the theories which employ such hypotheses. And it is also capable of explaining the basis on which we distinguish physical claims about theoretical domains from the claims of pure mathematics. The role of robust theory-mediated measurements has been largely missed by the models of empirical adequacy and predictive success that have dominated partial interpretation and other traditions in the philosophy of science; and it has also been missed by some of logical empiricism's severest critics.

Acknowledgments. Support from the Social Sciences and Humanities Research Council of Canada is gratefully acknowledged. I wish to thank Thomas Uebel for presenting this paper to the CLMPS on my behalf.

Bibliography

Carnap, R. (1956). The methodological character of theoretical concepts. In H. Feigl, & M. Scriven (Eds.), *The foundations of science and the concepts of psychology and psychoanalysis, Minnesota Studies in the Philosophy of Science Vol 1.* (pp. 38–76). Minneapolis: University of Minnesota Press.

Carnap, R. (1961). On the use of Hilbert's ε-operator in scientific theories. In Y. Bar-Hillel (Ed.), *Essays in the foundations of mathematics dedicated to A. A. Fraenkel* (pp. 156–164). Jerusalem: The Magnus Press of Hebrew University.

Carnap, R. (1963). Replies and systematic expositions. In P. A. Schilpp (Ed.), *The philosophy of Rudolf Carnap* (pp. 859–1013). La Salle IL: Open Court.

Demopoulos, W. (2007). Carnap on the rational reconstruction of scientific theories. In M. Friedman, & R. Creath (Eds.), *The Cambridge companion to Carnap* (pp. 248–272). Cambridge: Cambridge University Press.

Demopoulos, W. (2013). *Logicism and its philosophical legacy*. Cambridge: Cambridge University Press.

Gupta, A. (2009). Definition. In E. N. Zalta (Ed.), *Stanford Encyclopedia of Philosophy*. (Spring 2009 Edition). url: plato.stanford.edu.

Gupta, A. (2013). The relationship of experience to thought. *The Monist, 96*, 252–294.

Hempel, C. G. (1965). The theoretician's dilemma: A study in the logic of theory construction. In *Aspects of scientific explanation* (pp. 173–226). New York: The Free Press.

Lewis, D. (1970). How to define theoretical terms. *Journal of Philosophy, 67*, 427–445.

Lewis, D. (1983). New work for a theory of universals. *Australasian Journal of Philosophy, 61*, 343–377.

Lewis, D. (1984). Putnam's paradox. *Australasian Journal of Philosophy, 62*, 221–236.

Merrill, G. (1980). The model-theoretic argument against realism. *Philosophy of Science, 47*, 69–81.

Psillos, S. (1999). *Scientific realism: How science tracks truth*. London: Routledge.

Putnam, H. (1977). Realism and reason. *Proceedings and addresses of the American Philosophical Association, 50*, 483–498.

Putnam, H. (2012). A theorem of Craig's about Ramsey sentences. In M. DeCaro, & D. Macarthur (Eds.), *Philosophy in an age of science* (pp. 277–279). Cambridge MA: Harvard University Press.

Ramsey, F. P. (1929). Theories. In R. Braithwaite (Ed.), *The foundations of mathematics and other logical essays by Frank Plumpton Ramsey* (pp. 212–236). (Published 1931). Paterson NJ: Littlefield and Adams.

van Benthem, J. (1978). Ramsey eliminability. *Studia Logica, 37*, 321–336.

Winnie, J. (1967). The implicit definition of theoretical terms. *British Journal for the Philosophy of Science, 18*, 223–229.

Winnie, J. (1970). Theoretical analyticity. In R. Cohen, & M. Wartofsky (Eds.), *Boston Studies in the Philosophy of Science, Vol. VIII* (pp. 289–305). Dordrecht and Boston: Reidel.

Author biography. William Demopoulos is professor emeritus from The University of Western Ontario, where he taught from 1972 to 2012. He has held numerous visiting positions, among them visiting professorships at Harvard University and the University of California, Irvine. He was recently the recipient of a Killam Fellowship. Background to his contribution to these *Proceedings* can be found in his recently published collection of essays, *Logicism and its Philosophical Legacy*, Cambridge University Press. He is currently at work on a monograph entitled *On Theories*.

11 How values can influence science without threatening its integrity

SVEN OVE HANSSON *

Abstract. Science provides us with a common set of highly plausible factual beliefs, the scientific corpus. It consists of those statements in the domain of science that we currently see no reason to doubt. In most cases our practical purposes are well served by basing our decisions on the information that is available in the corpus. But the fit is not perfect. Two major types of conflict can arise between the criteria of corpus entry and those of practical decision-making. The first type of conflict arises when we have reasons to act as if some statement is true, even though the evidence is not strong enough for accepting it as scientifically valid. In such cases we do not typically adjust the requirements for scientific acceptance downwards, but instead distinguish between different criteria for scientific and practical purposes. The second, much less discussed, type of conflict arises when a higher level of evidence is required for acting as if something is true than for accepting it as scientifically valid. In such cases we typically adjust the requirements for scientific acceptance upwards so that they coincide with what is called for in the practical decision. Such adjustments are seldom explicitly discussed but they are common for instance in medical science. With this implicit, but rather sophisticated, two-branched strategy we can apply the appropriate criteria of evidence in practical decision-making while at the same time upholding the integrity of science and keeping down the conflicts between scientific and practical criteria to a minimum.

Keywords: epistemic values, practical decision, scientific corpus, value-neutrality.

---

*Department of Philosophy and History, Royal Institute of Technology (KTH), Brinellvägen 32, 100 44 Stockholm, Sweden

1 Introduction

In this contribution I will propose a new take on an old debate, namely whether practical (non-epistemic) values should be allowed to influence science.[1] There are of course many ways in which values can have an influence in science. For instance, they can have impact on the choice of issues to investigate and on what methods of inquiry are considered to be (ethically) acceptable. Following a common practice in this debate I will only consider the influence that values can have on what statements we assent to, given the evidence at hand. I will call this influence on *scientific assent*.

According to what is often called the "traditional view", judgments of scientific assent should be uninfluenced by our values and wishes, especially by ethical and political values. (Merton, [1942] 1973) In a famous paper, Richard Rudner (1953) claimed that in spite of its intuitive appeal, this viewpoint is untenable in scientific practice. His main argument was that a decision whether or not to accept a scientific hypothesis must take into account not only the available empirical evidence but also the seriousness of the two possible types of mistakes: accepting an incorrect hypothesis and rejecting a correct one.

> Thus, to take a crude but easily manag[e]able example, if the hypothesis under consideration were to the effect that a toxic ingredient of a drug was not present in lethal quantity, we would require a relatively high degree of confirmation or confidence before accepting the hypothesis – for the consequences of making a mistake here are exceedingly grave by our moral standards. On the other hand, if say, our hypothesis stated that, on the basis of a sample, a certain lot of machine stamped belt buckles was not defective, the degree of confidence we should require would be relatively not so high. *How sure we need to be before we accept a hypothesis will depend on how serious a mistake would be.* (Rudner, 1953, p. 2)

In a reply to Rudner, Isaac Levi (1960) conceded that scientists have to assign "minimum probabilities for accepting or rejecting hypotheses" but argued that these numbers can be part of the scientific standards of inference that scientists are committed to. He adjusted the value-neutrality thesis accordingly, saying that "the value-neutrality thesis does not maintain that the scientist *qua* scientist makes no value judgments but that given his commitment to the canons of inference he need make no further value judgments in order to decide which hypotheses to accept and which to reject." (Levi, 1960, p. 356) In a similar vein, Carl Hempel (1960) proposed that scientists should

---

[1] The word "science" will be used in an extended sense that includes the humanities (like the German "Wissenschaft"). For a clarification why this is a better demarcation than the conventional English one, see Hansson (2013). Science is formed by a community of knowledge disciplines that share the same overarching goal of providing us with a joint repository of reliable factual knowledge. They are also characterized by mutual respect for each other's results and methods.

be influenced by "the value or disvalue which the different outcomes have from the point of view of pure scientific research rather than the practical advantages or disadvantages that might result from the application of an accepted hypothesis". He called these "purely scientific, or epistemic, utilities" (Hempel, 1960, p. 465). Later he called them "epistemic values" and maintained that epistemic values should reflect the usefulness of truth, simplicity, explanatory power and other desiderata of scientific theories (Hempel, 1981, p. 398). This is the modern version of the value-neutral ideal. It has been defended at least in part for instance by Hugh Lacey (1999) and Gerald Doppelt (2007).

There is also a parallel debate in modern epistemology. It is couched in terms of individual rather than collective knowledge claims, and refers to knowledge in general rather than scientific knowledge. Discussions concern whether a person's values etc. can have a legitimate influence on her beliefs (Vahid, 2014) and whether the fact that someone knows something is sufficient reason for her to rely on it in practical reasoning. (Brown, 2008; Buckwalter, 2010; Fantl & McGrath, 2007; Hawthorne, 2004; Stanley, 2005)

In Sections 2-3 some preconditions will be presented that are essential for the argument. Section 4 investigates cases in which practical decisions have to be based on criteria of evidence that differ from those used for internal scientific purposes. In Section 5 this investigation is extended to cases in which different internal scientific purposes can give rise to divergent criteria for scientific assent. In Section 6 and the concluding Section 7 it will be argued that the pattern uncovered in the previous sections represents an implicit yet highly sophisticated practice that allows the criteria of scientific assent to be influenced by practical, non-scientific, considerations only in the cases when this can be done without damage to the integrity of science.

2 Separating facts and values

It is a basic fact about human reasoning that we separate, or make a distinction between, facts and values. As individuals we distinguish our factual beliefs from our other attitudes and reactions to what is happening around us. In some social contexts, including science, this separation is further refined. Such processes seem to be present in all types of human societies, including illiterate ones. For instance, !Kung hunters discussing animal behaviour take great care to distinguish facts (for instance tracks that they have seen) from theories and attitudes. (Blurton-Jones & Konner, 1976)

We do not know much about how members of other species "think". In particular, we do not know whether or not they have mental representations of factual circumstances that are separate from the mental representations of reactions or attitudes to these circumstances. For instance, when a python is in the process of digesting a large prey, it can let a rat pass within strike range without trying to snatch it. Does this inactivity depend on some factual belief ("this is not a tasty prey"), a judgment of value ("it is not good for me to take this tasty prey"), or some mode of thinking that is devoid of this distinction? The latter option is far from implausible. But why then do we humans make the fact/value distinction?

I have found no other credible answer to that question than that maintaining such a separate sphere of facts makes us more successful in our practical dealings with the world we are living in and with each other. Separating out the facts (or what we believe to be the facts) and reasoning about them in relative isolation from other components of our minds seems to have survival value for an organism like us, given the level and character of our cognitive abilities. In other words, this is a cognitive pattern that provides us with evolutionary advantages, and that is why we have developed it.[2]

The separation of facts from values operates in our individual thought processes and we can assume that it contributes to our cognitive abilities as individuals. However, its importance appears to be even greater when we communicate with each other. Organized collective decision-making for instance in a court of law, a board of directors or a parliament is based on ways of discussing that depend on this separation.

It is against this background that we can understand the practical successfulness of science. Beings like us, who isolate the factual components of our thoughts, can communicate, deliberate, and co-ordinate with others much more efficiently if we have a large common ground in the form of a joint set of factual beliefs. This is what science does for us; it provides a common repository of reliable factual beliefs that we can all share.

3 A scientific corpus

The statements that are accepted by the community of scientists form a continuously changing corpus of science, or repository of scientific knowledge.[3] Some authors, notably Richard Jeffrey, have opposed the idea of a scientific corpus. According to Jeffrey, scientists should not accept or reject hypotheses. Instead, "the activity proper to the scientist is the assignment of probabilities (with respect to currently available evidence) to the hypotheses which, on the usual view, he simply accepts or rejects."

---

[2]This conforms with David Papineau's (2003, p. 40) proposal that theoretical rationality "piggy-backs on the evolution of cognitive abilities for 'understanding of mind' and for means-ends thinking".

[3]The use of the word "knowledge" in this context is not quite accurate since we normally assume that knowledge implies truth. The corpus consists of "justified provisional knowledge claims", but no sufficiently short term with that meaning is available. – In public discussions the scientific consensus is often hidden from view. False scientific controversies have repeatedly been created by enemies of the scientific consensus, e.g. creationists, tobacco propagandists, and deniers of climate science. (Oreskes & Conway, 2010) This gives rise to what Wendy Wagner (1995) called a "science charade" in which divergences in interests and policy viewpoints are camouflaged as differences in scientific opinion.

(Jeffrey, 1956, p. 257) To illustrate his standpoint, Jeffrey pointed out that a physician who decides whether or not to inoculate a child with a polio vaccine may very well come to another conclusion than a veterinarian who decides whether or not to use the same vaccine on a monkey. To accept or reject an hypothesis about this vaccine once and for all is "to introduce an unnecessary conflict between the interests of the physician and the veterinarian." (Jeffrey, 1956, p. 245) It is better that the scientist "contents himself with providing them both with a single probability for the hypothesis (whereupon each makes his own decision based on the utilities peculiar to his problem)."

As Jeffrey himself noted, this is a far cry from how science is actually conducted. What is worse, it does not even seem to correspond to a feasible way to conduct science. If we followed this recipe, science would be a vast, growing web of intricately interconnected hypotheses with probabilities greater than 0 but smaller than 1. This would make scientific reasoning and argumentation unmanageably complex. For instance, as was pointed out by McLaughlin (1970, p. 121), no reasonably simple account of measurement procedures would be available. Human cognitive powers are not sufficient to handle such a mass of uncertainty. Therefore, we have to take sufficiently plausible knowledge claims provisionally for certain.

We will have use for three important observations about the scientific corpus. First, *membership in the corpus is always provisional*. When we accept or reject empirical statements, we do not do so once and for all, come whatever may. We do not classify them as incorrigible, only as provisionally not to be doubted. This means that although we do not actively doubt them, we reserve the right to do so at a later point in time, should we learn something new that gives us reason for a reappraisal. The provisional nature of the corpus is one of the major reasons why science is more successful than its competitors. For instance, homeopathy is still based on principles from the late 18th century that were thoroughly refuted by discoveries in chemistry in the early 19th century. In contrast, scientific pharmacology has repeatedly been improved and revised throughout this period, largely in response to a long series of discoveries in chemistry.

Due to its provisional nature, the scientific corpus cannot easily be modelled with the tools of standard probability theory. In that theory the only way to express that a sentence is no longer doubted is to assign to it the probability 1. But once a statement has unit probability, it is immune to change and can never be put to doubt again. In this respect, belief revision models are more versatile since they allow us to retract full beliefs. (Hansson, 2010)

Secondly, *the scientific corpus is a general-purpose repository of knowledge*, not limited to any particular usage or application. This means that it is used both for all scientific purposes and for all practical purposes for which it is relevant. For example, we assent to the same statements about the chemistry of DNA irrespectively of whether we are studying human evolution, investigating ongoing speciation in the apple maggot fly (Xie, X. et al., 2007), performing laboratory diagnosis of hereditary diseases, or working with forensic DNA profiling. The general-purpose nature of the corpus is of course necessary due to our cognitive limitations. There is no chance that we would be able to deal efficiently with a multitude of corpora, one for each knowledge area.

Thirdly, *the scientific corpus has high entry requirements.* In science, the burden of proof falls to those proposing a new theory or claiming the existence of a previously unknown phenomenon. The high entry requirements are justified by the use of the corpus in further scientific investigations. Scientific progress can be seriously halted by the acceptance of false information, and therefore we should only accept new claims if we have strong reasons to believe them to be true. The high entry requirements are matched by fairly low exit requirements. If good reasons come up to doubt some element of the corpus, it should no longer be taken for granted but instead be subject to investigations aimed at clarifying its status.

The question that we started with can now be reformulated in a much more precise way: *Can the entry criteria for the scientific corpus be legitimately influenced by practical considerations?*

4 Conflicts between scientific and practical criteria

I hope to have shown that given our cognitive constitution, the use of a scientific corpus makes it possible to manage our knowledge in a much more efficient way than what we could otherwise have done. However, this does not make the corpus perfectly fitted for all practical decisions that we have to make. There are occasions when we want to take action based on a rather weak indication that something is the case. For instance, suppose that bacteriologists tell us that our drinking water contains bacteria that may be of either of two types. One of these types is quite toxic, whereas the other is innocuous. Presumably, most of us would want the authorities to respond to this information in the same way as they would if it were known with certainty that the bacteria are toxic – namely provide everyone concerned with safe water and strongly recommend them not to drink the tap water until definite information is available. However, in this situation a statement to the effect that these bacteria are harmful would not satisfy the requirements for corpus entry as delineated above. We therefore have a conflict between scientific and practical criteria for the evaluation of that statement.

The following examples can be used to investigate conflicts between scientific and practical criteria of acceptance.

> *The baby food additive*
> New scientific evidence indicates that a common preservative agent in baby food may have a small negative effect on the child's brain development. According to the best available scientific expertise, the question is far from settled but the evidence weighs somewhat in the direction of there being such an effect. A committee of respected scientists have unanimously stated that although the evidence is not conclusive it is more

probable that the effect exists than that it does not. The food safety agency has received a petition whose signatories request the immediate prohibition of the substance.

Ski lift inspections
A new variant of stainless steel has been shown in laboratory tests to be amazingly resistant to the type of wear that is the main reason why the ropes of aerial lifts such as ski lifts are subject to frequent safety inspections. Articles about the new alloy have been published in the best journals, and its high wear resistance is considered by the best experts to be scientifically established. The first ski lift that uses wires of the new material is going to be built. Its owner asked the safety inspectorate for permission to reduce the frequency of rope inspections from once a month to once a year. His argument was that it has been scientifically established that the wire will be less abraded in a year than a traditional wire in a month. But the safety inspectorate did not grant this request.

In the baby food case, we have two crucial questions. First, should the food safety agency act as if it is known that the additive is harmful, i.e. prohibit or otherwise restrict its use? And secondly, should the scientific community make an exception from the standard requirements of corpus entry, and consider the toxicity of this additive to be scientifically established?

In my view, there are convincing reasons to answer the first question in the affirmative.[4] However, there are equally convincing reasons to answer the second question negatively. Taking it for settled that the substance is harmful can have various kinds of negative consequences. It would presumably lead us to refrain from further investigations aimed at finding out whether the substance really is toxic. This could potentially prevent us from discovering that the substance is harmless (and perhaps useful as a substitute for some other, harmful preservative in a product where a preservative is needed). It could also lead us astray in attempts to understand the relationships between neurotoxicity and chemical structure (thereby preventing us from making correct predictions about other substances). As these examples show, lowering the requirements for corpus entry could have negative consequences, not only for purely scientific investigations, but also for future practical decisions. To avoid this we should retain the high evidential requirements for scientific assent in spite of adopting lower requirements for the practical decision.

As far as I can see, this is also how such a case would normally be treated in practice. It is almost unthinkable for a scientist or a government official to say: "We have conclusive scientific evidence that the substance has these toxic effects, but we admit that it is almost equally likely that it does not." Instead it would be recognized

---

[4]Obviously, for the argument to go through it is only required that there are cases exhibiting this pattern, not that this particular example is such a case.

that even if the evidence for neurotoxicity is not strong enough to be scientifically conclusive it may be strong enough for a decision-maker to act *as if* the substance is neurotoxic.

In the ski lift example we have, correspondingly, the following two questions to answer: First, should the safety inspectorate reduce the frequency of safety inspections, or should a possible such decision be postponed until further evidence has been collected? And secondly, if the latter option is chosen, should the scientific community follow the authority in raising the standard of evidence and thus make the wear-resistance of the new material an open issue also scientifically, or should they maintain their previous assessment?

In answer to the first question I propose that the safety agency would be justified in requiring additional evidence although the standard scientific criteria of corpus entry have been met.[5] My answer to the second question is that it would be justified to raise the demands for corpus entry so that they coincide with those applied in the safety-critical practical decision. A distinction between two levels would be confusing in a case like this, since the lower of these levels would be the one usually indicating that there are no reasonable doubts in the matter. And importantly, raising the level of evidence required for corpus entry would not have the severe negative consequences that lowering the level could have. Therefore, it seems sensible to adjust the requirements for corpus entry upwards in this case.

What would happen in practice in a case like this? At any rate we should not expect safety inspectors to say: "There is conclusive scientific evidence that the wire has these properties, but we do not believe it." They would simply say that the evidence is not sufficient. This exemplifies a more general pattern: A claim that something is scientifically established but still has to be doubted has a distinctly paradoxical ring. This is of course because that which is scientifically established (included in the scientific corpus) is assumed not to be subject (at present) to reasonable doubt. Therefore we tend to raise the demands for corpus entry rather than introduce separate, higher demands of evidence for practical purposes.

The picture emerging from these two examples is a two-branched strategy for the influence of practical considerations on the requirements of corpus entry: When a practical decision demands a higher level of evidence than what is required for purely scientific purposes, then we tend to raise the demands for corpus entry accordingly. However, when a practical decision demands a lower level of evidence than what science *per se* would require, then we do not lower the demands for corpus entry. Instead, we maintain a distinction between two levels, one for the practical decision and one for corpus entry. If there is more than one practical decision to be made, then they may require different levels of evidence.

---

[5] Again, the argument only requires that there are cases exhibiting this pattern, not that this particular example does so.

In my view, this two-branched strategy provides a much better account of the influence of practical values on scientific assent than approaches that do not distinguish between upwards and downwards adjustments of the requirements of corpus entry. The reader may well take this to be the main message of this contribution. However, to complete the picture some further complications need to be taken into account, mainly concerning the "purely" scientific requirements for corpus entry.

5 Conflicts between different scientific criteria

As noted above, the "purely scientific" requirements of corpus entry are high, primarily because much damage can be done to our scientific endeavours by the inclusion of false statements into the corpus. However, the extent of that damage to science is not the same for all statements that we may consider for inclusion in the corpus. It is useful to distinguish between two types of effects on science that may follow from the mistaken inclusion of a statement into the corpus.[6] First there are *direct epistemic effects* on scientific deliberations and other decisions on scientific assent. For instance, if we accept an incorrect statement about the strength of the jaw muscles of prehistoric predators, then this may lead us to draw incorrect conclusions about their diet and their role in the ecological system. Misdating an ancient text can induce us to draw erroneous conclusions about language development.

Secondly there are *indirect epistemic effects* on our scientific deliberations, by which is meant effects that are mediated by impacts on scientific investigations.[7] For example, if we conclude incorrectly that a certain island has always been uninhabited, then we will not perform archaeological investigations there. If we misidentify the conditions under which a phenomenon in particle physics will appear, then that may lead to a massive waste of money on experiments that provide no new information.

Let us now consider some examples of internally scientific considerations that at least seemingly call for adjustments in the criteria of corpus entry.

> *A new type of glass*
> A new type of glass makes it possible to produce thinner mirrors with the same quality and the same strength as traditional ones. Results showing that the new glass has these properties have been presented at conferences in materials science, and experts in glass science take this information to

---

[6]The distinction is related to that between theoretical and epistemic rationality. (Papineau, 2003; Kelly, 2003) Epistemic rationality is the rationality we exert when believing or disbelieving various statements, given the evidence that we have. Theoretical rationality is a wider concept that also includes other parts of our pursuance of cognitive goals, such as our decisions to perform actions that will provide us with information.

[7]The distinction between direct and indirect effects is not knife-sharp; indeed it cannot be more precise than the distinction between scientific deliberation and scientific investigation.

be scientifically conclusive. A proposal has been made to use the new type of glass in a new space telescope predicted to cost several billion dollars. Due to the high economic stakes, the scientists planning its construction decide not to rely on the available scientific evidence that the glass has the desired properties, and therefore order new investigations.

This is a clear case of an indirect epistemic effect. For reasons paralleling those pertaining to the ski lift in the previous section, we can assume that the decision to reconsider the issue will not be expressed in terms such as "There is conclusive scientific evidence that the glass has these properties, but we are still not sure that it has". Instead, we can expect the decision to be justified by assertions that there are still reasonable scientific doubts in the issue.

Neutrinos travelling faster than light
In 2011 measurements were reported according to which neutrinos travelling from Cern to Gran Sasso in Italy had reached speeds higher than that of light in a vacuum. This result was met with much scepticism since if correct, it would overturn special relativity that is a cornerstone of physics and supported by an immense amount of empirical evidence. Much higher demands of evidence were applied to the validation of this experiment than what would have been applied to an experiment confirming that neutrinos do not exceed the speed of light. Arguably, the demands were indeed higher than those that would have been placed on almost any other physics experiment. (Overbye, 2011)[8]

Just as in the previous case, the scientific demands of evidence (i.e. the requirements for corpus entry) were raised. (It would have been almost unthinkable for physicists to say: "There is sufficient scientific evidence that these neutrinos travelled faster than light but we still doubt it." Legitimate doubt excludes scientific assent.) However, contrary to the previous case, this one represents a direct epistemic effect. The high demands on evidence in this case can be explained with reference to the common dictum that extraordinary claims require extraordinary evidence. (The common attribution of this phrase to David Hume is unsubstantiated, but what it says is compatible with his standpoints.)

A substance with uncertain effects
At a recent conference, a group of biochemists announced preliminary results indicating that a newly synthesized substance binds strongly to an enzyme, thereby deactivating it. These results were criticized by other researchers, and they are not considered to be sufficiently validated. Another research group is much in need of a substance that deactivates the enzyme in question. They decide to perform experiments with the new

---

[8] The result was soon shown to be an anomaly depending on deficiencies in the equipment. (Reich, 2012; Patrizii, L. et al., 2012)

substance in spite of the remaining uncertainty whether it actually deactivates the enzyme.

This is another example of an indirect epistemic effect. It is quite similar in structure to the example of the baby food additive in the previous section. The second group decides to act as if the substance in question is an inhibitor although there is insufficient scientific evidence that this is so. The value of performing exciting new experiments plays the same role here as the value of not exposing babies to a poison in the baby food case. The comparatively small disvalue of performing a few unsuccessful experiments plays the same role as the comparatively small disvalue of prohibiting an innocuous preservative in baby food. More importantly, in both cases the standard requirement of corpus entry is left unchanged. Instead of lowering it, a distinction is set up between the requirements for scientific assent and those for a non-epistemic decision (using the potential inhibitor respectively prohibiting the preservative).

These three examples can be classified as follows:

> indirect effect, incentive to raise requirements (*A new type of glass*)
> direct effect, incentive to raise requirements (*Neutrinos travelling faster than light*)
> indirect effect, incentive to lower requirements (*An enzyme with uncertain effects*)

The three examples exhibit the same pattern that we saw in Section 4 for the influence of practical considerations: When there is a justification for raising the "standard" requirements for corpus entry, then this will be done. However, when there is a justification to lower them, then this is not done, but instead a separate lower level of requirement is established for the purpose at hand.

What about the fourth combination, "direct effect, incentive to lower requirements"? The standard requirements for corpus entry represent a level of evidence that we do not, for epistemic reasons, wish ever to go below. It is difficult to see how we could have epistemic reasons to lower that limit. Therefore, this fourth form of conflict between scientific criteria does not seem to exist.

6 A common but mostly implicit practice

In the previous two sections we have seen that the level of evidence required for scientific assent can differ significantly between statements that are candidates for such assent. This can also be exemplified with the levels of evidence required for claims on the effects of pharmaceutical drugs. Relatively little evidence is required before it is considered to be known that a drug has a serious side effect. Usually, a single, well-conducted study on patients taking the drug is sufficient. Much more evidence is required for a statement that a drug has no serious side effects.[9] For that purpose, extensive evidence from each of the phases of drug testing must be available, and this usually includes several independent, large clinical trials. Only if such evidence is available will the medical community consider the claim to be beyond reasonable doubt for both scientific and clinical purposes.

Notably, no distinction is made here between the scientific and the clinical criteria. This is a general pattern in medical science: The high demands of evidence that are justified in clinical medicine are assimilated into medical science, leading to correspondingly high demands for corpus entry in the medical part of the corpus of science. This is the reason why we never hear statements such as "From a scientific point of view we are confident that this diagnostic test gives the right answer, but we are still not sure that it does" or "We are not sure that this drug has the desired effect, but the scientific evidence shows conclusively that it does".

Decision-making in clinical medicine is imbued with values: values concerning the desirable state of the human body, the seriousness of various conditions, what risks it is justified to take in order to achieve various medical objectives, etc. In consequence, the criteria of scientific assent in medical science are also highly value-laden. However, this has usually not been seen as a problem since the values in question are with few exceptions uncontroversial. Most of the recent discussion on the impact of values in science has focused on controversial values.[10] Arguably, the largest part of that impact derives from uncontroversial values such as the value of health in medicine, functionality and safety in technology, efficient learning in pedagogical research, crime reduction in criminology, etc. We have a tendency to take these values for granted, and sometimes we may not even think of them as values. If uncontroversial values are taken into account, then the impact of practical values in science will be seen to be much larger than what has usually been assumed. (Hansson, 2009)

7 Conclusion

We have investigated the impact of values on the criteria for scientific assent or, in other words, the evidential requirements for a statement to be included in the scientific corpus. The picture that has emerged is somewhat more complex than most previous proposals, but I hope to have shown that it is also more accurate. It can be summarized as follows:

1. There is a bottom line for the evidential requirements for corpus entry, a bottom line that we never go below. Its level is fairly high, and it places the burden

---

[9] Strictly speaking, such a statement is inaccessible to empirical investigation. For purely statistical reasons, the best that we can obtain is evidence strongly indicating that the frequency of serious side effect is below some low but non-zero number. (Hansson, 1995, 1997)

[10] Note however, that the early discussion referred to uncontroversial values (Jeffrey, 1956; Rudner, 1953).

of evidence on those proposing a new theory or claiming the existence of a previously unknown phenomenon. The primary reason why the bottom line is high is that the introduction of false statements into the corpus can severely debilitate future research. If there are no other considerations at play then the actual evidential requirements will coincide with the bottom line.

2. We often have reasons to choose an even higher level than the bottom line. Such reasons may be either practical or theoretical. Including a false statement in the corpus may have negative consequences on our practical doings or it may be particularly damaging for our continued pursuit of science (over and above what the bottom line is intended to protect us against). In both cases, we solve this by raising the requirements for corpus entry to the higher level that is required.

3. There can also be both practical and theoretical reasons to apply, for some particular purpose, evidential requirements that are lower than the bottom line. In such cases, we introduce a separate standard for the purpose in question, and distinguish it from the requirements for corpus entry. We do not move the requirements for corpus entry below the bottom line.

I propose that this is an idealized but on the whole largely correct picture of how values in practice influence scientific assent. The epistemic strategy that it represents has the advantage of maintaining the *integrity* of science as a source of reliable knowledge. This is because we never go below a bottom line constructed to keep the influx of false statements into the corpus at a manageable, very low level. It also has the advantage of preserving the *unity* of science, since we have only one corpus, and all deviations from its requirements are dedicated to the solution of some particular decision problem.

The most important feature that distinguishes this model from previous ones is the distinction that it makes between movements upwards and downwards from the basic criteria of scientific assent. Another important feature is its uniform treatment of scientific and practical reasons for deviations from the bottom line requirements on corpus entry.[11]

---

[11] Heather Douglas (2009) also gave up the usual assumption that epistemic and non-epistemic values should be treated differently. Her criterion for acceptable influence of values on scientific assent is that values should only have an indirect role, and not directly influence the acceptance or rejection of hypotheses. The present approach allows values to have both direct and indirect influence, but in both cases they are only allowed to raise, not lower, the evidential requirements for scientific assent.

Bibliography

Blurton-Jones, N., & Konner, M. (1976). !Kung knowledge of animal behaviour. In R. Lee, & I. DeVore (Eds.), *Kalahari hunter-gatherers* (pp. 326–348). Cambridge: Harvard University Press.

Brown, J. (2008). Subject-sensitive invariantism and the knowledge norm for practical reasoning. *Nous, 42*, 167–189.

Buckwalter, W. (2010). Knowledge isn't closed on saturday: A study in ordinary language review of philosophy and psychology. *Review of Philosophy and Psychology, 1*, 395–406.

Doppelt, G. (2007). The value ladenness of scientific knowledge. In H. Kincaid, J. Dupré, & A. Wylie (Eds.), *Value-free science: Ideals and illusions?* (pp. 188–217). New York: Oxford University Press.

Douglas, H. E. (2009). *Science, policy and the value-free ideal*. Pittsburgh: University of Pittsburgh Press.

Fantl, J., & McGrath, M. (2007). On pragmatic encroachment in epistemology. *Philosophy and Phenomenological Research, 75*, 558–589.

Hansson, S. O. (1995). The detection level. *Regulatory Toxicology and Pharmacology, 22*, 103–109.

Hansson, S. O. (1997). Can we reverse the burden of proof? *Toxicology Letters, 90*, 223–228.

Hansson, S. O. (2009). An agenda for the ethics of risk. In L. Asveld, & S. Roeser (Eds.), *The ethics of technological risk* (pp. 11–23). London: Earthscan.

Hansson, S. O. (2010). Changing the scientific corpus. In E. J. Olsson, & S. Enqvist (Eds.), *Belief revision meets philosophy of science* (pp. 43–58). Dordrecht: Springer.

Hansson, S. O. (2013). Defining pseudoscience – and science. In M. Pigliucci, & M. Boudry (Eds.), *The philosophy of pseudoscience* (pp. 61–77). Chicago: Chicago University Press.

Hawthorne, J. (2004). *Knowledge and lotteries*. Oxford: Clarendon Press.

Hempel, C. G. (1960). Inductive inconsistencies. *Synthese, 12*, 439–469.

Hempel, C. G. (1981). Turns in the evolution of the problem of induction. *Synthese, 46*, 389–404.

Jeffrey, R. C. (1956). Valuation and acceptance of scientific hypotheses. *Philosophy of Science, 23*, 237–246.

Kelly, T. (2003). Epistemic rationality as instrumental rationality: A critique. *Philosophy and Phenomenological Research, 66*, 612–640.

Lacey, H. (1999). *Is science value free?: Values and scientific understanding*. London: Routledge.

Levi, I. (1960). Must the scientist make value judgments? *Journal of Philosophy, 57*, 345–357.

McLaughlin, A. (1970). Science, reason and value. *Theory and Decision, 1*, 121–137.

Merton, R. K. ([1942] 1973). Science and technology in a democratic order. *Journal of Legal and Political Sociology, 1*, 115–126. Reprinted as: The normative structure of science. In Merton, R. K. (1973) *The Sociology of Science. Theoretical and Empirical Investigations* (pp. 267-278). Chicago: University of Chicago Press.

Oreskes, N., & Conway, E. M. (2010). *Merchants of doubt: how a handful of scientists obscured the truth on issues from tobacco smoke to global warming*. New York: Bloomsbury Press.

Overbye, D. (2011). Scientists report second sighting of faster-than-light neutrinos. *New York Times, November 18*.

Papineau, D. (2003). *The roots of reason. Philosophical essays on rationality, evolution and probability*. Oxford: Clarendon Press.

Patrizii, L. et al. (2012). Measurement of the neutrino velocity with the OPERA detector in the CNGS beam. *Journal of High Energy Physics*, *10*(093).

Reich, E. S. (2012). Flaws found in faster-than-light neutrino measurement. Two possible sources of error uncovered. *Nature News*. February 22.

Rudner, R. (1953). The scientist qua scientist makes value judgments. *Philosophy of Science*, *20*, 1–6.

Stanley, J. (2005). *Knowledge and practical interest*. Oxford: Clarendon Press.

Vahid, H. (2014). Varieties of pragmatic encroachment. *Acta Analytica*, *29*, 25–41.

Wagner, W. E. (1995). The science charade in toxic risk regulation. *Columbia Law Review*, *95*, 1613–1723.

Xie, X. et al. (2007). Hawthorn-infesting populations of Rhagoletis pomonella in Mexico and speciation mode plurality. *Evolution*, *61*, 1091–1105.

Author biography. Sven Ove Hansson is professor in philosophy at the Royal Institute of Technology, Stockholm. He is editor-in-chief of *Theoria* and the two book series *Philosophy, Technology and Society* and *Outstanding Contributions to Logic*. He is member of the editorial boards of the journals *Synthese, Philosophy & Technology, Techné, Studia Logica*, and *Journal of Philosophical Logic*, and the book series *Logic, Argumentation & Reasoning*. His research areas include philosophy of science and technology, epistemology, logic, ethics, and value theory. He is the author of well over 300 articles in international refereed journals. His most recent books are *The Ethics of Risk* (2013), *Norms in Technology* (2013, edited with Marc J. de Vries and Anthonie W.M. Meijers), *David Makinson on Classical Methods for Non-Classical Problems* (edited, 2014), *The Role of Technology in Science: Philosophical Perspectives* (edited, 2015) and *The Argumentative Turn in Policy Analysis* (2016, edited with Gertrude Hirsch Hadorn).

Part C

Philosophical Issues of Particular Disciplines

12 Prospects for an integrated history and philosophy of composition

HASOK CHANG [*]

Abstract. I propose a new line of argument against metaphysical reductionism, focusing on the actual practices of composition and decomposition in chemistry and physics. Such a focus exposes significant problems with the common assumption that every material entity is made up of smaller units in a straightforward way. In the early days of chemical analysis there were worries that the analytical techniques might be altering the substances being analyzed, rather than simply taking them apart. In those reactions where molecules seemed to be cleanly dissociated into smaller units, the latter most often turned out not to be atoms. Such worries are reproduced and amplified when we consider the practices of modern nuclear and particle physics. "Atom-smashing" has never been Lego-like disassembly, since mass is not conserved in nuclear disintegrations. Generally, experiments in high-energy physics do not produce results that are in line with the naïve philosophical view of reductive levels, as illustrated very well in proton–proton collisions and pair-creation/pair-annihilation events. Compositionism, namely the notion that things are made up of stable units that persist through reactions, has been strongly discredited by developments in chemistry and physics. This fact of scientific practice argues against a simple-minded application of mereological principles to physical composition. But why are there such widespread intuitions in favor of metaphysical reductionism in the manner of mereology? I argue that the source of such intuitions is our predilection for compositionist analysis, for

---

[*]Department of History and Philosophy of Science, University of Cambridge.

the conduct of which the compositionist principles hold a quasi-Kantian conceptual necessity. We do apply compositionist analysis successfully in many areas of mathematics and some areas of science and everyday life. However, it is unwise to let ourselves be guided by these intuitions in situations in which attempts at compositionist analysis show little evidence of empirical success.

Keywords: reductionism, microreductionism, composition, analysis (chemical), ontological principles, conservation.

1 Introduction

In this paper I would like to propose a new line of argument against metaphysical (or ontological) reductionism. Following John Dupré, Nancy Cartwright and others, my approach is based on a commitment to respect the best scientific practices and their outcomes, while not renouncing philosophical judgment. There is nothing particularly new in taking a naturalistic approach, but the particular kind of focus I seek to give is not on the verdict of theory on what the world is like, but on the nature of the actual laboratory operations through which entities are (presumably) broken down into their components and synthesized again from their components. If we want to talk seriously about the composition of objects and substances, I think we need to pay close attention to what happens when we actually break things down and put them together.

The gist of my argument will be that successful analytic and synthetic practices in chemistry and physics do not fit the Lego-like picture of decomposition as clean dissection, and composition as simple assembly of unchangeable units. If my argument is correct, we must ask where the widespread reductionist intuitions come from, and to what extent they can be justified or trusted, if at all.

A brief comment about how I came to think about this topic might be helpful, self-indulgent as it may be. In my previous work I have made various arguments concerning reductionism and realism, but only from an epistemic point of view. However, recently I have come to concede that it is unavoidable to enter into some metaphysics if I want to make my epistemic positions truly coherent. This paper is one of my first attempts to grapple more seriously with metaphysical issues.[1]

My thinking on reductionism began by following one line from John Dupré's early arguments: even if we grant ontological reduction, epistemic reduction does not follow. With an eye on some classic reductionist positions, Dupré focused on "compositional reductionism" or "microreductionism", which he understood as "the view that the ultimate scientific understanding of a range of phenomena is to be gained exclusively from looking at the constituents of those phenomena and their properties." (Dupré,

---

[1] Another attempt is Chang (2016). The epistemological arguments that have led me to the issues discussed in the present paper are contained in Chang (2015).

1993, p. 88.) There are many ways to argue against such a view, even while we accept that things are made up of smaller things and everything can be smashed up into elementary particles ultimately. There are different observational terms at different levels, and different abstractions and idealizations that are effective at different levels; if one smashed up a plate and smoothed out the rough edges on the pieces, one could not reconstitute the original plate by putting the smoothed-out pieces together. It is also important to note, following Dupré again, the general mismatch between functional and structural categories, and the lack of a guarantee that structural analysis will be informative of the nature of functionally defined entities (e.g. chair, hemoglobin).

I used to think that the above points spelled an effective end to the reductionism debate: we can accept ontological microreduction, but epistemic microreduction does not follow. But I now think that it is equally important to subject ontological microreductionism to full critical scrutiny. Dupré gives some arguments against the hierarchical compositional ontology assumed in reductionist scheme, given by Paul Oppenheim and Hilary Putnam for example (1958, p. 9): (1) Elementary particles; (2) Atoms; (3) Molecules; (4) Cells; (5) (Multicellular) living things; (6) Social groups. I want to push those arguments further, and also take them to the heart of the presumed reductive base, namely to the realm of chemistry and then physics. I want to question seriously the commonly accepted assumption that everything can be smashed up into smaller and smaller parts, ultimately down to elementary particles, in any straightforward sense. My motivation for questioning that ontological microreductionist assumption comes not from any metaphysical investigations, but from my studies in the history of chemistry and physics.

2 The history of decomposition: Analytical chemistry

Let's start with chemistry. The practice of decomposing substances into their presumed components began in analytical chemistry centuries ago, long before physicists were attempting any such thing (perhaps with the exception of Newton's optical experiments). But a careful look at the history of chemistry reveals that decomposition has always been a contentious sort of practice. In the early days of chemical analysis, there were worries that the processes of alleged decomposition might be altering the substances being analyzed or even creating new ones. There was a very long tradition of "fire analysis" in alchemy and chemistry. It was a common observation that the application of strong heat tended to break things down, and a common chemical practice to study the composition of complex substances by dry-distillation or other techniques. But the cogency of fire-analysis was questioned by the likes of Robert Boyle (see Debus, 1967). How could one be sure that the application of fire was merely breaking things up into their constituents, rather than altering their very nature – or having the particles of fire sticking to them? Similar kinds of doubts were raised even more obviously concerning other methods of chemical analysis, such as the dissolution of substances by the application of acids.

Overcoming these doubts enabled the establishment of what I have called "compositionism" in chemistry, starting from the late 18th century, most notably with the line of development going from Lavoisier to Dalton. Compositionism, as an ontological doctrine, is the notion that chemical substances are made up of stable units that persist through chemical reactions (see Chang, 2012, chapter 1, and for more detail, Chang, 2011). This idea also grounded the understanding of chemical reactions as the rearrangement of distinct and stable building-blocks which retain their identity (even when their properties are not manifest in a state of combination). This idea can also be extended beyond the realm of chemistry, as we will see later.

An important part of the compositionist tradition of chemistry was Lavoisier's operational definition of chemical elements:

> ...if, by the term *elements*, we mean to express those simple and indivisible atoms of which matter is composed, it is extremely probable we know nothing at all about them; but, if we apply the term *elements*... to express our idea of *the last point which analysis is capable of reaching*, we must admit, as elements, all the substances into which we are capable, by any means, to reduce bodies by decomposition. (Lavoisier, 1789/1965, p. xxiv.)

In Jean-Antoine Chaptal's more compact formulation, Lavoisier's definition amounted to taking "the endpoint of analysis" as elements.[2] The point of such a conception of elements was to identify the building-blocks of chemical substances that one could actually work with in the laboratory. Such a hands-on notion of units enabled the articulation of a practical ideal in the study of the constitution of matter, which consisted in one's ability to take things apart and put them back together — or, for analysis and synthesis, or decomposition and recomposition.

But this was hardly the end of ontological doubts. In order to apply Lavoisier's definition, we need to know how to tell which chemical operations are decompositional. That requires us to know whether the products of a chemical operation are simpler than the ingredients going into it. But the whole point of the analytical definition of chemical element was to define elementarity (or simplicity) in terms of decompositional operations, so we are caught in a tight circularity: what are elements (or, simple substances)? — those that cannot be decomposed further — what are decompositional operations? — those that produce simpler substances. This circularity threatens to make the analytical definition of elements operationally empty. The difficulty can be illustrated quickly through a couple of examples from the Chemical Revolution (see Chang, 2012, chapter 1, for a detailed treatment). Is sulfur an element? The phlogiston theorists didn't think so, because they thought (and showed by experiment)

---

[2] J. A. Chaptal, *Elemens de Chymie*, 3rd ed., 1796, vol. 1, p. 55 (quoted in Siegfried & Dobbs, 1968, p. 283).

that it could be decomposed into sulfuric acid and phlogiston. The Lavoisierians disagreed, arguing instead that sulfuric acid was a composite substance, which they could decompose into sulfur and oxygen. Are metals elements? The phlogistonists didn't think so, as they knew how to decompose them into calxes and phlogiston. Again, the Lavoisierians saw the same reaction as a synthetic one, in which the metals combined with oxygen and formed oxides.

Readers who know something about this history will be thinking that Lavoisier solved this problem by the weight criterion. When a metal oxidizes, the rust is heavier than the metal, and moreover the weight of the rust is equal to the sum of the weights of the metal and the oxygen. Isn't it clear from these facts that the rust is composed of the metal and the oxygen? That is indeed how the circularity was broken, at least to Lavoisier's own satisfaction. But the matter was not so simple. The precision of Lavoisier's chemical accounting was much exaggerated (for example, witness his utter confidence that the weight-ratio of hydrogen and oxygen making up water was 15:85 precisely, rather than roughly 1:8). More importantly for the purpose of the present paper: who decided that weight was the magic criterion for telling the composition of substances? Phlogiston was generally considered weightless, and Lavoisier's own chemistry was based on the notion of caloric, namely the substance of heat, which was weightless. Even if we disregard the weightless substances, weight as the criterion of composition only works if we assume that weight (or mass) is conserved – we now know it is not – but now I am getting ahead of the story.

Set aside all of those questions and doubts. What is even more important is the fact that, as compositionist chemistry unfolded, it became clear that most of the useful analytical techniques did not involve simple, clean decompositions. Take water, the highlight of Lavoisier's triumph in the Chemical Revolution. Lavoisier never actually came up with a method of simply decomposing water into hydrogen and oxygen. What he did do was to pass steam over very hot iron (a gun barrel), which resulted in the rusting of the iron and the production of hydrogen gas; he presumed that the oxygen from the water (steam) must have combined with the iron, but he was never able to get the oxygen in pure form from water. In fact most decompositional operations in chemistry at that time was of this "sticky" form: not $[AB] \to [A] + [B]$, but $[AB] + [C] \to [AC] + [B]$, or worse.

In the case of water, the first clean separation of it into its constituents (i.e., hydrogen and oxygen) was only achieved in 1800, several years after Lavoisier's death, thanks to Volta's invention of the battery, which quickly led to the first electrolysis of water by William Nicholson (with Anthony Carlisle).[3] This is represented in a modern rendition in **Figure 1**.[4] But when this clean decomposition was achieved, the earlier

---

[3]There was a previous decomposition of water by electric sparks, but in that case the hydrogen and oxygen gases were produced in a mixture.

[4]I take the figure from Pauling and Pauling (1975, p. 357). Note an error in the volumes of hydrogen and oxygen in the figure, which should be in a 2:1 ratio. (Lesson: always check your page proofs.)

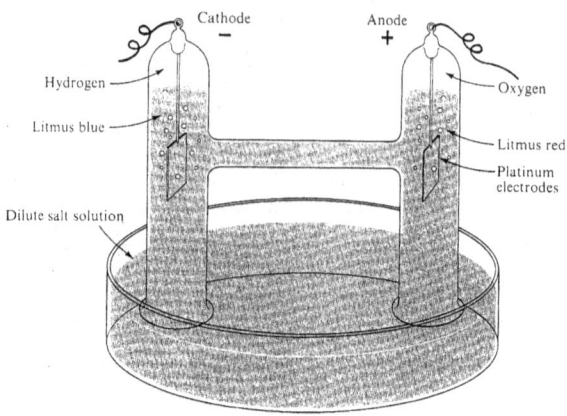

Figure 1. *A modern view of the electrolysis of water.*

circularity between simplicity and composition came back. First, note a quandary that Nicholson himself admitted right from the start: if electrolysis was the decomposition of each water molecule by the application of electricity, why did the oxygen and hydrogen not originate from the same spot where the water molecule would have been? This difficulty, which I have called the "distance problem", threatened to overturn the basic Lavoisierian interpretation of the composition of water. Johann Wilhelm Ritter understood the same electrolysis experiment as a pair of *syntheses*: hydrogen from water and negative electricity at the negative electrode, and oxygen from water and positive electricity at the positive electrode. And this interpretation, annoyingly for Lavoisierians, lined up exactly with Henry Cavendish's latter-day phlogistonist interpretation, if one identified phlogiston with negative electricity (see **Table 1**). No one denied at the time that electricity was a substance, and what exactly happened ontologically in electrolysis was at best contentious at that time, and subsequently for most of the 19th century. What happened to the distance problem in the end is too long a story to include here (see Chang, 2012, chapter 2), but the point here is that the chemical reaction itself did not carry its own obvious compositional interpretation.

Let us just assume, for now, that the circularity was eventually overcome. That is to say, I'm now just going to stick with the story of what happens in these reactions as accepted in modern chemistry, in terms of atoms and molecules. But that modern picture still embodies some fundamental difficulties for ontological reductionism. When chemists did manage to make clean decompositions of molecules, the smaller units produced often turned out not to be atoms. That is not necessarily a problem for microreductionism in general, but at least inconvenient for the usual hierarchical picture as presented by Oppenheim and Putnam.

Table 1. Competing views of the composition of water.

We observe:	Inflammable air	and Vital air	combine, to make	Water
Lavoisier says:	Hydrogen	Oxygen	→	H-O [compound]
Cavendish says:	Phlogisticated water	Dephlogisticated water	→	Water [element]
Ritter says:	Negatively electrified water	Positively electrified water	→	Water [element]

Think about water again. What actually happens, in modern terms, when we break up a molecule of water? H_2O never breaks up into two atoms of hydrogen and one atom of oxygen. If the decomposition is done by ionic dissociation (which happens spontaneously even in pure water), H_2O breaks up into H^+ and OH^-. If it is done by electrolysis, $(2)H_2O$ breaks up into $(2)H_2$ and O_2. If actual decomposition is what tells us about the building-blocks, then hydrogen and oxygen atoms are not the building-blocks of water. As it turns out, the stable units in chemical reactions are often not atomic (e.g., H^+ and H_2, not H). If we ask what really happens in the electrolysis of water, the picture is not simple, and not completely agreed among 20th-century chemists, either. In Pauling and Pauling's view, shown in **Figure 2** (Pauling and Pauling, 1975, p. 356), H_2O molecules first gain or lose some electrons, ending up as H_2 and O_2 molecules, as well as H^+ and OH^- ions. In chemical formulas we have:

(cathode side) $4H_2O + 4e^- \rightarrow 2H_2 + 4OH^-$

(anode side) $2H_2O \rightarrow O_2 + 4H^+ + 4e^-$

From the resulting products, $4OH^-$ and $4H^+$ combine again to make $4 H_2O$ molecules. The $4e^-$ on the left-hand side of the first equation represents the input of electrons from the battery through the cathode, and the $4e^-$ on the right-hand side of the second equation represents the flow of electrons into the battery through the anode. In another well-known textbook, by David Oxtoby et al. (1999, p. 432), the story is similarly complex, but not quite the same as the one given by Pauling and Pauling:

(cathode) $2H_3O^+ + 2e^- \rightarrow H_2 + 2H_2O$

(anode) $3H_2O \rightarrow (1/2)O_2 + 2H_3O^+ + 2e^-$

Adding the two equations together, we can see the whole reaction as one measure of H_2O breaking up into one measure of H_2 and half a measure of O_2. In neither case do we have simple decomposition into atoms, or even simple re-constitution of atoms from pre-existing ions.

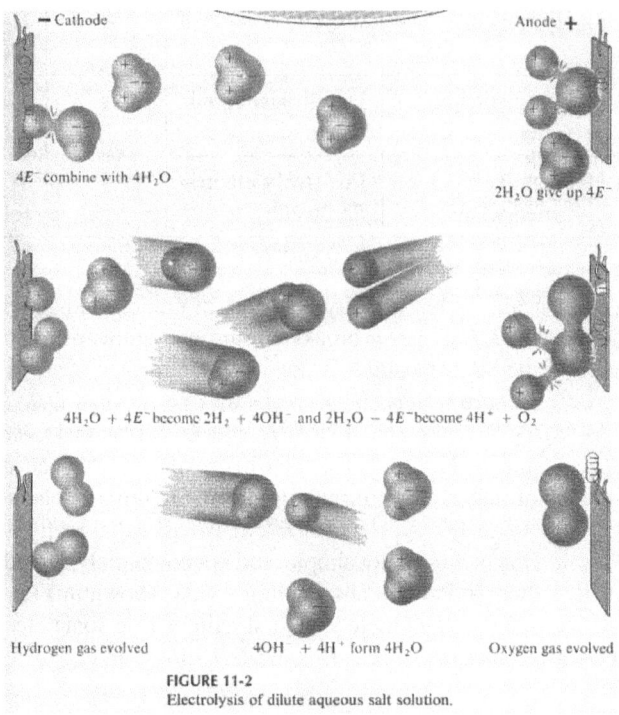

Figure 2. *The electrolysis of water, in Linus Pauling and Peter Pauling's interpretation.*

It may be easy to say, and it may have been almost universally agreed, that water is made up of oxygen and hydrogen. But it is quite difficult to translate that into a truly reductionist compositional statement, such as "H_2O is made up of H_2 and O_2" or "H_2O is made up of H and O." The former is not true in any rigorous metaphysical sense, and the latter is never realized in decompositional (or compositional) laboratory practice. Generally speaking, decomposition is not often instantiated in the practice of chemistry in any way that would satisfy the standard reductionist picture of composition. In the end, no laboratory practices in chemistry have been able to effect the kind of decomposition into atoms that Dalton's diagrams such as the one shown in **Figure 3** (Dalton, 1808, Plate 5, opposite p. 560) and other similar depictions lead us to imagine. In the 20th century, the advanced practices of analytical chemistry have moved on to spectroscopic techniques including nuclear magnetic resonance (NMR), in the employment of which no molecules are decomposed at all (see Reinhardt, 2006 for a detailed account of these developments).

Prospects for an integrated history and philosophy of composition 233

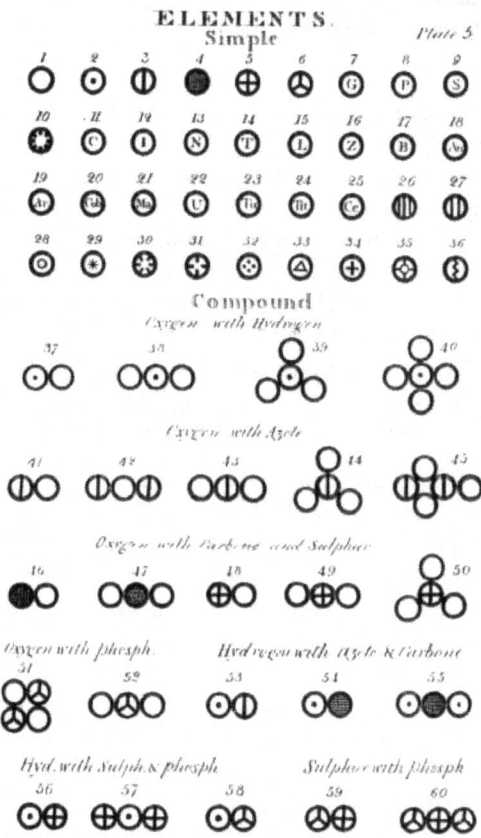

Figure 3. *John Dalton's atomic diagrams.*

3 The history of decomposition: Modern experimental physics

At this point I anticipate the response that of course chemistry is a mess, and reductionist science made real headway only with the advent of modern physics. If we look briefly again at the composition and decomposition of water, the presence of H^+ (proton) and e^- (electron) in the reactions clearly indicate that what we need here is elementary particles, not the outmoded notion of Daltonian atoms. So we must take a look at the atom-smashing practices of modern experimental physics, and see if they can provide some reassurance to the ontological reductionist. Here I am only going to rely on very rudimentary modern physics, the kind that has been very much taken for granted for many decades now, as that will be sufficient for my current purposes.

There should not be anything controversial about any of the physics that I will be discussing, and any physicists reading this may wonder why I am taking pains to state the obvious.

The philosophically interesting thing here is that the same kinds of worries about compositionist reductionism that we saw in the history of chemistry are reproduced anew, and even amplified further, when we consider the practices of modern nuclear and elementary particle physics. "Atom-smashing" has never been Lego-like disassembly: when atomic nuclei are broken up, energy is almost always added or subtracted, and according to modern physics mass is a form of energy. In nuclear reactions experimental physicists have taken careful note of the interconversion of mass and energy. It won't do to suggest that the Lego-like picture is approximately true since the amounts of mass lost or gained in these reactions are relatively small compared to the total mass of the reactants: no theory that has to dismiss nuclear bombs as merely inconvenient anomalies should be regarded as "approximately true" in the relevant sense.

Let's think a little bit further about the history of mass-energy equivalence. $E = mc^2$ came out of Einstein's special theory of relativity, which was accepted for very general theoretical reasons, but we might ask what experimental evidence compelled people to believe this particular aspect of the theory. In this context, what is invoked most often are particle-collision experiments in which the masses of the ingredients do not add up exactly to the sum of the masses of the products (see Fernflores, 2012, esp. Section 4). Especially famous is the 1932 experiment of John Cockcroft and Ernest Walton, who bombarded a lithium nucleus with a proton and obtained two helium nuclei (α-particles). Their measurements showed that the sum of the masses of the reactants was $1.0072 + 7.0104 = 8.0176$ amu,[5] but the masses of the products only added up to 8.0022 amu, indicating that 0.0154 amu has "disappeared." After over a century of dominance, Lavoisier's principle of the conservation of mass had been overturned in physical and chemical practice. An inspection of the periodic table of elements also easily shows that the masses of atoms are slightly different from the masses of the protons and neutrons (and electrons) that constitute them. These have been regarded as indisputable facts for many decades now by physicists and chemists, but the basic metaphysical implications of such facts have not got through to the sensibilities of reductionists. If we still regard mass as the primary indicator of the amount of matter, then it is clear that the amount of matter is not preserved in elementary-particle collisions or nuclear fusion or fission. That does not present a problem for the conservation of energy, of course, but it does destroy the naïve notion that atoms are simply put together from elementary particles. Atomic nuclei are not mereological sums of protons and neutrons; they are not made up of protons and neutrons in a straightforward sense.

Generally speaking, the results of experiments in high-energy physics do not support the naïve philosophical view of reductive levels in the Oppenheim–Putnam mode,

---

[5] 1 amu (atomic mass unit) is 1/12 of the mass of carbon-12 in its ground state.

Prospects for an integrated history and philosophy of composition   235

according to which atoms are made up of elementary particles, which are themselves unchangeable building-blocks. When two protons collide with each other in a particle accelerator, a whole host of other particles are created:[6] should we say that a proton already contained these particles? Should pair-creation and pair-annihilation lead us to conclude that a pair of photons consist of an electron and a positron, or *vice versa*?[7] When a photon is absorbed by an atom, it is annihilated and raises the energy level of the atom; so the photon is not an unchangeable unit, not even a persisting one. These are merely a handful of illustrative examples. Generally speaking, smashed-up pieces do not necessarily pre-exist in the whole. This recognition led Geoffrey Chew to advance his "bootstrapping" view of elementary particles, according to which no particles are fundamental and all elementary particles are made up of each other.[8] This view was sidelined with the advent of quarks and the Standard Model, but it may be worth revisiting, after all. The early modern chemists' mereological fears have turned into reality "with a vengeance", in much more serious ways than they could have imagined. There is no need for "the Tao of physics", but there are good reasons to have our metaphysical thinking informed by respect for actually viable practices.

Again, I must stress that all of what I have discussed in this section is based on *obvious* physics. I do not claim any special knowledge of modern physics, and I am not drawing from any controversial parts of physics. The experimental facts that I cite are very well-known and accepted in an unquestioning way by physicists. I am not even invoking the difficult ontological questions raised by the notion of quantum superposition and entanglement, or indistinguishability, or virtual particles and vacuum fluctuations, or quark confinement, which are bound to complicate the picture much further and in all likelihood in anti-reductionist directions.[9]

4 Scientific naturalism, everyday intuition, and conceptual habits

At this point some philosophers may say: "But what these physics experiments (or the theories created in order to explain them) seem to show can't really be the true metaphysical picture. There *must be* unchangeable basic building blocks, and everything must ultimately be made of them." Of course, no amount of scientific knowledge would *prove* that the metaphysical reality is not like Lego. I think that way of thinking

---

[6] See a friendly presentation of the basic facts and ideas in the "International Physics Masterclasses" section on proton collisions (http://atlas.physicsmasterclasses.org/en/zpath_protoncollisions.htm).

[7] Positron–electron pair-annihilation is now even in the realm of everyday technology as the basis of PET (positron-emission tomography) scans; it is not something we can afford to ignore in our thinking about the world.

[8] For more on the metaphysics of bootstrapping and the S-matrix theory, see McKenzie (2011).

[9] For a cutting-edge discussion, see Caulton (2015).

about reality is not productive, but that is not the argument I am making in this paper. For now I want to focus on a different question: where do people get the intuition that reality is Lego-like?

(1) Old science. In the discussion above I have implied that philosophical reductionists are quite out of step with modern science. So could it be that the reductionist intuitions come from some older phase of science that was once successful, from which we've inherited some now-outdated theoretical assumptions, which parade as metaphysical intuitions? This sort of situation has been common in the history of science. As Philipp Frank put it (1949, pp. 475, 478): "the Copernican system was declared to be 'mathematically true' but 'philosophically false.' And this severe judgment has been passed again and again by philosophers upon new physical theories." Such impulse is understandable, but it is something we need to learn to overcome: "We understand now very well that these 'established philosophic principles' are nothing else than physical hypotheses in a state of petrification."

In this case, however, I do not think that successful past science can be a full explanation of our intuitions. It is true that *some* Lego-like science has been successful: e.g., 19th century organic structural chemistry. I think such cases did have some effect on creating, or at least bolstering, compositionist intuitions. But there are some significant caveats that negate this point almost entirely. (a) A lot of other successful science (such as 20th-century physics) has not been compositionist. (b) A lot of compositionist science has been unsuccessful; for example, Alan Chalmers (2009) has shown persuasively that much of early atomic science was rather useless. (c) Compositionist ideas were widespread and popular long before there was successful compositionist science; Democritus, or even the 17th-century mechanical philosophers, would not have got their atomistic ideas from some successful previous science. (d) Even successful compositionist science often showed awareness that physical reality was not entirely compositionist. The "ball-and-stick" molecular models needed the sticks to connect the balls, and it was recognized that the sticks were exactly the mysterious part of the picture: the nature of the chemical bond remained enigmatic throughout the 19th-century except in clear cases of ionic bond.[10]

(2) Everyday life. It could be that the compositionist intuitions are rooted in our everyday life, rather than in successful previous sciences. Robert Northcott, in a serious joke, comments that we must have such intuitions because we all grew up playing with Lego. Lego itself was not invented till 1949 so it can't have been responsible for the advent of compositionism, but could it be that much of our everyday life, dealing with "medium-sized dry goods" is like playing with Lego, giving rise to compositionist intuitions? We can smash a plate and glue it back together, build a house out of bricks, and take apart a watch and put it back together.

But these practices of composition do not work in Lego-like ways. When we try to put

---

[10]Cohesion was a similar mystery. For a fascinating history of cohesion, see Rowlinson (2002).

things together in everyday life, it does not normally work out in the compositionist way. Medium-sized dry goods generally do not stick to each other. Bricks need mortar to make a wall or a house. What we know of as "Lego", first marketed as "*automatic binding blocks*", was such a commercial and cultural success precisely because it was a very clever arrangement, an ingenious combination of rigidity, elasticity and friction, in which bricks do stick together without the help of anything else![11] Almost nothing else in nature or human life behaves like Lego,[12] and that is the secret of its success. We do not live in Legoland. Where *do* we live? Our quotidian paradigm for sticking things together is the use of glue. How glue works is not at all like Lego, so our experience of gluing (or stapling, or clamping, or strapping) things together does not explain why we have Lego-ish intuitions.

(3) Quasi-Kantian conceptual necessity. I suggest that that our compositionist intuitions come not from practical experience, but from a quasi-Kantian conceptual necessity. I say quasi-Kantian because the necessity involved here is conditional (paradoxical as that may sound), being consequences of definitions that we happen to adopt. More generally, ontological principles (as I term them) are assumptions that we have to adopt about the presumed entities that we are dealing with, if we are to carry out certain kinds of epistemic activities. It is our choice to engage in a particular epistemic activity; once we have made that choice, however, there are some metaphysical assumptions we must subscribe to, because they are necessary for that activity to be successful or even intelligible. For a very simple example: if we want to engage in the activity of counting, then we have to presume that the things we are trying to count are discrete. In other words, the metaphysical principle of discreteness is required in the activity of counting, and enables that activity. What we have here is a distinct type of a transcendental argument — laying out the necessary conditions for something that we may or may not choose to do, rather than investigating the precondition of what is already and definitely given. So I think of it as a contingent transcendental argument in the form: "*if* we want to do X, *then* we must presume Y."[13]

In illustrating these ideas briefly, it is helpful to refer back to the work of the much-neglected Clarence Irving Lewis. In a move that is only apparently anti-Kantian, he denies that there are any synthetic *a priori* principles. For Lewis all *a priori* principles follow from the nature of the concepts we choose to craft and use:

> The paradigm of the *a priori* in general is the definition. It has always been clear that the simplest and most obvious case of truth which can be known in advance of experience is the explicative proposition and those consequences of definition which can be derived by purely logical anal-

---

[11] For the history of Lego as told by the company itself, see http://www.lego.com/en-us/aboutus/lego-group/the_lego_history/1940.

[12] Gretchen Siglar points out that Velcro is similar.

[13] The nature of a contingent transcendental argument is explained in full in Chang (2008) and Chang (2009).

ysis. These are necessarily true, true under all possible circumstances, because definition is legislative. (Lewis, 1929, pp. 239-240.)

And yet, Lewisian *a priori* principles in science are not mere tautologies; they do serve as significant laws of nature, in a way that I think is consonant with Michael Friedman's views on what constitutive principles do (Friedman, 2001). Lewis chose Einstein's definition of simultaneity in special relativity to help make his point about the role of definitions:

> As this example well illustrates, we cannot even ask the questions which discovered law would answer until we have first by *a priori* stipulation formulated definitive criteria. Such concepts are not verbal definitions nor classifications merely; they are themselves laws which prescribe a certain behavior to whatever is thus named. Such definitive laws are *a priori*; only so can we enter upon the investigation by which further laws are sought. (Lewis, 1929, p. 256.)

Finally, let me come back to the matter of composition. I submit that compositionist intuitions are ontological principles necessitated when we choose to carry out compositionist analysis, by which I mean the the activity of understanding an object as a mereological sum of its parts, each of which persists unless broken up into its own parts. So compositionism, now generalized from the sense I gave it in the context of chemistry, can be re-stated as the metaphysical doctrine that all objects are made up of persisting parts, whose identity is not affected by any combinations into which they enter. To this basic compositionist assumption, scientists and philosophers often add two further ideas, which are actually not necessary for compositionist analysis *per se*: (1) that there are fundamental parts that cannot be further decomposed, and (2) that there are only a small number of types of fundamental parts (units).

If we choose to carry out a compositionist analysis, it makes no sense not to adopt the basic compositionist ontology; that would render our activity incoherent and unintelligible. And if we do routinely carry out compositionist analysis, we may understandably form compositionist habits of the mind. We often do engage in some compositionist analyses in the activities of everyday life. Simple-minded accounting is a very good example; money is a wonderfully compositionist entity, which can be divided and added up with no loss or change of any presumed parts. We also engage in compositionist analysis in theoretical science, for example when we apply various conservation laws, or when we learn to divide and add up angles or lengths in geometry. Perhaps most fundamentally, the standard kind of arithmetic that most of us are taught in childhood is firmly founded on compositionist intuitions. The experience of such epistemic activities gives us familiarity with compositionist ontological principles, and disposes us to think in compositionist terms.

However, this is not to say that we *should* always engage in compositionist analysis. And it is incoherent to try to reason on the basis of compositionist intuitions when we are not in fact carrying out any compositionist analysis. Similarly, it is futile to attempt a compositionist analysis where we cannot find reliably persistent units out of which the objects of our interest can be said to be made. Whether or not there are such units is an empirical question, a contingent matter, which can only be clarified in an operational way, by devising and attempting to carry out actual operations of physical assembly and disassembly.

It is a mistake to make an unthinking application of mereology to physical combination. Judging from the findings of modern chemistry and physics, it seems that the mereological part–whole relation is an inappropriate framework for understanding physical combination. Physical fusion and disintegration may violate axioms that one considers reasonable or even indispensable in mereology, such as transitivity. If so, the most reasonable conclusion may be that physical composition is not a matter of part-whole relation. We need to make sure that quasi-Kantian conditional necessity does not degenerate into pseudo-Kantian metaphysical prejudice.

5 Conclusion

Attention to the actual practices of chemistry and physics reveals that there has never been unequivocal scientific warrant for ontological reductionism as it is commonly conceived. And if ontological reductionism doesn't work in chemistry and physics, then it is not likely to work in other sciences, partly because the usual reductionist strategy concerning those other sciences is to reduce entities in them to the entities of chemistry and physics. The main source of widespread intuitions in favor of ontological reductionism is neither everyday life nor successful past science. Rather, it is our predilection for compositionist analysis, which we apply successfully in many areas of mathematics and some areas of science and everyday life. But it is unwise to let ourselves be guided by these intuitions in other situations, especially those in which attempts at compositionist analysis show little evidence of empirical success.

Bibliography

Caulton, A. (2015). Is mereology empirical? Composition for fermions. In T. Bigai, & C. Wüthrich (Eds.), *Metaphysics in contemporary physics* (pp. 293–322). Leiden/Boston: Brill/Rodopi.

Chalmers, A. (2009). *The scientist's atom and the philosopher's stone: How science succeeded and philosophy failed to gain knowledge of atoms*. Dordrecht: Springer.

Chang, H. (2008). Contingent transcendental arguments for metaphysical principles. In M. Massimi (Ed.), *Kant and philosophy of science today* (pp. 113–133). Cambridge: Cambridge University Press.

Chang, H. (2009). Ontological principles and the intelligibility of epistemic activities. In H. De Regt, S. Leonelli, & K. Eigner (Eds.), *Scientific understanding: Philosophical perspectives* (pp. 64–82). Pittsburgh: Pittsburgh University Press.

Chang, H. (2011). Compositionism as a dominant way of knowing in modern chemistry. *History of Science, 49*, 247–268.

Chang, H. (2012). *Is water H_2O?: Evidence, realism and pluralism*. Dordrecht: Springer.

Chang, H. (2015). Reductionism and the relation between chemistry and physics. In T. Arabatzis, J. Renn, & A. Simões (Eds.), *Relocating the history of science: Essays in honor of Kostas Gavroglu* (pp. 193–209). Dordrecht: Springer.

Chang, H. (2016). The rising of chemical natural kinds through epistemic iteration. In C. Kendig (Ed.), *Natural kinds and classification in scientific practice* (pp. 33–46). London: Routledge.

Dalton, J. (1808). *A new system of chemical philosophy*. Manchester.

Debus, A. G. (1967). Fire analysis and the elements in the sixteenth and the seventeenth centuries. *Annals of Science, 23*(2), 127–147.

Dupré, J. (1993). *The disorder of things: Metaphysical foundations of the disunity of science*. Cambridge, MA: Harvard University Press.

Fernflores, F. (2012). The equivalence of mass and energy. In E. N. Zalta (Ed.), *Stanford Encyclopedia of Philosophy (Spring 2012 Edition)*. Retrieved from http://plato.stanford.edu/archives/spr2012/entries/equivME/

Frank, P. (1949). *Modern science and its philosophy*. Cambridge, MA: Harvard University Press.

Friedman, M. (2001). *Dynamics of reason*. Stanford, CA: CSLI Publications.

Lavoisier, A. L. (1965). *Elements of chemistry, in a new systematic order, containing all the modern discoveries* (R. Kerr, Trans.). New York: Dover. (Original work published 1789)

Lewis, C. I. (1929). *Mind and the world-order: Outline of a theory of knowledge*. New York: Dover.

McKenzie, K. (2011). Arguing against fundamentality. *Studies in History and Philosophy of Science Part B: Studies in History and Philosophy of Modern Physics, 42*(4), 244–255.

Oppenheim, P., & Putnam, H. (1958). Unity of science as a working hypothesis. In H. Feigl, M. Scriven, & G. Maxwell (Eds.), *Concepts, theories, and the mind-body problem* (pp. 3–36). Minneapolis, MN: University of Minnesota Press.

Oxtoby, D. W., Gillis, H. P., & Nachtrieb, N. H. (1999). *Principles of modern chemistry* (4th ed.). Philadelphia, PA: Saunders College Publishing.

Pauling, L., & Pauling, P. (1975). *Chemistry*. San Francisco, CA: W. H. Freeman and Co.

Reinhardt, C. (2006). *Shifting and rearranging: Physical methods and the transformation of modern chemistry*. Sagamore Beach, MA: Science History Publications.

Rowlinson, J. S. (2002). *Cohesion: A scientific history of intermolecular forces*. Cambridge: Cambridge University Press.

Siegfried, R., & Dobbs, B. J. (1968). Composition, a neglected aspect of the Chemical Revolution. *Annals of Science, 24*(4), 275–293.

Author biography. Hasok Chang is the Hans Rausing Professor of History and Philosophy of Science at the University of Cambridge. He received his degrees from Caltech and Stanford, and has taught at University College London. He is the author of *Is Water H_2O? Evidence, Realism and Pluralism* (2012), and *Inventing Temperature: Measurement and Scientific Progress* (2004). He is a co-founder of the Society for Philosophy of Science in Practice (SPSP), and the Committee for Integrated History and Philosophy of Science.

13 A new look at the history of science. The transnational orientation of the Genetics and Radiobiology Program in Mexico in the 1960s

ANA BARAHONA [*]

Abstract. The transnational approach in the history of science is very recent and has been influenced by the effects of globalisation, multiculturalism and the formation of circuits of practices, organizations, objects, goods, knowledge and people, in which scientific developments go beyond nation-state borders, collaborative networks being the units of historical analysis. Recent debates regarding global and local contexts have called attention to circulation networks that explore inter-regional exchanges and transnational circuits that allow quicker cross-border transmission of scientific practices and a faster flow of people, ideas and artefacts. In the case of science studies in Latin America, a great deal of research performed under this approach has indicated that despite their historiographical and epistemological importance, narratives on the national sciences perspective have revealed their analytical limitations. This research has expressed the need to reconstruct transnational stories that account for how the knowledge produced in developing countries forms part of international knowledge as it circulates via international networks of collaboration. A case in point is the cre-

---

[*]Department of Evolutionary Biology, School of Sciences, UNAM México 04510, D. F, Mexico.
Email: ana.barahona@ciencias.unan.mx.

ation of the first Genetics and Radiobiology Program of the National Commission of Nuclear Energy in Mexico in the 1960s wich was in line with the international trends for the peaceful uses of atomic energy, and with local needs for creating a program to study the effects of radiation in human populations in the country. Alfonso León de Garay's role in the development and establishment of radiobiology and human genetics in Mexico was fundamental.

Keywords: transnational science, circulation of knowledge, genetics and radiobiology, nuclear energy, Alfonso León de Garay.

1 Introduction

The transnational perspective in the history of science is very recent and can be traced to the end of WWII, when there was growing interest for international peace that gave rise to international cooperation programs and organizations that produced important changes on the international political landscape. In the aftermath of WWII, not only were cultural and social processes reconfigured, sea changes occurred in the area of science itself, where the global circulation of scientific instruments, workforce and ideas had come increasingly into focus. Nevertheless, some authors have highlighted the absence of the historian of science in this discussion (Turchetti, Herrán, & Boudia, 2012). They have explained this absence due to "the ways in which the discipline has changed while this approrach has emerged, epistemic univeralism (which understands science as a transcendent, truth-finding activity that, in principle, should not be affected by national, class or ethnic difference), or social interactions that define science as cross-border, transnational activity" (Turchetti et al., 2012, p. 323). According to them, what is needed is that scholars take the role of science in a global dimension seriously.

Recent debates regarding global and local contexts have called attention to circulation networks that explore inter-regional exchanges and transnational circuits that allow quicker cross-border transmission of scientific practices and a faster flow of people, ideas and artefacts. A great deal of research performed under this approach has indicated that despite their historiographical and epistemological importance, narratives on the national sciences perspective have revealed their analytical limitations. A case in point is the creation of the first Genetics and Radiobiology Program (Programa de Genética y Radiobiología, PGR) of the National Commission of Nuclear Energy (Comisión Nacional de Energía Nuclear, CNEN) in Mexico in the 1960s, which was in line with international trends for peaceful uses of atomic energy, and with local needs for creating a program to study the effects of radiation in human populations in the country. Alfonso León de Garay's role in the development and establishment of radiobiology and genetics in Mexico due to his belonging to international networks was fundamental.

2 A new look at the history of science, the transnatinal perspective

In the case of history of sience focused outside the United States and Europe, there have been two leading models to explain the flow of knowledge, diffusionism and dependency. Many historians, especially those focused on colonial science, referred to and often sought to implement the model proposed by George Basalla in 1967. According to this model, there are three phases in the diffusion of science from the metropolis towards the peripheries; that is, from Europe to non-European countries. These phases are lineal, sequential, progressive, and necessary for the transformation of science in local or peripheral contexts (Basalla, 1967); or have focused on the asymmetrical relationships between centre and periphery that implies the inability of the latter to develop an autonomous system of science and technology (see for example Cardoso & Faletto, 1979). During the 1970s and the 1980s, these patterns framed the development of historical studies seeking to account for the development of science particularly in Latin American countries (Sagasti & Guerrero, 1974; Stepan, 1981; Lafuente & Sala-Catalá, 1989; Chambers, 1987). None of these studies acknowledge on the global or reciprocal connections, or focus on the networks of collaboration that may help explain the construction of knowledge at both the local and the global level.

However, recent historiography of science has focused on how knowledge circulates in widely different contexts. In the last decades of the 20th Century, the historiography of science, imperialism and postcolonialism elaborated upon the dynamic interactions between what have been traditionally called centers and peripheries. For example, recent studies in the history of science in developing countries have confronted the accepted idea about the coexistence of the global circulation of people, practices and techniques, and the symmetries between scientists working in metropolitan centers and those working in less-developed ones, showing that not all science on the so-called periphery should be regarded as peripheral, and criticizing those perspectives that see the diffusion process as unilateral and promoted by metropolitan centers and adapted in the peripheries. Moreover, some historians have proposed the abandoning of the terms center and periphery, inasmuch as they do not reflect the dynamics or circulation of the elites from less-developed countries who made outstanding contributions and participate in international networks (Kreimer, 2010).

Debates regarding global and local contexts have called attention to circulation networks formed by shared interests through which exchange is negotiated and in which the circulation of knowledge, people and practices occurs (van der Vleuten, 2008; Sivasundaram, 2010; Safier, 2010; Birn & Necochea López, 2011; Hofmeyr, 2013; Druglitrø & Kirk, 2014). Sivasundaram, for example, emphasizes that in order for science to be successful it has to travel, "studying networks fit well with global history because networks cross empires, nations, and regions" (Sivasundaram, 2010, p. 158). Secord, for his part, has demonstrated that science can be understood as knowledge in transit, which has to cross national, temporal and disciplinary borders owing to its social nature (Secord, 2004). Even more recently, Finnegan proposed four distinct types of place in which scientists operate: sites, regions, territories, and boundaries or circulation, the latter used to understand the dynamics of knowledge that became universal by circulating (Finnegan, 2008); for Kohler, these categoriesries embrace the material and the social aspects of place, the local and socially constructed character of scientific knowledge (Kohler, 2012). The role played by the creation of networks in the stabilization of scientific facts has focused on how local knowledge becomes universally accepted. Important works that show the techniques by which scientists convinced their peers about their knowledge claims highlight the agency of local actors in the flow of knowledge (to name a few see Latour & Woolgar, 1979; Rudwick, 1984; Collins, 1985; Shapin & Schaffer, 1985; Latour, 1988; Daston & Galison, 1992; Shapin, 1994; Porter, 1995; Golinsky, 1998; Daston & Galison, 2007).

Recent studies have emphasized the need to write connected transnational narratives based on a reciprocal treatment of global and local contexts (Subrahmanyam, 1997). This transnational approach abandons the nation as a unit of analysis, Eurocentric narratives, cultural-diffusion interpretations, and the rigid opposition of the categories of "center" and "periphery" in order to explain the dynamics of transnational circuits and the global and local circulation of knowledge, people, artefacts and scientific practices. This rich approach problematizes the notion of "international science" and has pending issues such as the precise definition of notions such as circulation, reception, adaptation and creativity. However, this new perspective thoroughly emphasizes the interaction of experts from different countries and the transnational circulation of people, knowledge and practices as an intrinsic part of knowledge production (Birn & Necochea López, 2011). "Exploring the transnational circulation of knowledge thus becomes a key feature in the analysis of how hazardous trades have been reinterpreted, negotiated and relocated in undeveloped and less-developed countries" (Turchetti et al., 2012, p. 328). Many case studies looking through the transnational glasses of the recent historiographical turn have been produced (for Latin American studies see Cañizares, 2000; Cueto, 2006, 2007; Hochman, 2008; Soto Laveaga, 2009; Medina, 2011). It is under this perspective that I analysed the creation of the PGR of the CNEN in the 1960s in Mexico together with the pioneering work of the Mexican physician-turned-geneticist Alfonso León de Garay. In this narrative the term 'transnational' allowed us to understand the formation of the Program based on local needs and the aim of working beyond geographical limitations to thus facilitate the circulation of knowledge, practices and people from and to Mexico.

3 Nuclear physics in México during the Cold War[1]

In the early 1950s, an acceleration in the arms race and atomic tests became apparent as a result of the bombings of Hiroshima and Nagasaki that put an end to the Second World War (WWII). Within this international context, President Dwight D. Eisenhower gave the *Atoms for Peace* speech on December 8th 1953 before the UN General Assembly in which he stated his preference for halting the military and warfaring uses of nuclear energy and offering peaceful nuclear technology to humanity instead (Eisenhower, 1953/2007).[2] Eisenhower suggested the creation of an international organization to regulate the process for creation and use of atomic energy and proposed that countries with nuclear technology projects contribute economical and technical resources to the development of peaceful uses of nuclear energy. In 1957 the International Atomic Energy Agency (IAEA) was created with its headquarters in Vienna.[3] The member countries of the UN were invited to participate by creating their own national institutions to attend relevant meetings and benefit from the knowledge produced and establish control mechanisms. Several of these organisms were created before or concurrently with the formalization of the IAEA in 1957, as was the case for the CNEN. The creation of the IAEA increased funding sources for nuclear scientists and technicians with emphasis on the peaceful uses of radioisotopes in medical genetics and agriculture, among others.

The interest of Mexican scientists and politicians did not begin with the creation of the CNEN. According to Vélez Ocón, Mexican nuclear projects began as a result of the atomic bombings in Japan, when the Mexican government declared nuclear deposits of uranium, thorium and actinium, national reserves in 1945 (Vélez Ocón, 1997). At the end of 1945, General Ávila Camacho's government lobbied enthusiastically for the attendance of a Mexican representative at the atomic tests the United States would hold. This was how Harvard-educated Guggenheim Fellow soil engineer Nabor Carrillo, who was at the time the Scientific Research Coordinator at the National Autonomous University of Mexico (UNAM), and Soil Mechanics Research Section Head of the Commission for Promotion and Coordination of Scientific Research (Comisión Impulsora y Coordinadora de la Invetigación Científica, CICIC), was invited to attend the Bikini Atoll tests as an observer in 1946, along with Colonel Juan Loyo González. Most countries, Mexico included, declared themselves in favor of the peaceful use of atomic energy, and stated that the atomic bomb and other weapons of destruction would bring permanent peace to the world. At a national level, President Ávila Camacho had founded the CICIC in 1942. Its first director, even while he was living

---

[1] Sections 3 and 4 are a revised and shortened version of Barahona (2015).

[2] As Creager and Santesmases have shown, 'even before President Eisenhower's initiative, biology, agriculture, and medicine served to represent the peaceful face of atomic energy', but it was only after WWII that medical practices and biological knowledge changed drastically (Creager & Santesmases, 2006).

[3] Among the Mexican delegation that attended the Conference were Manuel Sandoval Vallarta, Nabor Carrillo and Carlos Graef. We will see the connection with the PGR below.

in the United States, was Dr. Manuel Sandoval Vallarta, the Mexican MIT-educated Guggenheim Fellow physicist. He had collaborated with MIT from 1923 to 1946, and traveled to Mexico several times to promote nuclear energy studies. In about 1943, the CICIC established the first radioactivity research laboratory, which had been promoted by Sandoval Vallarta with the main objective of developing radiochemistry. This was because one of his chief scientific concerns was acid rain. The head of the soils mechanics laboratory at the time was his colleague and friend, Nabor Carrillo (Bulbulian & Rivero-Espejel, 2012).

From this moment on, interest in the study of atomic energy grew in Mexico, and so the nuclear research program was set up. Dr. Carrillo, who was Rector of the UNAM at the time, visited the American company High Voltage Engineering Corporation in Cambridge, Massachusetts, thanks to Dr. William Buechner's invitation to attend the facilities and familiarize himself with the equipment. This allowed the UNAM to install a Van de Graaff accelerator with federal support in the 1950s (Domínguez-Martínez, 2012). The immediate effect of this was, on the one hand, the development of research, including the commencement of various projects and the training of technicians on radioisotope techniques, and on the other, an increase in and consolidation of institutional relationships and collaborations between Mexican and foreign scientists. Chief among these collaborations was that of Mexican Harvard-educated and Guggenheim Fellow physicist Dr. Carlos Graef –Sandoval's former student-, who was interested in cosmic radiation and relativity, with Harlow Shapley, the high-profile promoter of the United States National Academy of Sciences and director of the Harvard College Observatory.

The creation of the CNEN was created in 1956 and highlighted the efficiency of nuclear energy as an energy source, besides considering it a more economical source than oil, coal or hydropower. Since its inception, the Mexican nuclear program has had civilian and not military purposes, so the word 'atomic' was excluded because of its implications for war and the arms race. Instead, the word 'nuclear' was included, due to its peaceful connotations – energy and non-energy applications, as well as studies in nuclear science that included agronomy and genetics among others (Vélez Ocón, 1997). President Adolfo Ruiz Cortines appointed the lawyer and ex-President of the Supreme Court José María Ortiz Tirado as chairman of the Commission, and Doctors Nabor Carrillo and Manuel Sandoval Vallarta as members. The Advisory Board would be comprised of Doctors Alberto Barajas Celis, Fernando Alba Andrade and Carlos Graef Fernández, José Mireles Malpica (M.Sc.) and the engineers Eduardo Díaz Losada and Jorge Suárez Díaz.

The CNEN was founded upon two general fields of interest: energy and non-energy applications; and nuclear science studies. It was created with a complement of nine programs: nuclear physics, education and training, seminars, reactors, radioisotopes, industrial afflictions from radiation, agronomy, genetics and radiological protection. Thanks to the efforts of Sandoval Vallarta, Nabor Carrillo and Carlos Graef, who prompted the study of nuclear physics in Mexico, land and resources were obtained to develop a nuclear center. In 1964, construction began on the 'Nabor Carrillo' Nuclear

Center as part of the CNEN in the town of Salazar, State of Mexico, and both a Triga Mark III reactor and a Tandem Van de Graaff positive ion accelerator were acquired; these facilities would house all the CNEN's programs and laboratories that were scattered throughout the city, and give support to other institutions such as the UNAM and the National Polytechnic Institute (Instituto Politécnico Nacional, IPN).

It is through the CNEN that nuclear physics and genetics and radiobiology were connected in Mexico with the foundation of the PGR, opening new contexts for the development of genetics and radiobiology.

4 The Genetics and Radiobiology Program and the pioneering work of Alfonso León de Garay

After having spent two years in the Galton Laboratory at University College London, under the supervision of British medical geneticist Lionel Penrose, the Mexican physician Alfonso León de Garay founded the PGR under the CNEN. As we shall see, de Garay was part of an international network in human genetics in which the circulation of researchers, knowledge and practices enabled the assimilation and modification of genetic practices in Mexico in the early 1960s. It was in these years, when global trends in human genetics were reshaping the field of biomedicine in the aftermath of WWII, and when growing international interest in understanding the effects of radiation on human beings led to the formation of institutions and a proliferation of multi-centered clinical trials and inter-laboratory studies (Cambrosio, Keatins, & Bourret, 2006).

Alfonso León de Garay (1920-2002) studied medicine in the 1940s at the Autonomous University of Puebla, and served for many years as a neurologist. In 1957, he decided to study population genetics with Lionel Penrose, the Head of the Galton Laboratory. Thanks to an agreement between the IAEA and some European universities, de Garay obtained a grant and went to London. The relationships he established in Europe were extremely influential on the PGR he founded upon his return to Mexico in 1960, and had a profound and lasting importance in his later career and on his thinking (Barahona, 2009).

When de Garay was in England, he attended the 1957 general assembly of the IAEA as a companion to the English representatives, and there he met the Mexican delegation composed of José María Ortiz Tirado, Nabor Carrillo and Manuel Sandoval Vallarta, and the Secretary General of the CNEN, Mr. Salvador Carmona. They urged him to complete his studies and return to Mexico in order to found a program where he could begin studies into the effects of radiation on health (Barahona, 2009).

As a result of this meeting, and due to the support of the Mexican delegation, de Garay went on to study the mutagenic effect of radiation, in line with projects that were being developed in other parts of the world. In 1957, with the support of both

Ortiz Tirado and Dr. Alexander Hollaender, head of the Division of Biology at the Oak Ridge (Tennessee) National Laboratory of the US Atomic Energy Commission, de Garay was able to go to Oak Ridge and become well-grounded in the field of radiobiology (Villalobos-Pietrini, Guzmán, & Levine, 2005). His return to Mexico was hastened because the CNEN insisted on the immediate set-up of a laboratory, due to international pressures that Mexico should develop its own research in genetics and radiobiology.

The PGR was established in 1960 in order to 'contribute to the conservation of health, physical and mental improvement, and sickness prevention, through the investigation of the factors which intervene (favorably or unfavorably) in the biological inheritance of the population' (de Garay, 1960).

De Garay's agenda was two-fold. On the one hand, he intended to open up spaces for the development of genetics and radiobiology in Mexico, and on the other, to promote the PGR on the international stage. This interplay between de Garay, the PGR, the CNEN and the IAEA fostered the circulation of ideas and practices beyond transnational borders. To achieve this agenda, de Garay's objectives were the study of the effects of radiation on human health, and the study of diverse specific aspects of the heredity process, from the molecular level up to population genetics.

The PGR consisted at first of a small staff composed of six researchers, including de Garay as director, Rodolfo Félix Estrada, chief of the *Drosophila* section, and María Cristina Cortina Durán, María Teresa Zenzes Eisembach, Víctor Manuel Salceda Sacanelles, and Claudina Berlanga Siller, who obtained their B.A. degrees in biology in 1960, as well as a technician, a secretary, and a service assistant. By the end of the 1960s, the number of sections increased to include, for example, human genetics and molecular genetics sections (de Garay, 1970; Barahona, Pinar, & Ayala, 2003).

The Program consisted of six sections: tissue culture (where cytology and genetic analysis were practiced), photography (microphotography and autoradiography), biochemistry (biochemistry and radiochemistry), education (preparation of educational materials and training of personnel), *Drosophila* labs (conventional experimentation and computing of mutations, including irradiated stocks), and statistics and social work (population genetics and family studies).

The three main lines of research conducted in the years after its foundation that were in line with international projects while responding to the national context were, first, cytogenetic studies of certain abnormalities; second, the study of the effects of radiation on hereditary material, and third, the study of population genetics in *Drosophila* and in Mexican indigenous groups.

4.1 Cytogenetic studies of chromosome abnormalities

The 1960s was a decade of intensive work in the area of human genetics; it was transformed into an enticing frontier for medical research, and the production of technological knowledge from the study of chromosomes held a great deal of hope for progress in the study of human heredity. The technical analysis of the karyotypes and practices that accompanied it became popular worldwide as it allowed human chromosomes to be analyzed and visualized.[4]

De Garay had come into contact with these techniques in England. Upon his return, he brought the first human chromosome preparations that have been studied in the Galton laboratory and donated by Penrose. These studies also led to the standardization of laboratory tissue culture methods, and in general, those practices enabling visualization of chromosomes. To do so, the PGR was assisted by Dr. David A. Hungerford, the American scientist and co-discoverer of the chromosomal abnormalities in cancer cells who worked at the Institute for Cancer Research in Philadelphia, United States. He came to Mexico in 1963 to work on the protocols for the tissue culture from peripheral blood technique that he and his colleagues had developed in the United States, which was originally designed for the study of chromosomes in patients with cancer and in laboratory animals with radiation-induced leukemia. As a result of this collaboration, Hungerford agreed to receive in Philadelphia, PGR technicians Tayde García and Pilar Quijano to specialize in tissue culture and other cytological techniques that had been standardized at Hungerford's laboratory, who upon their return were incorporated to the cytogenetic section of the PGR.

As part of his personal agenda and in response to national needs, de Garay encouraged young people to learn cytogenetic techniques that were being standardized in other laboratories, always with the support of the IAEA through the CNEN. Maria Cristina Cortinas de Nava went to the Hospital des Enfants Malades to learn techniques of human cytogenetics with the French geneticist Jean de Grouchy's group in France (De Grouchy, Cortinas de Nava, & Bilski-Pasquier, 1965); and Maria Teresa Zenses Eisenbach left for Texas University to work with Chinese geneticist T. C. Hsu the techniques for human and animal cytogenetics.

4.2 The study of the effects of radiation on the hereditary material

After WWII, the international call from institutions such as the IAEA, to study the effects of radiation on humans, led to the establishment of local programs, such as the PGR, where de Garay introduced the study on the effects of radiation in the hereditary

---

[4] Another research group was being formed in the 1960s at the Mexican Institute of Social Security. See Barahona (2009, 2015).

material. According to de Garay, the increasing uses of nuclear energy in electricity production and medical applications, agriculture and industry, forced him 'to increase research and concentrate scientific, technical and legal efforts intended to prevent any damage that may arise from nuclear development' (de Garay, 1962). After that, new prospects for the peaceful application of nuclear energy and research in general opened up gradually within the PGR.

The *Drosophila* section studied the mutagenic effects of diverse radiation-emitting sources by using tracer isotopes. As part of the staff training, annual courses about radioisotopes and nuclear instrumentation were organized in order to obtain and analyze these mutants. These were offered in the Radioisotope Laboratory at the Physics Department of the School of Sciences of the UNAM in close collaboration with the CNEN, by himself and Augusto Moreno y Moreno, whom he had met at the IAEA meeting in Vienna in 1957. The first course was held from March-July in 1960. These courses works were compulsory for the personnel who worked in the Program (Olvera & Guzmán, 2001). By 1962 de Garay and collaborators had obtained 57 *Drosophila* mutants that were used for research and educational purposes.

In the ensuing years, and to be competitive on a worldwide scale, the PGR included the Aquatic Invertebrate Radiobiology section and the Molecular Genetics Laboratory in 1965 and 1966. The idea was to work, respectively, on protective compounds against radiation effects in the flatworm *Planaria*, and to investigate the genetic mechanisms operating in microorganisms, as well as the radiation effects and consequences of radioprotective substances at the molecular level. In parallel, de Garay encouraged Magadalena Carrillo Santín to go to the Atomic Energy Institute at Sao Paulo, Brazil, to study radiochemistry in 1965, and he invited the then-Director of the Oak Ridge National Laboratory of the US, Alexander Hollaender, whom he had met in 1957 at Oak Ridge, to see the facilities and to work on future collaborations in 1966.

4.3 Population genetics studies

Although population genetics had been developed in the first decades of the 20th Century, it was in the post-war years that serological studies of the distribution of blood groups were developed with the intention of measuring the intra-specific variability in human populations. This was possible thanks to the introduction of new techniques such as gel electrophoresis and paper chromatography, which were quickly built to track immune reactions, mainly thanks to the work of Harry Harris at the Galton Laboratory (and whom de Garay had met in England) with human isoenzymes.

In Mexico, human population studies started in the 1940s with the work of physician Mario Salazar Mallén and his students at the General Hospital on the detection of hemoglobin abnormalities in rural communities. Between 1944 and 1949 he published several papers on the blood agglutinogens of the Mexicans, and blood type surveys of urban and indigenous populations (Salazar-Mallén, 1949; Barahona, 2009). This work was followed by Adolfo Karl at the IPN, who studied the distribution of abnormal hemoglobin in a Mazatecan group of the Papaloapan basin using the horizontal electrophoresis technique (Karl, 1957). At that time, the Mexican populations were investigated for ethnographic, anthropological or economic purposes, but rarely from the genetic point of view, so these studies were important in the Mexican local context.

These studies, although aligned with the work of other laboratories in other parts of the world, did not impact Mexican genetics, until other groups such as those of de Garay and Mexican physician Rubén Lisker began to use molecular tracers and more up-to-date electrophoresis techniques to measure the genetic variability of Mexican indigenous populations.[5]

To start these kinds of investigations at the PGR, de Garay invited the physician and biologist Hans Kalmus, the prewar refugee from Czechoslovakia, whom he had met in the Galton Laboratory, to design the study of population genetics of *Drosophila* and Mexican indigenous populations.

Kalmus' first visit thanks to the support of the IAEA was in 1962, but only for a three-month period; these visits became more common and longer in the ensuing years. Under his supervisión, gathering expeditions were made to Chiapas and Oaxaca to collect data on *Drosophila*. Similarly, thanks to this collaboration they began the study of population genetics in indigenous populations using tracer genes of Tzeltal and Tzotzil groups in Chiapas and Mixtec groups near the coast of Oaxaca. Later on, the Lacandon populations in Chiapas and the Otomí group in Hidalgo were investigated. The results of these studies showed that in some communities, the frequencies of certain genetic characteristics increased by prolonged isolation in a small geographical area (de Garay, 1963; Barahona, 2009).

Finger and palm prints were also taken, and tests were performed for phenylthiocarbamide tasting, enzyme deficiencies in the school population, color blindness and the chemical composition of earwax (Kalmus, de Garay, Rodarte, & Cobo, 1964). In order to obtain blood samples and analyze them later in the laboratory, several gathering expeditions were necessary owing to the geographical conditions of the regions studied and the intervention of other Mexican institutions such as the National Indigenist Institute and the Summer Institute of Linguistics.

In order to strengthen this line of research, de Garay arranged for Víctor Salceda to be a visiting scholar at the Rockefeller University in New York, at the laboratory of the Russian-born American geneticist and evolutionist Theodosius Dobzhansky to study genetic load in irradiated populations. This occurred from 1965 to 1967, and enabled the PGR to collaborate with Dobzhansky, who visited Mexico in 1974 at the invitation of de Garay.

---

[5] In 1960, another research group headed by Mexican physician Rubén Lisker at the Nutrition Diseases Hospital (later the Salvador Zubirán National Institute of Nutrition) was working on the genetic characterization of Mexican indigenous populations. See Barahona (2009).

The IAEA provided Salceda with a scholarship to go to New York, and he joined Dobzhansky's laboratory in November 1965 to do research on the genetic load of irradiated *Drosophila* flies. When Salceda went back to Mexico in 1967, he tried to initiate collections of *D. pseudoobscura* in natural populations under Dobzhansky's advice, in order to study the geographical distribution of the chromosomal inversions that characterize this species. Because of Dobzhansky's illness and because de Garay's group was involved in the organization of the Olympic Games held in Mexico in 1968, this project did't take off at the time (Barahona & Ayala, 2005b, 2005a).

The context and the conditions in which this research was performed, the commencement of international collaborations, and the role of de Garay and his group in the circulation of knowledge were key pieces in the development and establishment of genetics in Mexico and its positioning at an international level.

5 Conclusions

As Turchetti and colleagues have said,

> in recent years historians of science have focused on the production of scientific knowledge in its social milieu, shedding new light on such key determinants as (among others) the transmission of new paradigms through the training of new generations, the orienation of belief in trading zones, the functioning of laboratories and the circulation of knowledge....They also have challenged narratives built around the celebration of scientific discoveries by presenti the production of knowledge as a complex and dynamic process in which the meaning of new scientific theories are negotiated by a number of different actors (Turchetti et al., 2012).

In this work, the idea of circulation allowed us to understand in greater detail how it is that knowledge and scientific practices have travelled across geographical and temporal spaces, crossing nations and passing through borders. This paper attempted to show how the PGR was created from a simultaneous dialog between the local context responding to national needs and concerns, and a transnational approach owing to its need for international networks through which scientific resources were mobilized in order to enter into a transnational material culture. From the start, de Garay was as committed to the construction of shared languages and practices as he was to networks of collaboration in order to guarantee the necessary conditions for establishing genetics in Mexico. This study also allowed us to place Mexican science within a global context in which interconnected narratives describe the interplay between global trends and national contexts.

I have tried to show how the creation of the CNEN in its local and international contexts, influenced the foundation of the PGR. The training of Alfonso León de Garay at the Galton Laboratory of University College in London was crucial to the establishment of the PGR. The members of the program were important pieces in the movement of knowledge across borders due to their standing as experts, which was thanks to the moral and epistemic authority they had acquired by studying abroad, and through their personal relationships. I tried to show how his subsequent membership of international networks facilitated the flow of knowledge and the circulation of scientific practices and people, which included the training of young geneticists in foreign academic institutions, thus enabling spaces to be opened for the development of genetics in Mexico. This historical reconstruction tried to show the relations between different actors in different countries and at different times, and the scientific programs and institutions that were created in the 1950s and 1960s in Mexico.

Acknowledgments. I would like to thank M.C. Alicia Villela González for expert research assistance; to Nuria Gutiérrez and Marco Ornelas, for their help and patience during this investigation; to David Bevis for his careful copyediting. I would also like to thank Hannes Leitgeb, Ilkka Niiniluoto, Elliot Sober, and Päivi Seppälä for putting this volume together. This research was supported by project INTEGRA CONACyT and the Bioethics University Program of the UNAM.

Bibliography

Barahona, A. (2009). *Historia de la genética humana en México. 1971-1970*. México: UNAM.
Barahona, A. (2015). Transnational science and collaborative networks. The case of genetics and radiobiology in Mexico. *Dynamis, 35*, 333–358.
Barahona, A., & Ayala, F. J. (2005a). History of genetics in México. *Nature Reviews / Genetics, 6*(11), 860–866.
Barahona, A., & Ayala, F. J. (2005b). The role played by Theodosius Dobzhansky in the emergence and institutionalization of genetics in Mexico. *Genetics, 170*, 981–987.
Barahona, A., Pinar, S., & Ayala, F. J. (2003). *La genética en México. Institucionalización de una disciplina*. México: Coordinación de Humanidades, UNAM.
Basalla, G. (1967). The spread of Western science. *Science, 156*, 611–622.
Birn, A.-E., & Necochea López, R. (2011). Footprints on the future: looking forward to the history of health and medicine in Latin America in the Twenty-First Century. *Hispanic American Historical Review, 91*, 503–527.
Bulbulian, S., & Rivero-Espejel, I. A. (2012). Historia de la radioactividad en el Instituto de Investigaciones Nucleares. *Boletín de la Sociedad Química de México, 6*(1), 15–26.
Cambrosio, A., Keatins, P., & Bourret, P. (2006). Objetividad regulatoria y sistemas de pruebas en medicina: El caso de la cancerología. *Convergencia, 13*(42), 135–152.
Cañizares, J. (2000). *Nature, empire, and nation: Explorations of the history of science in the Iberian World*. Stanford, CA: Stanford University Press.
Cardoso, F. H., & Faletto, E. (1979). *Dependency and development in Latin America*. Berkeley, CA: University of California Press.
Chambers, D. W. (1987). Period and process in colonial and national science. In N. Reingold, & M. Rotherberg (Eds.), *Scientific colonialism: A cross cultural comparison* (pp. 297–391). Washington: Smithsonian Institution Press.
Collins, H. (1985). *Changing order: Replication and induction in scientific practice*. Beverly Hills: Sage.
Creager, A. N. H., & Santesmases, M. J. (2006). Radiobiology in the Atomic Age: Changing research practices and politics in comparative perspective. *Journal of the History of Biology, 39*, 637–647.
Cueto, M. (2006). Excellence in twentieth century biomedical science. In J. J. Saldaña (Ed.), *Science in Latin America* (pp. 231–240). Austin: University of Texas Press.
Cueto, M. (2007). *Cold War, deadly fevers: Malaria eradication in Mexico, 1955-1975*. Washington, DC / Baltimore: Woodrow Wilson Center / Johns Hopkins University Press.
Daston, L., & Galison, P. (1992). The image of objectivity. *Representations, 40*, 81–128.
Daston, L., & Galison, P. (2007). *Objectivity*. Brooklyn, NY: Zone Books.

de Garay, A. L. (1960). *Programa de genética y radiobiología. Informe de labores 1960*. México: Comisión Nacional de Energía Nuclear, Archivo de Información, Biblioteca ININ.
de Garay, A. L. (1962). *Programa de genética y radiobiología. Informe de labores 1962*. México: Comisión Nacional de Energía Nuclear, Archivo de Información, Biblioteca ININ.
de Garay, A. L. (1963). *Programa de genética y radiobiología. Informe de labores 1963*. México: Comisión Nacional de Energía Nuclear, Archivo de Información, Biblioteca ININ.
de Garay, A. L. (1970). *Programa de genética y radiobiología. Informe de labores 1970*. México: Comisión Nacional de Energía Nuclear, Archivo de Información, Biblioteca ININ.
De Grouchy, J., Cortinas de Nava, C., & Bilski-Pasquier, G. (1965). Duplication d'un Ph1 et suggestion d'une evolution clonale dans une leucémie myéloïde chronique en transformation aiguë. *Nouvelle Revue Francaise d'Hematologie, 5*(69).
Domínguez-Martínez, R. (2012). Los orígenes de la física nuclear en México. *Revista CTS,* (7), 95–112.
Druglitrø, T., & Kirk, R. G. W. (2014). Building transnational bodies: Norway and the international development of laboratory animal science, ca. 1956-1980. *Science in Context, 27*(2), 177–186.
Eisenhower, D. D. (2007). Atomic power of peace. In J. Pilat (Ed.), *Atoms for peace. A future after fifty years?* (pp. 239–246). Washington, DC: Woodrow Wilson Center Press. (Original work published 1953)
Finnegan, D. A. (2008). The spatial turn: Geographical approaches in the history of science. *Journal of the History of Biology, 41*, 369–388.
Golinsky, J. (1998). *Making natural knowledge. Constructivism and the history of science.* Cambridge: Cambridge University Press.
Hochman, G. (2008). From autonomy to partial alignment: National malaria programs in the time of global eradication, Brazil, 1941-1961. *Canadian Bulletin of Medical History, 25*(1), 161–192.
Hofmeyr, I. (2013). African history and global studies: A view from South Africa. *The Journal of African History, 54*, 341–349.
Kalmus, H., de Garay, A. L., Rodarte, U., & Cobo, L. (1964). The frequency of PTC tasting, hard ear wax, color blindness and other genetical characters in urban and rural Mexican populations. *Human Biology, 36*(134).
Karl, A. (1957). Estudio electroforético de la hemoglobina de los indígenas mazatecos de la cuenca del Papaloapan. *Ciencia, 17*, 85–86.
Kohler, R. E. (2012). Practice and place in Twentieth-Century field biology: A comment. *Journal of the History of Biology, 45*, 579–586.
Kreimer, P. (2010). *Ciencia y periferia. Nacimiento, muerte y resurrección de la biología molecular en la Argentina: aspectos sociales, políticos y cognitivos.* Buenos Aires, AR: Eudeba.
Lafuente, A., & Sala-Catalá, J. (1989). Ciencia colonial y roles profesionales en la América Española del Siglo XVIII. *Quipu, 6*, 387–403.

Latour, B. (1988). *The Pasteurization of France*. Cambridge, MA: Harvard Univeristy Press.
Latour, B., & Woolgar, S. (1979). *Laboratory life: The social construction of scientific facts*. Bevery Hills: Sage.
Medina, E. (2011). *Cybernetic revolutionaries: Technology and politics in Allende's Chile*. Cambridge, MA: The MIT Press.
Olvera, O., & Guzmán, J. (2001). Interview with author, Laboratorio de Genética, ININ. Carretera México-Toluca, Km. 36.5, Salazar, estado de México.
Porter, T. M. (1995). *Trust in numbers: The pursuit of objectivity in science and public life*. Princeton, NJ: Princeton University Press.
Rudwick, M. J. S. (1984). *The great Devonian controversy. The shaping of scientific knowledge among gentlemanly specialists*. Chicago: The University of Chicago Press.
Safier, N. (2010). Global knowledge on the move. Itineraries, Amerindian narratives, and deep histories of science. *ISIS, 101*, 133–145.
Sagasti, F., & Guerrero, M. (1974). *El desarrollo científico y tecnológico en América Latina*. Buenos Aires, AR: Instituto para la Integración de América Latina.
Salazar-Mallén, M. (1949). El aglutinógeno Lewis en la sangre de los mexicanos. *Boletín del Instituto Médico Biológico, 7*, 25–30.
Secord, J. A. (2004). Knowledge in transit. *ISIS, 94*, 654–672.
Shapin, S. (1994). *A social history of truth: Civility and science in Seventeenth-Century England*. Chicago: University of Chicago Press.
Shapin, S., & Schaffer, S. (1985). *Leviathan and the air pump: Hobbes, Boyle and experimental life*. Princeton: Princeton University Press.
Sivasundaram, S. (2010). Sciences and the global: On methods, questions and theory. *ISIS, 101*, 146–158.
Soto Laveaga, G. (2009). *Jungle laboratories; Mexican peasants, national projects, and the making of the Pill*. Durham: Duke University Press.
Stepan, N. (1981). *Beginnings of Brazilian science: Oswaldo Cruz, medical research and policy, 1890-1920*. New York: Science History Publications.
Subrahmanyam, S. (1997). Connected histories: Notes towards a reconfiguration of early modern Eurasia. *Modern Asian Studies, 31*, 735–762.
Turchetti, S., Herrán, N., & Boudia, S. (2012). Introduction: Have we ever been 'transnational'? Towards a history of science across and beyond borders. *The British Journal for the History of Science, 45*(3), 319–336.
Vélez Ocón, C. (1997). *Cincuenta años de energía nuclear en México 1945-1995*. México: Programa Universitario de Energiá, UNAM.
Villalobos-Pietrini, R., Guzmán, J., & Levine, L. (2005). Homenaje a Alfonso León de Garay. *Revista Internacional de Contaminacíon Ambiental, 21*, 5–1.
van der Vleuten, E. (2008). Towards a transnational history of technology: Meanings, promises, pitfalls. *Technology and Culture, 49*(4), 974–994.

Author biography. Ana Barahona studied biology and graduate studies at the National Autonomous University of Mexico (UNAM). She is Full Time Professor in the Department of Evolutionary Biology of the School of Sciences at the UNAM. She did posdoctoral studies at the University of California, Irvine, with Francisco J. Ayala, and has done research stays at Harvard University with Professor Everett Mendelsohn and Janet Browne, at the American Philosophical Society, at the Max Planck Institute for the History of Science, and at the Florida State University. Pioneer in the science and technology studies (STS) since 1980, she founded the area of Social Studies of Science and Technology in the School of Sciences. She was President (2009-2011) of the *International Society for the History, Philosophy and Social Studies of Biology*, member of the International Committee of the *Scientific Society Sigma Xi* (2003-2005), and Council Member of the Division of History of Science and Technology of the International Union of History and Philosophy of Science (DHST/IUHPS, 2009-2013). She has been recently acknowledged as a Corresponding Member of the International Academy for the History of Science.

14 Comparing causes – an information-theoretic approach to specificity, proportionality and stability

ARNAUD POCHEVILLE [*], PAUL E. GRIFFITHS [†] AND KAROLA STOTZ [‡]

Abstract. It would be useful if the interventionist account of causation, in addition to distinguishing causes from non-causes, could define various desirable properties of causal relationships. Amongst these properties are specificity, proportionality and stability. In earlier work we offered an information theoretic analysis of causal specificity, using an approach which parallels existing work in complex systems science. Here we extend this approach to proportionality and stability. First, we show that the interventionist criterion of causation, 'minimal invariance', is formally equivalent to non-zero specificity. We then show that there are natural, information theoretic ways to explicate the distinction between potential and actual causal influence. With these foundations in place we show that there is a natural information-theoretic approach to describing causal variables that explicates the idea that causes should be proportional to their effects. Then we draw a clear distinction between two ideas in the existing literature, the range of invariance of a causal relationship and its stability. Range of invariance is simply specificity. Stability concerns the effect of additional variables on the relationship between some focal pair of cause and effect variables. We show

---

[*] Department of Philosophy and Charles Perkins Centre, The University of Sydney, NSW2006, Australia
[†] Department of Philosophy and Charles Perkins Centre, The University of Sydney, NSW2006, Australia
[‡] Department of Philosophy, Macquarie University, NSW 2109, Australia

that in an information theoretic framework there is an important distinction between the extent to which these additional variables influence the effect and the extent to which they influence the relationship between the focal cause and effect variable. We show how to measure the influence of additional variables in both these respects. The overall result of this work is to provide precise explications of a whole family of intuitive notions associated with the interactionist account of causation. In principle, these properties can now all be measured on a causal graph. The information theoretic approach has substantial technical limitations, however, as we discuss towards the end of the paper. The real value of our work lies as much in the way it reveals the ambiguity and equivocation in earlier, qualitative discussions as in the actual measures we construct.

Keywords: causality, intervention, invariance, specificity, stability.

1 Invariance and causal explanation

The interventionist approach to causal explanation is based on the insight that "causal relationships are relationships that are potentially exploitable for purposes of manipulation and control" (Woodward, 2010, p. 314). Interventionists approach causation via the relationships between the variables that characterise an organised system. These relationships can be represented by an acyclic directed graph. In such a graph, variable C is a cause of variable E when a suitably isolated manipulation of C would change the value of E. With suitable restrictions on the idea of 'manipulation' this test provides a criterion of causation, distinguishing causal relationships between variables from merely correlational relationships (Woodward, 2003, pp. 94–107).

The interventionist account only applies to 'change-relating' generalisations, where at least one intervention upon C will produce some change in E. Generalisations which are not change-relating are not candidates to provide causal explanations. Non-change-relating generalizations may state the impossibility of certain affairs: nothing can be accelerated past the speed of light. Or they may relate an outcome to a reliable but irrelevant antecedent: men who take birth control pills will never become pregnant (Woodward 2000, 206f).

Change-relating generalisations provide causal explanations in virtue of being invariant under interventions rather than because they hold widely in nature, or have nomological force as traditionally conceived (Woodward, 2003, p. 16):

> [E]xplanation has to do with the exhibition of patterns of counterfactual dependence describing how the systems whose behavior we wish to explain would change under various conditions. ... Explanatory generalizations allow us to answer what-if-things-had-been different questions: they show us what the value of the explanandum variable depends upon. (Hitchcock & Woodward, 2003, pp. 182–183)

Invariance under intervention simply means that the relationship between variables C and E continues to hold when interventions are made on C.

> I will say that a generalization is invariant simpliciter if and only if (i) the notion of an intervention is applicable to or well-defined in connection with the variables figuring in the generalization (...) and (ii) the generalization is invariant under at least some interventions on such variables. ... To count as invariant it is not required that a generalization be invariant under all interventions. (Woodward, 2000, p. 206)

The idea of invariance is sometimes expressed in terms of the 'stability' of the generalization:

> A generalization is invariant if (i) it is ... change-relating and (ii) it is stable or robust in the sense that it would continue to hold under a special sort of change called an intervention. (Woodward, 2000, p. 198)

However, as we will shortly see, it is more convenient to reserve the term 'stability' for a different idea associated with the interventionist account.

Woodward makes a clear distinction between the actual criterion of causation and various desirable properties of causal relationships. The criterion of causation is 'minimal invariance' – invariance in the face of at least one possible intervention. A wider range of invariance is a desirable property of causal relationships: a relationship that holds for more values of C and E is a more powerful means of intervention. However, while a minimally invariant relationship may be less useful, it is not less causal.

'Specificity' is another desirable property of causal relationships. The intuitive idea behind specificity is that interventions on C can be used to produce any one of a large number of values of E, providing what Woodward terms "fine-grained influence" over the effect variable (Woodward, 2010, p. 302).

'Proportionality' is a further desirable feature of causal relationships, or, more accurately, of how causal relationships are described:

> ... causal description/explanation can be either inappropriately broad or general, including irrelevant detail, or overly narrow, failing to include relevant detail. (Woodward, 2010, pp. 296–7)

Woodward provides several striking example where a causal explanation is weakened because the choice of variables suffers from one of these vices. Saying that one person went bungy-jumping whilst another did not because only one has a 'gene for bungyjumping' is less explanatory than saying that only one has a gene associated with risk-seeking behavior. The former explanation excludes important information that the latter provides.

'Stability' is a final desirable property of causal relationships. Whilst invariance concerns the relationship between C and E, stability concerns the relationship between other variables and that relationship. Intuitively, C is a stable cause of E if it continues to cause E across some range of values of other variables Z, W, etc. These other variables are sometimes referred to as 'background' variables. There is much more to be said (and settled) about stability and its relationship to invariance, as we will see below.

In earlier work with other collaborators we have developed an information-theoretic approach to measuring the specificity of causal relationships within the interventionist framework (Griffiths et al., 2015). In this paper we extend that approach to (1) explore the relationship between invariance and specificity, (2) distinguish between potential and actual causal influence, (3) explicate the idea of proportionality, (4) distinguish invariance from stability, (5) draw a further distinction between the stability of an effect the stability of the relationship between cause and effect, and (6) show how to measure both forms of stability. We conclude by discussing the limitations of an information-theoretic approach.

2 Minimal invariance and specificity

In earlier work we noted that specificity is not entirely independent of the criterion of causation (Griffiths et al., 2015). This is a straightforward consequence of our measure of specificity, which formalises the idea that, other things being equal, the more a cause specifies a given effect, the more knowing the value set for the cause variable will inform us about the value of the effect variable. This idea led us to propose a simple measure:

> Spec: the specificity of a causal variable is obtained by measuring how much mutual information interventions on that variable carry about the effect variable.[1]

The mutual information of two variables is simply the redundant information present in both variables. Where $H(X)$ is the entropy of X (see Appendix), the mutual information of X with another variable Y, or $I(X;Y)$, is given by:

$$I(X;Y) = H(X) - H(X|Y)$$

Mutual information is not in itself a suitable measure of causal influence. It is symmetrical, that is $I(X;Y) = I(Y;X)$, and variables can share mutual information without being related in the manner required by the interventionist criterion of causation. However, any variables that satisfy the interventionist criterion of causation will show some degree of mutual information between *interventions* and effects. If $C \to E$ is minimally invariant, that is, invariant under at least one intervention on C, then $I(do(C);E) > 0$, where $do(C)$ means that the value of C results from an intervention

---

[1]This measure has been independently proposed in cognitive sciences by Tononi et al. (1999) and in computational sciences by Korb et al. (2009). For related measures see also Ay & Polani (2008), Janzing et al. (2013). Ay and Polani's measure captures what we call SAD below.

on C (Pearl, 2009). To simplify writing, we will from now on represent the $do(\ )$ operator by a hat on the variable: $do(X) \equiv \widehat{X}$.[2] So our measure of specificity does not simply measure the mutual information between variables C and E. Instead, it measures the mutual information between interventions on the variable C and the variable E. This is not a symmetrical measure because the fact that interventions on C change E does not imply that interventions on E will change C: in general, $I(\widehat{C};E) \neq I(\widehat{E};C)$.[3] Furthermore, if any pair of variables $\{C,E\}$ satisfies the interventionist criterion of causation, with C being a cause of E, then C will have some degree of specificity for E. So minimal invariance is equivalent to non-zero specificity.

This raises the obvious further question of how the *degree* of specificity of a causal relationship relates to its *range* of invariance – the range of values of the variables across which a causal relationship holds. Marcel Weber has argued in qualitative terms that the degree of specificity is just the same thing as its range of invariance (Weber, 2006). Woodward questioned Weber's proposed equivalence because a causal relationship might hold across a large range of invariance but fail to be bijective, and thus to offer the sort of fine-grained control associated with the idea of specificity: "a functional relationship might be invariant and involve discrete variables but not be 1–1 [injective] or onto [surjective] " – that is, it might fail to be bijective (2010, p. 305 fn 17). In our earlier paper we argued that measuring the mutual information between two variables is a good way to formalize Woodward's idea that the mapping between the cause and effect may 'approximate a bijection'. We then showed that with a slight correction corresponding to Woodward's caveat, Weber is correct. He is correct because the mutual information between cause and effect variables will typically be greater when these variables have more values, simply because the entropy of both variables is higher. Woodward's caveat corresponds to the fact that it is not enough to increase the number of values of a cause variable unless the additional values of the cause map onto distinct values of the effect. Our measure of specificity captures both points. Increasing the entropy of the cause variable will not increase mutual information when no additional entropy in the effect variable is captured. We can see this by contrasting the cases in Figures 14.1 and 14.2.

In fact, we would argue, it does not really make sense to say that the relationship $C \to E$ in Figure 14.2 has a greater range of invariance than $C \to E$ in Figure 14.1. The variable C merely has a large number of nominal values. The appropriate way to divide a causal variable into discrete states for the purposes of an interventionist account of causal explanation is to group together states that make the same difference. A description of the variable that does not respect this constraint is effectively

---

[2] We take this convention from related work in computer sciences applying information theory to causal modeling (see fn above, Ay & Polani, 2008; Lizier & Prokopenko, 2010).

[3] These quantities can be equal if and only if the two variables are not causally connected. Indeed, at least one of these quantities is null since C and E are variables in a causal graph: if C causes E, E can't feed back on C (causal graphs are acyclic, see Pearl (2009)). Thus the two quantities can be equal if and only if they are both null.

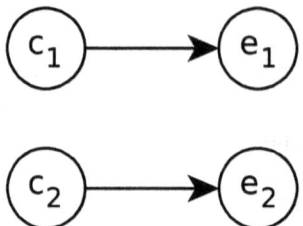

Figure 14.1: Causal mapping showing a bijection between causal values and effect values. Complete ignorance (maximum entropy) obtains when each value of the effect has a probability of $\frac{1}{2}$ before intervening on the value of the cause: $H(E) = -\sum_{j=1}^{2} p(e_j) \log_2 p(e_j) = -\sum_{j=1}^{2} \frac{1}{2} \log_2(\frac{1}{2}) = 1$ bit. After knowing the value set for the cause (c_1 or c_2), the effect is fully specified and the conditional entropy is: $H(E|\widehat{C}) = -\sum_{i=1}^{2} p(\widehat{c}_i) \sum_{j=1}^{2} p(e_j|\widehat{c}_i) \log_2 p(e_j|\widehat{c}_i) = -\sum_{i=1}^{2} \frac{1}{2} \sum_{j=1}^{2} 1 \log_2(1) = 0$ bit. The information gained by knowing the cause can be obtained by measuring the difference between the entropy before and the entropy after intervening to set the value of the cause. This quantity is the mutual information between E and $\widehat{C}$: $I(E;\widehat{C}) = H(E) - H(E|\widehat{C}) = 1$ bit.

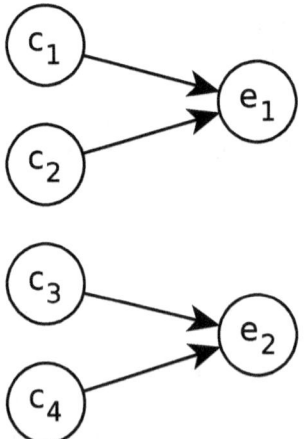

Figure 14.2: Here, different values of the cause lead to the same outcome. As in Figure 14.1, $H(E) = 1$ bit. Although here two values of the cause can lead to the same effect, intervening to set the value of the cause fully specifies the value of the effect just as effectively as it does in Figure 14.1. Therefore, the difference in uncertainty about the effect between before and after intervening to set the value of the cause is the same: $I(E;\widehat{C}) = H(E) - H(E|\widehat{C}) = 1 - 0 = 1$ bit.

a gerrymandered description, as we discuss in Section 4. So, in line with Weber's proposal, the range of invariance of a causal relationship is simply the specificity of that relationship (assuming for simplicity an equal weighting of values).

In this section we have argued that both the interventionist criterion of causation – minimal invariance – and the desirable property of having a greater range of invariance can be assessed using our measure of specificity (Spec). The information-theoretic framework we have adopted allows a quantitative formulation of these two key elements of the interventionist framework. However, the information theoretic approach requires us to specify a probability distribution over the cause, something that earlier, qualitative discussions seemed to be able to do without.

Our measure of specificity (Spec) depends on what probability distribution we choose to impose on the the causal variable C, as well as on the mapping from C to E. In our earlier paper we showed that this is very much a feature and not a bug of our measure. As we will now discuss, specificity measured with different distributions over C corresponds to different properties that are of interest to the philosophy of causation (Griffiths et al., 2015).

3 Actual and potential difference-making

Measuring Spec with different probability distributions over $\widehat{C}$ corresponds to different views of causal specificity in the existing, qualitative literature.

One way to measure specificity corresponds to Woodward's (2010) characterisation of fine-grained influence (INF). In his presentation the value of C depends only on interventions by an idealised agent. Since the aim is to characterise how one variable causally depends on another, we assume that this agent does not favour one value over another, so that every value is equiprobable. The distribution of values of C is therefore the maximum entropy distribution:

INF: $I(\widehat{C};E)$, where the distribution of $\widehat{C}$ has maximum entropy.

In our earlier work we suggested that INF was a good measure of the potential of C to causally affect E (Griffiths et al., 2015). However, another non-arbitrary choice is to construct a distribution which maximizes specificity. Such a distribution does not necessarily maximize the entropy of the cause variable (see Korb, Hope, & Nyberg, 2009).

MaxSpec: $I(\widehat{C};E)$, where the distribution of $\widehat{C}$ maximises Spec.

One formal advantage of MaxSpec is that it is insensitive to finer redescription of the variables. MaxSpec is unaffected if we divide C or E into a greater number of nominal values.

Whereas INF measures how much influence C exerts on E in an unbiased set of intervention experiments, MaxSpec measures how much influence C exerts on E under ideal conditions. This is the 'causal power' of C with respect to E (Korb et al., 2009) and can also be thought of as a measure of C's *potential* influence on E. We are inclined to think MaxSpec a better explication than INF of the intuitive idea that a system has an intrinsic causal structure and that this structure is independent of how the system operates on some particular occasion. Understanding causal connections in this sense is a central aim of science – seeking to understand how the parts of a mechanism are connected to one another, rather than how often each connection is used or whether they are used on a particular occasion.

A different view of causal specificity has been advocated by Kenneth Waters (2007). Waters draws attention to contexts in which scientists are only interested in the actual causes of differences in some population, situations in which, he argues, they seek to characterise the causes which are 'specific actual difference makers' in that population (SADs). In earlier work we argued that this amounts to measuring *Spec* when C takes the distribution it has in the actual population. Although Water's stresses the *observed* distribution of properties in a population, his discussion makes it clear that he intends SAD to be a conception of causation, not merely of correlation, so rather than measuring the mutual information between the actual distributions of C and E, we need to imagine a set of interventions that create the same distribution of values of C that we see in the population, hence:

SAD: $I(\widehat{C};E)$ where the distribution of $\widehat{C}$ is identical to the actual distribution of C in some population.[4]

If we take Spec to correspond to SAD rather than MaxSpec we get a rather different picture of causation from the one we sketched in Section 2. First, it will no longer be true that all causes have some degree of specificity for their effects. It may simply be that the range of C within which there is a relationship between C and E does not occur in the population from which we derive the distribution of C. In other words, the 'experiment of nature' does not include the experiment that reveals how E depends on C. For the same reason, under the SAD interpretation, the degree of specificity of C for E may not correspond to the range of invariance of the relationship $C \rightarrow E$.

We interpret SAD as a measure of a complementary idea to potential causal influence, namely actual causal influence – how much difference a cause *actually* makes to an effect. For example, in a causal graph representing a firing squad, the potential causal influence of the variable SHOOT with respect to the variable DIE, as measured by MaxSpec, will be greater than that of the variable SAY BOO, but SAY BOO will

---

[4] In addition, Marcel Weber (2013) has argued that in the biological sciences specificity should be assessed using a wider range of values of C than actually occur in any given population, but not all possible values of C. He suggests we should restrict ourselves to 'biologically normal' values of C. We interpret this to mean that C should be restricted to the range of variation that could be produced by known mechanisms operating on the timescale of whatever process we are trying to study. We have suggested that within that range, $\widehat{C}$ should conform to the maximum entropy distribution and named this additional flavour of specificity REL for relevant specificity (Griffiths et al., 2015). But it is also possible construct a version of relevant specificity based on the MaxSpec measure.

have greater actual causal influence on DIE than SHOOT does in a population where more prisoners die from fright than from bullets. The same idea has been termed 'information flow' (Ay & Polani, 2008). Ay and Polani explicitly conditionalise on a set of background variables S. We intervene to set S equal to what we observe in nature, derive a distribution for A and then ask what part of the correlation between A and B results from a causal relationship between them. To do this we need to import causal information derived from intervening on A, but the distribution over A whose effect we are assessing is observed, not imposed by intervention. Information flow is intended to measure the causal impact of variables on one another in a specific set of data, or what we have called actual causal influence (Lizier & Prokopenko, 2010).

In Section 2 we analysed the relationship between specificity and invariance. In this section we showed that specificity can be used to measure both potential and actual causal influence. In Section 4 we move on from our examination of specificity to examine a second desirable property of causal relationships, proportionality.

4 Proportionality

The proportionality of cause to effect is a matter of "whether the cause and effect are characterized in a way that contains irrelevant detail" (Woodward, 2010, p. 287) This idea has been discussed extensively in the philosophy of causation, where it has been explained via examples and qualitative characterisations:

> Yablo suggests that causes should "fit with" or be "proportional" to their effects—proportional in the sense that they should be just "enough" for their effects, neither omitting too much relevant detail nor containing too much irrelevant detail. (Woodward, 2010, p. 297)

In an effort to characterise the idea more precisely, Woodward has characterised it as a 'proportionality constraint' on the mapping between value of the cause and values of the effect.

> (P) There is a pattern of systematic counterfactual dependence (with the dependence understood along interventionist lines) between different possible states of the cause and the different possible states of the effect, where this pattern of dependence at least approximates to the following ideal: the dependence (and the associated characterization of the cause) should be such that (a) it explicitly or implicitly conveys accurate information about the conditions under which alternative states of the effect will be realized and (b) it conveys only such information – that is, the cause is not characterized in such a way that alternative states of it fail to be associated with changes in the effect. (Woodward, 2010, p. 298)

We stress that Woodward is not adding an additional condition to his criterion of causation. Like specificity, proportionality is meant to enrich the theory of causation by capturing why some causal facts may legitimately be of more interest to us than others, and thus may be highlighted in our explanations whilst other causal facts are omitted. Highly specific causes provide more precise control over an effect, and explain outcomes with greater precision. Proportional descriptions of causes provide us with all and only the information relevant to intervening or explaining with those causes.

We are now in a position to spell out the relationship between proportionality and specificity, and by doing so to define proportionality more precisely. If we choose a set of values for a causal variable, and a probability distribution over those values, which maximizes specificity, then, by definition, we cannot have omitted any relevant detail, since we have explained as much of the differences in the effect variable as possible. How can we make sure not to include any irrelevant detail? This is performed by minimizing the entropy of the cause variable by aggregating values which make the same difference, whilst maintaining its specificity: the less the entropy of the cause, the less information about the cause we have included in our explanation. Ideal proportionality is thus achieved when the cause is described in a way which minimizes its entropy and still maximizes specificity.

We can see how this works with Yablo's original example (1992, p. 4). A pigeon called Sophie has been trained to peck in response to any stimulus which is some shade of red. Yablo contrasts two explanations:

1. Sophie pecked because she was exposed to a red stimulus

2. Sophie pecked because she was exposed to a scarlet stimulus

Yablo suggests that 1 is a better causal explanation than 2. Like many philosophical thought experiments, this one is underspecified. We have two variables, P, with the values 'peck' and '~peck', and S. What values should S take? The combined probability of all values of a random variable must sum to one, so let us take the values of S to be the actual colour chips available in the laboratory, which neatly avoids the problem that birds do not perceive human spectral colours like red and scarlet. We stipulate that there are colour chips of more than one shade of red, and of some non-red shades. Finally, we stipulate that Sophie has been trained to peck at each of the colour chips that is a shade of red, giving us a causal graph in which P has the value 'peck' if and only if S has one of the values which is a shade of red.

We now construct the maximum specificity distribution, in this case making the combined weight of probablity on the red values equal to that on the non-red values. The graph we have described resembles that in Figure 14.2 above, and is exactly that graph if there are just two red and two non-red values. If we coarse-grain the values of our variable, so that S now has just two values, red and $\sim$ red, then we get the graph in Figure 14.1. S now has the same specificity as before, but the entropy of S has been reduced from 2 bits to 1 bit. This is is the optimally proportional way to divide the variable S into discrete values. No more specificity can be obtained by fine-graining and any further coarse-graining will reduce the specificity.

The artificiality of the example produces some problems. Whilst this is the optimal way to discretise the variable S for this single experiment with Sophie, it is not optimal for a wider experimental program! A better example of proportionality might be an experimentalist who sets her values for S to correspond to the distinctions in the pigeon's own tetrachromatic spectrum, since this would make S express only the 'differences that can make a difference' to the pigeon's behavior.

Woodward's other example of a failure of proportionality is taken from psychiatric geneticist Kenneth Kendler:

> To illustrate how this issue of the appropriateness of level of explanation may apply to our evaluation of the concept of "a gene for..." consider these two "thought experiments":
>
> Defects in gene X produce such profound mental retardation that affected individuals never develop speech. Is X is a gene for language?
>
> A research group has localized a gene that controls development of perfect pitch ... Assuming that individuals with perfect pitch tend to particularly appreciate the music of Mozart, should they declare that they have found a gene for liking Mozart?
>
> For the first scenario, the answer to the query is clearly "No." Although gene X is associated with an absence of language development, its phenotypic effects are best understood at the level of mental retardation, with muteness as a nonspecific consequence. X might be a "gene for" mental retardation but not language.
>
> Although the second scenario is subtler, if the causal pathway is truly gene variant → pitch perception → liking Mozart, then it is better science to conclude that this is a gene that influences pitch perception, one of the many effects of which might be to alter the pleasure of listening to Mozart. It is better science because it is more parsimonious (this gene is likely to have other effects such as influencing the pleasure of listening to Haydn, Beethoven, and Brahms) and because it has greater explanatory power. (Kendler 2005 p. 1249–50, quoted in Woodward 2010 p. 300–301)

The grain of description of the cause variable in these cases is fixed by the technology used to detect the genetic variant. The failure of proportionality is supposedly the result of describing the *effect* in too fine-grained a manner. But 'proportionality' here is not the same phenomena that we identified in the pigeon-pecking case, and nor is it really a matter of fine- versus coarse-graining. There are two alternatives to saying that the genetic variant is a gene for language or a gene for liking Mozart. The first is to say that it is a gene for an intervening variable, a variable which is linked to a host of behavioral and cognitive deficits in the former case or a host of musical preferences and abilities in the later. The second is to that say it is a pleiotropic gene with effects on many phenotypes. The first option corresponds to redrawing the causal graph by inserting an intervening variable, not to redescribing the effect variable. 'Perfect pitch' and 'Liking Mozart' do not stand to one another as variable and value but as cause and effect. The second alternative, pleitropy, also amounts to adding connections between the cause and additional variables, not coarse-graining the original variable: liking Mozart is not an instance of liking Haydn.

In these two later examples it is the causal graph itself that is too 'coarse grained' rather than one of the variables. That is consistent with Woodward's original characterisation of proportionality as 'neither omitting too much relevant detail nor containing too much irrelevant detail', but it reveals an ambiguity in that description. Choosing which variables to include in the graph and choosing how finely or coarsely to discretize a variable are clearly very different problems and it is better to keep them distinct. We therefore prefer to define 'Proportionality' more narrowly[5]:

> Proportionality constraint: Given an effect variable E that is a target of intervention or causal explanation, a causal variable C should be discretised so as to minimise the entropy of C whilst maximising specificity for E.

The main philosophical dispute in which the notion of proportionality has figured concerns whether lower-level, reductive explanations of phenomena are always superior to high-level explanations of the same phenomena. Carl Craver, for example, has argued that in some cases the lower-level explanation merely recognises additional differences that make no difference (e.g. Craver, 2007, Ch 6). Whether this argument is successful or not, our version of the proportionality constraint seems suitable to capture the intention behind it.

In this section we have added an account of proportionality to our earlier account of specificity.[6] In the remaining sections we tackle the concept of 'stability'.

5 Stability: Some clarifications

The interventionist account of causation aims to identify causes that "are likely to be more useful for many purposes associated with manipulation and control" (Woodward, 2010, p. 315). One aspect of this is the 'stability' of causal relationships.

> Among change-relating generalizations, it is useful to distinguish several sorts of changes that are relevant to the assessment of invariance. First, there are changes in the background conditions to the generalization. These are changes that affect other variables besides those that fig-

---

[5]The other issues being discussed under the heading of 'proportionality' seem to concern what statisticians call 'model selection'.

[6]Our discussion in this section is indebted to conversations with Jun Otsuka, Pierrick Bourrat and Brett Calcott

ure in the generalization itself. ... Second, there are changes in those variables that figure explicitly in the generalization itself... (Woodward, 2003, p. 248)

At this point some terminological stipulation is needed. We will reserve the term 'invariance' strictly for the properties of Woodward's "variables that figure explicitly in the generalization itself." Invariance characterizes the relationship between two variables, one of which can be used to intervene on the other. The invariance of that relationship is the range of values of those two, focal variables across which one can be used to intervene on the other. We showed above that the range of invariance can be captured by the degree of specificity. We will refrain from using the term 'stability' in connection with the relationship between those two focal variables. Instead, we use it strictly to describe how the relationship between those two variables is related to other variables. Stability is about of whether a causal relationship continues to hold across a range of background conditions.

Hitchcock and Woodward distinguish between two senses in which a causal generalization may be said to hold against a 'background' of other factors. In their first sense the 'background' to a causal generalization is simply everything not mentioned in the generalization. Most of the background, in this sense, is causally irrelevant. In their second sense, the 'background' consists of variables that are causally relevant to the effect but not explicitly represented in the model (Hitchcock & Woodward, 2003, p. 187). In our terms, causally relevant background conditions are additional variables that have some degree of specificity for the effect variable.

Sandra Mitchell has also made extensive use of something she calls 'stability' in an account of causal generalisations. Using 'invariance' in a broader sense, rather than in the restricted sense we have stipulated, she writes that:

> Stability for me is a measure of the range of conditions that are required for the relationship described by the law to hold, which I take to include the domain of Woodward's invariance. ... Stability does just the same work [as Woodward's invariance], however it is weaker and includes what might turn out to be correlations due to a non-direct causal relationship. But for there to be a distinction between stability and invariance, then we would have to already know the causal structure producing the correlation. (Mitchell, 2002, pp. 346–347))

Mitchell's 'stability' is a matter of whether a generalization holds across a range of values of other variables that are statistically relevant to the effect, either because they are causally relevant to it or due to confounding factors. Her treatment of stability is thus very different from Woodward's, and from ours. Mitchell's work is centrally concerned with complex systems for which there may be no practical way to reliably and fully document their causal structure. Hence she emphasises the scientific and pragmatic value of generalisations that are stable in her sense irrespective of what other, more stringent requirements they may satisfy. She also doubts the value, in her chosen context, of the distinction between the range of invariance of a relationship and its stability.

Despite the different foci of their work, there is real disagreement between Woodward and Mitchell about what distinguishes causally explanatory relationships between variables from mere correlations. Mitchell argues that causal generalisations are explanatory to the extent that they are stable. Woodward's criterion of causation was outlined above – causally explanatory generalisations need to be minimally invariant. Nothing more is needed to make them causally explanatory, and without this property no amount of stability in Mitchell's sense will make a generalization causally explanatory. The role of stability in Woodward's account is not to provide a criterion of causation, but to identify more *useful* causal relationships. Hitchcock and Woodward remark, using invariance in the same, wider sense as Mitchell, that,

> Invariance under changes in background conditions does not render a generalization explanatory; yet *greater* invariance [our stability] under changes in background conditions can render one generalization *more* explanatory than another. ... Briefly, if G is sensitive to changes in background conditions, that is because it has left out some variable(s) upon which the explanandum variable depends. (Hitchcock & Woodward, 2003, p. 187, italics in original)

As Hitchcock and Woodward emphasise, genuine background conditions are factors that could, and often should, be explicitly represented in a causal model:

> [C]laims about the invariance of a relationship under changes in *background conditions* are transformed into claims about invariance under interventions *on variables figuring in the relationship* through the device of explicitly incorporating additional variables into the relationship. (Hitchcock & Woodward, 2003, p. 188, italics in original)

One further distinction is needed to think clearly about the relationship between causal generalisations and background conditions. There are two different things with respect to which a variable may act as a (genuine) background condition. A variable may be a background condition with respect to the outcome of a causal process – the effect variable. Or it may be a background condition with respect to a causal relationship in the model. So if C is a cause of E, a third variable Z may affect the value of E, but it may also affect the way in which C is related to E. These two effects of Z are closely connected, and it may not be clear to the reader that they are distinct, but they are in fact importantly different (see Section 6 and Appendix).

6 Stability of causal relationships

When we speak of the 'stability' of some relationship $C \to E$ we often have in mind, not the influence of background variables on E, but whether the relationship $C \to E$ itself changes across a range of background conditions. For example, alternative

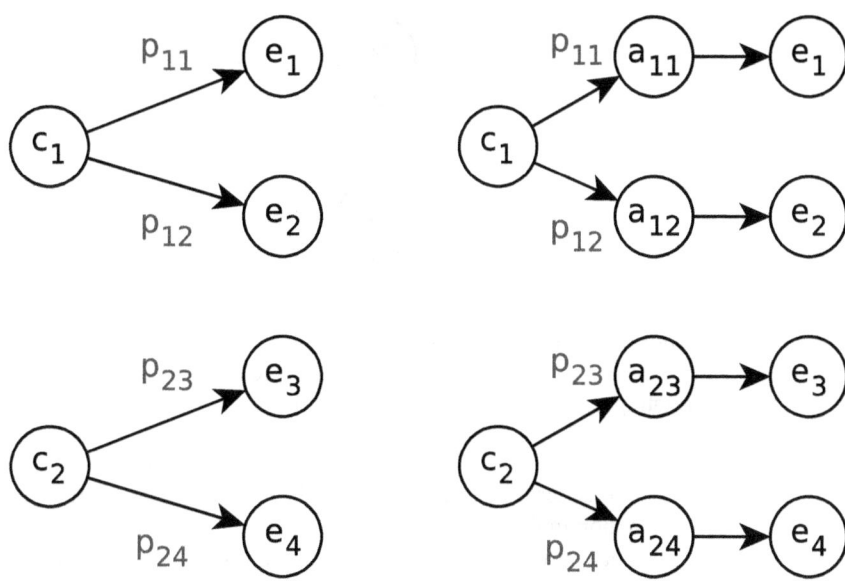

Figure 14.3: On the left, a causal mapping relates values of a nominal causal variable to values of a nominal effect variable. In this example, each causal value can lead to two proper (and incompatible) effect values, each arrow being associated with a probability $p_{ij} = p(e_j|\hat{c}_i)$. On the right, 'arrows' are now explicitly represented as values of a new variable, A, which represents the mapping between C and E.

splicing of genes depends on splicing regulatory elements (SREs), short nucleotide sequences in the pre-mRNA that bind protein factors that either activate or repress the use of adjacent splice sites. The causal relationship between the presence of an SRE and binding of its protein can be affected by the surrounding RNA sequence, because the shape of the whole RNA molecule can render the SRE more or less accessible to the factors for which it has an intrinsic binding affinity. Hence the same sequence can act as an SRE in one organism, but not in the orthologous gene of another organism, due to changes elsewhere in the gene (Wang & Burge, 2008). The molecular facts in these cases are very naturally represented as a focal causal relationship in which C is the sequence of the SRE and E is whether the protein binds or not, plus one or more background variables representing the structure of rest of the gene, which can interfere with that focal causal relationship.

It is stability and instability in this sense that we now proceed to analyse. Our aim in this section is not to come up with a definitive measure of causal stability for every purpose, but rather to show how to relate the idea of stability of causal relationships to our measure of causal specificity.

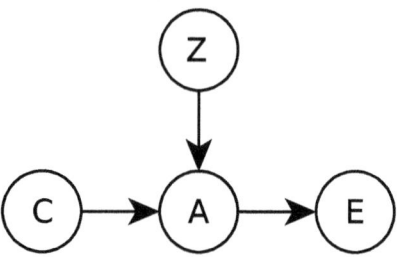

Figure 14.4: Causal graph with a variable representing the arrows A mapping C to E as they are affected by Z. (We draw reader's attention to the fact that this diagram is a causal graph relating variables, not a mapping relating values.)

To start with, let's consider a causal relationship $C \to E$ represented by a mapping between values of a (nominal) causal variable C to values of a (nominal) effect variable E (Fig. 14.3). Each causal value c_i can lead to one or several effect values e_j. To look at how the mapping can be influenced by a third variable, we will focus on the arrows connecting the values c_i and e_j. Each arrow a_{ij} can be defined as a couple of one causal value and one effect value. In formal terms, $a_{ij} \equiv (\widehat{c}_i, e_j)$.

When an intervention which sets C to c_i leads to e_j, we will say that the causal arrow a_{ij} has been instantiated, or that the variable A (for arrow – Figure 14.4) has taken the value a_{ij}.[7] The mapping between the values of the causal variable and the values of the effect variable is the set of these causal arrows, together with their associated conditional probabilities.

Now, let's consider that the mapping between C and E is somehow unstable with respect to a background variable Z. That is, Z makes the instantiation of some arrows more or less probable than it would be otherwise. We now treat the instantiations of the arrows a_{ij} as the events that are to be explained, and Z as the variable explaining them.

We first consider the arrows stemming from one causal value. Let's intervene on C to set it to value $\widehat{c}_1$. Given $\widehat{c}_1$, we look at how intervening on Z changes the probability of the arrows $a_{1j} : \widehat{c}_1 \to e_j$ that will be instantiated.[8] The amount of change can be measured by the mutual information between $\widehat{Z}$ and the variable A given c_1, that is, in formal terms, $I(A; \widehat{Z} | \widehat{c}_1)$.[9]

Figure 14.5 illustrates this idea. An intervention on Z has no effect on the mapping when the causal probabilities are unchanged, in which case $I(A; \widehat{Z} | \widehat{c}_1) = 0$ bit. The

---

[7]Because both C and E are sets of alternative events, it is axiomatic that one and only one arrow is instantiated in every intervention on the cause C. Also, because C and E are nominal variables, the composite

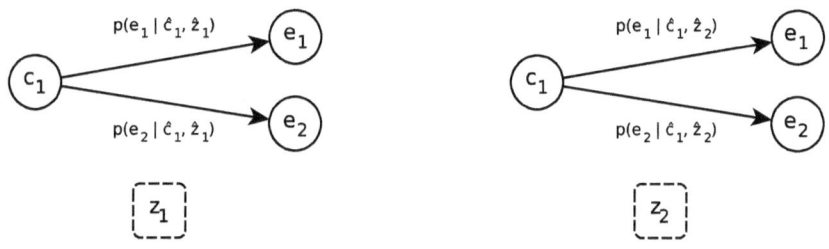

Figure 14.5: Diagram showing how interventions on Z can modify the mapping from C to E. For simplicity, only a single value of the causal variable is considered.

mapping between C and E is then maximally stable with respect to $\widehat{Z}$ (in this limiting case, Z is an irrelevant background condition, see Section 5). An intervention on Z has a maximum effect when it completely specifies the causal arrows being instantiated, that is, when one of the causal arrows is always instantiated when the intervention results in $\widehat{z}_1$ and reciprocally when $\widehat{z}_2$. In this case $I(A; \widehat{Z}|\widehat{c}_1) = 1$ bit, which is the maximum possible instability for this mapping. In between these two limiting cases stability will come in degrees.

When more than one value of C is considered (which is a necessary condition to be able to speak of C as a putative cause of E), it is reasonable to average the conditional mutual information $I(A; \widehat{Z}|\widehat{c}_i)$ over all the values of the causal variable C. The rationale for this is that causal arrows stemming from causal values that are themselves improbable (or impossible) should count less in characterizing the properties of the mapping. Calculating this average is equivalent to computing the conditional mutual information $I(A; \widehat{Z}|\widehat{C})$. This quantity characterizes how much Z affects the mapping between C and E when C is given in the background, or, in other words, the instability of the mapping with respect to Z.

However, not all mappings between C and E represent causal relationships. If C is not a causally relevant variable with respect to E, then the mapping between them is one where any value of C maps to all values of E (Fig. 14.6). The method we just outlined may nevertheless detect an effect of Z on the arrows being instantiated, but this will be due solely to the direct effect of Z on E. What we are after, however, is not just the mere effect of the variable Z on the effect E, it is rather how much the cause C and the background Z interact when they are both causes of E (Fig. 14.7). This, in our view,

---

variable A is also a nominal variable.

[8] Given $\widehat{c}_1$, the probabilities $p(a_{1j}|\widehat{c}_1)$ sum to 1.

[9] We condition on $\widehat{c}_1$ for pedagogical reasons, but it also makes philosophical sense. If $\widehat{Z}$ and $\widehat{C}$ are not independent, then we want to control for $\widehat{C}$ before assessing any effect of $\widehat{Z}$ on the arrows, as Z can be a cause C. If they are independent, conditioning makes no difference (see Appendix).

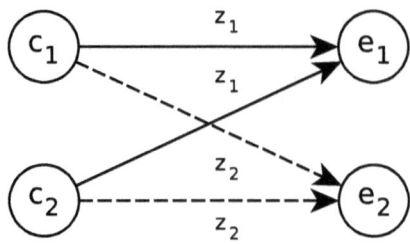

Figure 14.6: Causal mapping where Z is the only cause of E.

is what it means to talk of the causal relationship $C \to E$ depending on Z.

To measure our real target, the extent to which C and Z interact when they are both causes of C, we can use a bit of calculus and remark that the (conditional) specificity of Z for the mapping is equal to the conditional specificity of Z for the effect. In formal terms, $I(A;\widehat{Z}|\widehat{C}) = I(E;\widehat{Z}|\widehat{C})$ (see Appendix). This term embeds both the proper information coming from $\widehat{Z}$ alone, which is here equal to $I(E;\widehat{Z})$, and the information coming from the interaction between $\widehat{Z}$ and $\widehat{C}$, which is what we are after (Fig. 14.7).[10] To measure this interaction we compute the quantity $I(E;\widehat{Z};\widehat{C}) = I(E;\widehat{Z}|\widehat{C}) - I(E;\widehat{Z})$. This quantity is called the interaction information between the three variables.[11] The interaction information represents the portion of the effect of Z on the relationship between C and E that is not merely a consequence of the direct effect of Z on E.

7 Stability: Some conclusions

In Section 5, we remarked that the stability of the causal relationship $C \to E$ with respect to a background variable Z must be distinguished from the stability of E with respect to Z. We can now make this point more precise. The stability of the relationship $C \to E$ in response to changes in Z can only be reduced to the effect of Z on E if C and Z have entirely non-interactive effects on E, that is, if the interaction information is zero (see Appendix). Another way to look at this condition is that there is no inter-

---

[10]These components are often referred to as the unique information and the synergistic information, respectively. Another component of information is often considered: the redundant information (e.g. Williams and Beer, 2010). Decomposing multivariate information into such components is a currently debated topic (e.g. Bertschinger et al., 2013a, 2013b; Rauh et al., 2014). Here we assume that C and Z are independently manipulated and do not share any redundant information with respect to E.

[11]The interaction information is symmetrical: $I(E;\widehat{Z};\widehat{C}) = I(E;\widehat{C}|\widehat{Z}) - I(E;\widehat{C}) = I(\widehat{Z};\widehat{C}|E) - I(\widehat{Z};\widehat{C})$. In philosophical terms, there is parity, in our framework, between the causal variable C and the background variable Z: both C and Z are causal variables in the mapping from $\{C,Z\}$ to E.

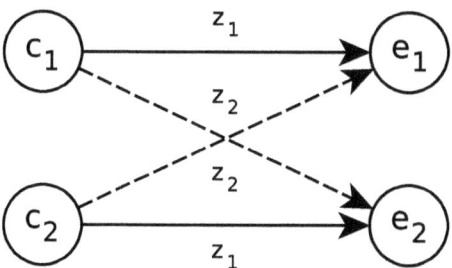

Figure 14.7: Example of interacting causes C and Z with respect to E. If the background Z is not controlled, the cause C is entirely not specific (assuming, for the ease of presentation, equiprobability between $\widehat{z}_1$ and $\widehat{z}_2$). Indeed, any intervention $\widehat{c}_1$ or $\widehat{c}_2$ can equiprobably lead to e_1 or e_2. Thus, $I(\widehat{C};E) = 0$ bit. However, once we know the background, C is entirely specific: $I(\widehat{C};E|\widehat{Z}) = 1$ bit (assuming, for the ease of presentation, equiprobability between $\widehat{c}_1$ or $\widehat{c}_2$). The interaction information in this case is $I(\widehat{C};E;\widehat{Z}) = I(\widehat{C};E|\widehat{Z}) - I(\widehat{C};E) = 1$ bit. (By design, the same holds when Z is the focal cause variable and C is as a background variable.)

action if and only if, given that we know which value E has taken, learning the value of the background variable Z gives us no additional information about which causal arrow from C has been instantiated: interventions on Z do not cause the same result in E to be produced in a different way. This is the case for instance in Figure 14.3 but not in Figure 14.7.

It is probably worth emphasizing that a relationship can be unstable with respect to a background variable but nevertheless have a stable conditional specificity under each background condition. This comes from the fact that the background variable affects the mapping between C and E but not necessarily the properties of the mapping, of which specificity is one. In other words, changing the background may produce a new mapping, but one that is exactly as specific as the original (e.g. Fig. 14.7).

The measures of stability described in Section 6 can be reconfigured in line with the 'specific actual difference making' (SAD) version of specificity favoured by Waters (2007) using the procedure described in Section 3. The intervention distribution of a background condition, $\widehat{Z}$, is forced to correspond to the actual distribution of Z in some population. The cost will be the same as for other applications of SAD any conclusions about stability will be relevant only for the population from which the actual distribution is derived.

In Section 6 we quantified how much a causal relationship depends on interactions of

Concepts	Relationship	Measure
Specificity, Stability	$C \to E$	$I(\widehat{C};E)$
Interaction	$Z \to (C \to E)$	$I(\widehat{Z};E\|\widehat{C}) - I(\widehat{Z};E)$

Table 14.1: Information theoretic measures for specificity, stability, and interaction.

the cause with background variables.[12] But what about the reverse of this – how much influence the cause can exert irrespective of the background background variables? In information theoretic terms, the answer is quite simple. The amount of causal information which is independent from the background variables is just the mutual information between interventions on the cause and the effect, without controlling for the background variables. In other terms, a positive notion of stability is automatically captured by the idea of specificity (Table 14.1).

It goes without saying that the strengths of the information theoretic formalism, its simplicity, and the way it helps us clarify our concepts, come with limitations. The fact that we can quantify specificity and stability by the same measure in part reflects the conceptual overlap between the two: both deal with how much a cause can affect an effect across a range of background conditions. In part, however, this formal homogeneity is a by-product of using a highly constrained theoretical framework, a theme we return to in our conclusion.

8 Conclusions

Our information-theoretic framework was developed for thinking about causal specificity within the interventionist approach to causation (Griffiths et al., 2015). In this paper we have used it to analyse several other key elements of the interventionist account. In Section 2 we showed that the property of 'minimal invariance', which provides the criterion of causation in Woodward's (2003) interventionist account, is equivalent to a non-zero degree of specificity in the relationship between a cause and its effect. The 'range of invariance' of a causal relationship can be measured by the degree of specificity of the cause for its effect. Our proposed measure of specificity is the mutual information between interventions on a causal variable and observations of an effect variable:

$$\text{Spec} = I(\widehat{C};E)$$

In our earlier work we suggested that the potential of C to causally influence E should be measured by Spec with a maximum entropy distribution over $\widehat{C}$. This seems to be

---

[12] Z can be any set of several variables.

the natural interpretation of Woodward's conception of fine-grained influence (INF), ultimately derived from David Lewis. Here, however, we have argued that that the potential causal influence of $\widehat{C}$ on E, considered in the abstract, is better measured by constructing the distribution over $\widehat{C}$ that maximises the value of Spec (MaxSpec, see also Korb et al. (2009)).

In Section 3 we examined Water's proposal to assess specificity using only the actual variation in a cause in some population, or 'specific actual difference making' (SAD). This conception of specificity can also be expressed information-theoretically and has useful applications, as we have argued elsewhere (Griffiths et al., 2015). We showed that SAD specificity corresponds to Spec when we intervene on $\widehat{C}$ to mimic the actual distribution of $\widetilde{C}$ in some population. It is instructive that different qualitative discussions of specificity correspond to different probability distributions over the causal variable. However, SAD behaves very differently from INF or MaxSpec and we interpret it as a measure of the *actual* influence of $\widehat{C}$ on E in some population, rather than of the *potential* influence of $\widehat{C}$ on E. We also suggested that another information-theoretic measure, information flow (Ay & Polani, 2008), is an alternative way to measure actual causal influence that has some advantages over SAD.

In Section 4 we argued that the controversial idea that causes should be described in a more or less fine-grained way so as to render the description of the cause 'proportional' to its effects could be made more precise in our framework. Ideal proportionality is achieved by simultaneously minimising the entropy of $\widehat{C}$ whilst maximising the specificity $I(\widehat{C};E)$. This amounts to discretising the variable $\widehat{C}$ so as to mark all and only differences that make a difference to E. We suggested that some features referred to in the qualitative literature as 'proportionality' but not captured by our proposal concern which variables to include in a causal graph in the first place, rather than the grain of description of a given variable.

In Sections 5 we suggested that the 'stability' of a causal relationship is the extent to which that relationship is not affected by additional variables, often termed background variables. We distinguished two ways in which a background variable could have an effect on a causal relationship. It might affect the value of the effect variable, or it might affect the relationship between the causal variable and the effect variable. To the best of our knowledge, this distinction has not been clearly drawn in any earlier discussions.

In Section 6 we offered an information-theoretic analysis of the instability of causal relationships. The effect of a third variable, Z on the causal relationship $C \to E$ is the effect of interventions on Z on the mapping from C to E. The greater this effect, the more unstable $C \to E$ is relative to the background variable Z. The amount of instability can be measured by the interaction information between C, Z and E. We showed that the impact of Z on $C \to E$ needs to be distinguished from the impact of Z on the value of E. The causal mutual information between Z and these two will only be equal under special conditions. The opposite of instability, the *in*sensitivity of the relationship $C \to E$ to background variables, is simply specificity of that relationship.

We believe that the work presented here adds precision to some important elements of the interventionist approach to causation and opens up many potential lines for further research. However, our use of information theory as a formal tool introduces some very severe limitations. Most importantly, we are restricted to using nominal variables. Individual values are different from one another, but not different by any amount. We are thus unable to capture the idea that highly specific relationships are 'smooth'. This might mean that the size of changes in the cause corresponds to the size of changes in the effect, for which we would need metric variables. Alternatively, it might mean that adjacent values of causes produce adjacent values of the effect, for which we would need at least ordinal variables. A related blind-spot for our approach to stability is whether changes to background variables have large, small, or negligible, impacts on a causal relationship. We can only measure *how many* changes in a background variable have *an* impact.

There are two possible responses to the intrinsic limitations of some formal framework. One is to return to a qualitative approach which can encompass the full richness of the relevant concepts, but at the price of being less clear about what constitutes that richness. That strikes us as a very high price. The other is to seek to approach different aspects of the topic using different formalisms. The interventionist framework would benefit very greatly from being given a treatment in an entirely different formalism, such as dynamical systems theory, but that is a project for another day.

9 Appendix

Here we provide a quick primer in information theory, proofs of equations cited in the text and expand on some of the ideas in Section 6.

9.1 Entropy, conditional entropy, and mutual information

We recall basic formulas of information theory. For a primer on information theory, see (Cover & Thomas, 2006). The Shannon entropy of a variable X is defined as:

$$H(X) \equiv -\sum_i p(x_i) \log_2 p(x_i)$$

The conditional entropy of a variable X knowing Y is defined as:

$$H(X|Y) \equiv -\sum_j p(y_j) \sum_i p(x_i|y_j) \log_2 p(x_i|y_j).$$

The mutual information of two variables X and Y can be computed as:

$$I(X;Y) = H(X) - H(X|Y) = \sum_i \sum_j p(x_i, y_j) \log_2 \left(\frac{p(x_i, y_j)}{p(x_i) p(y_j)} \right).$$

The conditional mutual information of two variables X and Y knowing a third variable Z can be computed as:

$$I(X;Y|Z) = \sum_k p(z_k) \sum_i \sum_j p(x_i, y_j | z_k) \log_2 \left(\frac{p(x_i, y_j | z_k)}{p(x_i | z_k) p(y_j | z_k)} \right).$$

Our measure of specificity of C to E is defined as the mutual information between $\widehat{C}$ and E: $Spec(C \to E) \equiv I(\widehat{C}; E)$. The formula of $I(\widehat{C}; E)$ reads (see Griffiths et al., 2015):

$$I(\widehat{C}; E) = \sum_i \sum_j p(\widehat{c}_i, e_j) \log_2 \left(\frac{p(\widehat{c}_i, e_j)}{p(\widehat{c}_i) p(e_j)} \right).$$

9.2 $I(A; \widehat{Z} | \widehat{C}) = I(A; \widehat{Z})$ when $\widehat{C}$ and $\widehat{Z}$ are independent:

We start with the plain formula for $I(A; \widehat{Z} | \widehat{C})$:

$$\begin{aligned} I(A; \widehat{Z} | \widehat{C}) &= \sum_i p(\widehat{c}_i) I(A; \widehat{Z} | \widehat{c}_i) \\ &= \sum_i p(\widehat{c}_i) \sum_j \sum_k p(a_{ij}, \widehat{z}_k | \widehat{c}_i) \log_2 \left(\frac{p(a_{ij}, \widehat{z}_k | \widehat{c}_i)}{p(a_{ij} | \widehat{c}_i) p(\widehat{z}_k | \widehat{c}_i)} \right) \\ &= \sum_i \sum_j \sum_k p(\widehat{c}_i) p(a_{ij}, \widehat{z}_k | \widehat{c}_i) \log_2 \left(\frac{p(\widehat{c}_i) p(a_{ij}, \widehat{z}_k | \widehat{c}_i)}{p(\widehat{c}_i) p(a_{ij} | \widehat{c}_i) p(\widehat{z}_k | \widehat{c}_i)} \right) \\ &= \sum_i \sum_j \sum_k p(a_{ij}, \widehat{z}_k, \widehat{c}_i) \log_2 \left(\frac{p(a_{ij}, \widehat{z}_k, \widehat{c}_i)}{p(a_{ij}, \widehat{c}_i) p(\widehat{z}_k | \widehat{c}_i)} \right) \end{aligned}$$

Now we use $p(a_{ij}, \widehat{c}_i) = p(a_{ij})$ and $p(a_{ij}, \widehat{z}_k, \widehat{c}_i) = p(a_{ij}, \widehat{z}_k)$ ($\widehat{c}_i$ is necessary to obtain a_{ij}), as well as $p(\widehat{z}_k | \widehat{c}_i) = p(\widehat{z}_k)$ (independence of $\widehat{C}$ and $\widehat{Z}$). We obtain:

$$I(A; \widehat{Z} | \widehat{C}) = \sum_i \sum_j \sum_k p(a_{ij}, \widehat{z}_k) \log_2 \left(\frac{p(a_{ij}, \widehat{z}_k)}{p(a_{ij}) p(\widehat{z}_k)} \right) = I(A; \widehat{Z})$$

9.3 Conditional specificity about the mapping is conditional specificity about the effect

We can transform $I(A; \widehat{Z} | \widehat{C})$, using the bijection (by construction) between the events (a_{ij}) and $(\widehat{c}_i, e_j)$:

$$I(A; \widehat{Z} | \widehat{C}) = I\left((\widehat{C}, E); \widehat{Z} | \widehat{C} \right) = I(E; \widehat{Z} | \widehat{C}).$$

Curious readers might wonder what would yield a reciprocal approach to computing $I(A; \widehat{Z} | \widehat{C})$, which would be to look at how C influences the mapping A, holding Z in the background. This actually amounts to computing the entropy of the cause:

$$I(A;\widehat{C}|\widehat{Z}) = I\left((E,\widehat{C});\widehat{C}|\widehat{Z}\right) = H(\widehat{C}|\widehat{Z}) = H(\widehat{C}).$$

The last equality obtains by hypothesis of independence between $\widehat{C}$ and $\widehat{Z}$. This reduction to the entropy of the cause comes from the fact that, by construction $\widehat{c}_i$ is necessary to obtain a_{ij} (recall that by definition $a_{ij} \equiv (\widehat{c}_i, e_j)$), while there is no such condition with respect to Z.

Acknowledgments. We thank Maël Montévil for reading a previous version of the manuscript. This publication was made possible through the support of a grant from the Templeton World Charity Foundation. The opinions expressed in this publication are those of the authors and do not necessarily reflect the views of the Templeton World Charity Foundation.

Bibliography

Ay, N., & Polani, D. (2008). Information flows in causal networks. *Advances in Complex Systems, 11*(01), 17–41.

Cover, T. M., & Thomas, J. A. (2006). *Elements of information theory* (2nd ed.). John Wiley & Sons.

Craver, C. F. (2007). *Explaining the brain.* New York and Oxford: Oxford University Press.

Griffiths, P. E., Pocheville, A., Calcott, B., Stotz, K., Kim, H., & Knight, R. (2015). Measuring causal specificity. *Philosophy of Science, 82*(4), 529–555.

Hitchcock, C., & Woodward, J. (2003). Explanatory generalizations, part II: Plumbing explanatory depth. *Nos, 37*(2), 181–199.

Janzing, D., Balduzzi, D., Grosse-Wentrup, M., & Schölkopf, B. (2013). Quantifying causal influences. *The Annals of Statistics, 41*, 2324–2358.

Kendler, K. S. (2005). 'A gene for...': The nature of gene action in psychiatric disorders. *American Journal of Psychiatry, 162*(7), 1243–1252.

Korb, K., Hope, L., & Nyberg, E. (2009). Information-theoretic causal power. In F. Emmert-Streib, & M. Dehmer (Eds.), *Information theory and statistical learning* (pp. 231–265). Boston, MA: Springer US.

Lizier, J. T., & Prokopenko, M. (2010). Differentiating information transfer and causal effect. *The European Physical Journal B, 73*(4), 605–615. doi:10.1140/epjb/e2010-00034-5

Mitchell, S. D. (2002). Ceteris paribus – an inadequate representation for biological contingency. *Erkenntnis, 57*(3), 329–350.

Pearl, J. (2009). *Causality; models, reasoning and inference.* New York: Cambridge University Press.

Tononi, G., Sporns, O., & Edelman, G. M. (1999). Measures of degeneracy and redundancy in biological networks. *Proceedings of the National Academy of Sciences, 96*(6), 3257–3262.

Wang, Z., & Burge, C. B. (2008). Splicing regulation: From a parts list of regulatory elements to an integrated splicing code. *RNA, 14*(5), 802–813. doi:10.1261/rna.876308

Waters, C. K. (2007). Causes that make a difference. *The Journal of Philosophy, 104*(11), 551–579.

Weber, M. (2006). The central dogma as a thesis of causal specificity. *History and Philosophy of the Life Sciences*, 595–609.

Weber, M. (2013). Causal selection versus causal parity in biology: Relevant counterfactuals and biologically normal interventions. In *What if? On the meaning, relevance and epistemology of counterfactual claims and thought experiments* (pp. 1–44). Konstanz: University of Konstanz.

Williams, P. L., & Beer, R. D. (2010). *Nonnegative decomposition of multivariate information.* arXiv Preprint arXiv:1004.2515. Retrieved from http://arxiv.org/abs/1004.2515

Woodward, J. (2000). Explanation and invariance in the special sciences. *The British Journal for the Philosophy of Science*, *51*(2), 197–254.

Woodward, J. (2003). *Making things happen: A theory of causal explanation.* New York: Oxford University Press.

Woodward, J. (2010). Causation in biology: stability, specificity, and the choice of levels of explanation. *Biology & Philosophy*, *25*(3), 287–318.

Yablo, S. (1992). Mental causation. *The Philosophical Review*, *101*(2), 245–280.

Author biography. Arnaud Pocheville is a theoretical biologist and a philosopher of biology. He received his PhD thesis in theoretical biology from the Ecole Normale Supérieure Paris , and pursued his research as a postdoctoral fellow at the Center for Philosophy of Science, University of Pittsburgh. Arnaud is now a research fellow in the Theory and Methods in Biosciences group, University of Sydney, funded by the Templeton World Charity Foundation project "Causal foundations of biological information."

Paul E. Griffiths is a philosopher of science, in the Department of Philosophy, University of Sydney. He is a Fellow of the American Association for the Advancement of Science, of the Australian Academy of the Humanities and former President of the International Society for History, Philosophy and Social Studies of Biology. He leads the Theory and Methods in Bioscience group and the Templeton World Charity Foundation project "Causal foundations of biological information."

Karola Stotz is senior lecturer and a Templeton World Charity Foundation Fellow at the Department of Philosophy, Macquarie University. She has worked at the Konrad Lorenz Institute for Evolution and Cognition Research in Austria, the University of Pittsburgh, Indiana University, and the University of Sydney. She is the co-author of Griffiths, P.E. and K. Stotz (2013) Genetics and Philosophy: An Introduction. CUP.

15 What is action-oriented perception?

ZOE DRAYSON

Abstract. Contemporary scientific and philosophical literature on perception often focuses on the relationship between perception and action, emphasizing the ways in which perception can be understood as geared towards action or 'action-oriented'. In this paper I provide a framework within which to classify approaches to action-oriented perception, and I highlight important differences between the distinct approaches. I show how talk of perception as action-oriented can be applied to the evolutionary history of perception, neural or psychological perceptual mechanisms, the semantic content or phenomenal character of perceptual states, or to the metaphysical nature of perception. I argue that there are no straightforward inferences from one kind of action-oriented perception to another. Using this framework and its insights, I then explore the notion of action-oriented perceptual representation which plays a key role in some approaches to embodied cognitive science. I argue that the concept of action-oriented representation proposed by Clark and Wheeler is less straightforward than it might seem, because it seems to require both that the mechanisms of perceptual representation are action-oriented and that the contents of these perceptual representations are action-oriented. Given that neither of these claims can be derived from the other, proponents of action-oriented representation owe us separate justification for each claim. I will argue that such justifications are not forthcoming in the literature, and that attempts to reconstruct them run into trouble: the sorts of arguments offered for the representational mechanisms being action-oriented seem to undermine the sorts of arguments offered for the representational content being action-oriented, and vice-versa.

Keywords: perception, action, representation, affordances, enactivism, embodiment.

1 Introduction

Contemporary scientific and philosophical literature on perception often focuses on the relationship between perception and action, emphasizing the ways in which perception can be understood as geared towards action. Research in psychology, for example, sometimes characterizes perception as "action-specific" (Cañal-Bruland & van der Kamp, 2009; Witt, 2011), and talk of perception as being "of affordances for action" is found in both psychology and philosophy (Richardson, Shockley, Fajen, Riley, & Turvey, 2008; Chemero, 2009). Descriptions of perception as "active" are found in vision science (Whitehead & Ballard, 1990), artificial intelligence (Weyns, Steegmans, & Holvoet, 2004), psychology (Aloimonos, Y. (Ed.), 2013), and philosophy (Noë, 2004); and as "enactive" in robotics (Morse & Ziemke, 2007), cognitive science (Froese & Spiers, 2007), and philosophy (Thompson, 2007). In psychology (Fajen, 2005) and also in sports science (Pijpers, Oudejans, & Bakker, 2007), perception is described as being "for action"; while the term "action-oriented" is applied to perception in philosophy (Gallagher & Zahavi, 2014), computer science (Gora & Wasilewski, 2014), and neuroscience (Ridderinkhof, 2014).

In what follows, I'll use the term 'action-oriented' as an umbrella-term to include approaches like these that share the idea of perception as oriented or geared towards action, in some sense. My first aim in this paper is to provide a framework within which to classify approaches to action-oriented perception, and to highlight important differences between distinct approaches.

My framework classifies action-oriented approaches to perception into one of five categories:

1. The evolution of perception as action-oriented: these approaches claim that perception evolved to guide action, or that perception was selected for its action-oriented capacities.

2. The mechanisms of perception as action-oriented: these approaches claim that the mechanisms of perception overlap with, or are closely coupled to, the mechanisms of action.

3. The contents of perception as action-oriented: these approaches claim that the contents of perception present the world to the perceiver in terms of possible actions.

4. The phenomenal character of perception as action-oriented: these approaches claim that a perceptual state is qualitatively experienced as encouraging or demanding a certain action.

5. The nature of perception as action-oriented: these approaches claim that perception has a necessary connection to action, that to be a perceiver is essentially to have certain agentive capacities or skills.

Part of my motivation for providing this framework is to clarify the discussions of action-oriented perception across philosophy and the sciences. I will show that each approach within the framework is logically independent of the others: none of the approaches can be derived from the others without further justificatory assumptions. We should thus not conclude from the fact that perception is action-oriented in one respect that it is action-oriented in any of the other respects.

This paper also has a second aim, which is to use this framework to explore the role of action-oriented perception in embodied cognitive science. In particular, I will focus on the claim that perceptual representations can be action-oriented. I will show that Clark (1997) and Wheeler (2005) put forward a concept of action-oriented perceptual representations which is committed to both the representational mechanisms of perception being action-oriented, and to their contents being action-oriented. But given that neither of these claims can be derived from the other, proponents of action-oriented representation owe us separate justification for each claim. I will argue that such justifications are not forthcoming in the literature, and that attempts to reconstruct them run into trouble: the sorts of arguments offered for the representational mechanisms being action-oriented seem to undermine the sorts of arguments offered for the representational content being action-oriented, and vice-versa.

In Sections 2-6, I introduce each of the distinct respects in which perception can be said to be action-oriented. Section 7 explores the use of action-oriented perceptual representations in embodied cognitive science.

2 The evolution of perception as action-oriented

There is an evolutionary sense in which perception can be described as action-oriented, that is often expressed with the claim that perception is "for action". Proponents of this view emphasise that our perceptual capacities evolved to guide our bodily interactions with the world, specifically those interactions which increased our adaptive fitness. Talking about visual perception, for example, Briscoe claims that:

> "[f]rom an biological or evolutionary standpoint, it is reasonable to think that vision is *for action*, that its preeminent biological function is to adapt an animal's bodily movements to the properties of the environment that it inhabits" (Briscoe, 2014, p. 202, my italics).

On one interpretation of such claims, the idea that perception is "for action" in an evolutionary sense seems trivially true, and it's not clear whether any believer in evolution would deny it. But the evolutionary claim is often put forward in a stronger way: the assumption is sometimes that perception evolved directly to guide action and *not* (or at least not directly) to present the perceiver with an action-neutral description of the objective world. If perception evolved for its action-guiding properties, so this line of thought goes, then it did not evolve to enable us to have beliefs about how the world is, independently of our own actions. Proponents of this view either conclude that we do not perceive the objective world in an action-neutral way, or that if we do, this is not the primary function of our perceptual systems: perception did not evolve in order to provide inputs to the human capacity to think and reason about the world. Representative versions of this view can be found in the works of philosophers like Patricia Churchland and Kathleen Akins:

> "looked at from an evolutionary point of view, the principle function of nervous systems is to get the body parts where they should be in order that the organism may survive. [...] Truth, whatever that is, definitely takes the hindmost." (Churchland, 1987, p. 548)

> "evolution will favor sensory solutions that package the information in efficient and quickly accessible formats, in ways that match the particular physical form of the motor system, its motor tasks, and hence informational requirements. [...] the elegant solutions that evolution eventually selects need not involve any straightforward (to our eyes) 'veridical' encoding of sensory information." (Akins, 1996, p. 353)

Notice that this involves a strong commitment to a particular evolutionary story. Proponents of this view have to rule out the possibility that the development of amodal cognition could have been more adaptive than a faster but less flexible system in which specific sensory inputs drove specific motor outputs. Some scientists claim exactly this, arguing that game theory demonstrates that selection pressures would make objective representations of the world an unlikely outcome (Hoffman, Singh, & Mark, 2013), but this is often assumed rather than argued for. Notice that even if one accepts that evolution selected for action-guiding perception, this is consistent with the claim that evolution *also* selected for the kind of perception that can yield reasoning capacities and abstract thought. It might be the case, in other words, that we have two forms of perception. This is a possibility raised by the dual-visual system theory of perception, discussed in the next section.

3 The mechanisms of perception as action-oriented

According to a traditional picture of the mind, perceiving and acting are distinct mechanisms which are separate from each other and from our thinking mechanisms: perception provides input for thought, and action emerges as the output of thought. (This is what Susan Hurley (1998) terms the 'classical sandwich' picture of the mind.) This picture is challenged by empirical work which suggests that the mechanisms of perceiving and acting are closely intertwined, overlapping, or even co-constituting. If this is true, then there is a sense in which the mechanisms of perception can be action-oriented regardless of their evolutionary history.

One example of this comes from the literature on so-called 'mirror neurons': in macaque monkeys, neurons known to control hand and mouth movements fire both when the monkey is manipulating an object (e.g. reaching for a piece of food) and when the monkey is watching someone perform a similar manipulation. In creatures like humans, where single-cell recordings are not possible, there is similar data showing that areas of the brain associated with movement are active during perceptual processing: one and the same neural mechanism seems to be involved in both perception and action. Perception is action-oriented in the sense that a perceptual mechanism seems to overlap with a mechanism for action.

In visual psychology, Milner and Goodale (1995) have advanced the 'dual visual systems' hypothesis, which concerns the way that visual perception builds up information about the world. They have demonstrated that sensory input to the visual system of primate brains can be processed by two independent pathways. The processes in the *ventral* pathway reflect the traditional picture of perception as the input to thought: they produce conscious perceptual states that we can categorize, memorize, and combine with thoughts to produce a broad range of actions. The processes in the *dorsal* pathway, however, do not seem to be the inputs to thought: these visual inputs instead lead only to the specific motor outputs involved in reaching and grasping objects with our hands. The dorsal processes seem to be action-oriented in the sense that visual input results in the appropriate motor output without the intermediary of conscious thought.

This sort of 'close coupling' between sensory input and motor output is also found in robotics and artificial intelligence. The traditional view of perception and action as distinct mechanisms separated by central thought processes creates engineering problems for designers of intelligent systems by causing bottlenecks to occur: the sorts of thinking required to update plans and amend instructions slows down the system's ability to respond to sensory stimuli. Roboticists like Rodney Brooks produced systems that could interact with their environments in real-time, by cutting out the central 'mind' and focusing instead on using specific input sensors to generate specific outputs. Brooks (1991) proposes that arranging these sensory-motor couplings in 'layers' in the appropriate way, surprisingly intelligent behavior can be produced. One layer might govern general locomotion, such that a robot will move around at random until it hits an obstacle. Another layer might then take over and turn the robot until the way is clear, before control reverts to the first layer. A third layer might sense red light and keep the robot on track to reach the light, thus overriding the first layer's random movement. Each sensory process in these robotic architectures is thus action-oriented in the sense that that each layer has its own sensors which operate exclusively for one kind of behavior. There is no amodal cognition or control: the communication between the layers is minimal, and amounts to just switching each other on and off. Such architectures can be used to create systems that display a remarkable amount of seemingly intelligent flexible behavior from purely reflex-like action-oriented sensory-motor couplings. It remains an empirical question whether such architectures can simulate higher-level behaviours, such as deciding between actions, without the addition of central thought mechanisms.

The three examples I've considered – mirror neurons, the dorsal visual stream, and sensory-motor architectures – are examples of ways in which the mechanisms of perception could be said to be action-oriented. Notice that claims about action-oriented perceptual mechanisms are independent from the previously considered claim that perception is action-oriented in the sense of having evolved to guide action. First, consider the robotic architectures just described. The fact that an optimal engineering solution uses action-oriented perceptual mechanisms (in the sense of reflex-like couplings between sensory inputs and motor outputs) doesn't tell us that human perceptual systems actually evolved such couplings: evolutionary solutions are often the satisfactory but suboptimal ones, rather than the optimal solutions. Second, while the dorsal visual stream's use of sensory input to directly guide action seems to support the evolutionary claim, it does so only with respect to a certain subset of perceptual processing: the guiding of fine-grained motor control, rather than action more generally. And the ventral visual stream seems to have evolved to generate the kind of perceptual states that can input into thought, which is exactly what the proponents of the stronger evolutionary claims want to deny. Third, despite many claims about the evolutionary purpose mirror neurons might have served, it's not clear that we can draw any conclusions about the action-oriented nature of perceptual evolution from the empirical data. Cecilia Heyes (2009), for example, has persuasively argued that mirror neurons are a byproduct of our general capacity for associative learning and not the result of evolutionary adaption.

4 The contents of perception as action-oriented

Many of the claims about action-oriented perception in philosophy concern the contents of perception: how the world is presented to us in experience. It is traditionally assumed that perception presents the world to us in a way that is entirely neutral with respect to the actions one might perform: while the contents of *beliefs* might present the world to us in terms of how we can act on it (presenting food as edible or stairs as climbable, for example), such contents are not available in perception, according to the traditional view of perceptual content as action-neutral. Proponents of action-oriented perceptual content, on the other hand, propose that action-involving properties like edibility or climbability can be presented to us in the contents of perception: "we *see* objects as edible, and do not just believe that they are" (Nanay, 2012, p. 430). Following Gibson (1979), we can use the term 'affordance' to describe properties like edibility and climbability, and characterize the action-oriented view of perceptual content as the claim that we perceive affordances. Action-oriented views of perceptual content come in different strengths, depending on how they characterize the idea of affordances. To perceive the world as affording a certain action might, on a weak reading, mean that we perceive the possibilities for action: the sorts of actions we are capable of performing. A stronger reading would claim that action-oriented perceptual content presents the world to us in a way that solicits or encourages us to act in a certain way; and an even stronger reading views affordance properties in perceptual content as mandating or demanding a particular action. Action-oriented perceptual content need not be consciously experienced by the perceiver, but there is a way of extending the idea of action-oriented perceptual content to include the phenomenal character of perception, which I'll discuss in the next section.

As I am using the notion of perceptual content here, talk of perceptual content merely commits us to the claim that the world is presented to the perceiver in experience. Such talk is not committed to the stronger claim that perceptual experience fundamentally consists in the subject perceptually representing their environment as being a certain way. The weaker notion of perceptual content is therefore not committed to representationalism about perception, but is compatible with at least some forms of relationalism (Siegel, 2010). Furthermore, proponents of action-oriented perceptual content might conceive of content as propositional or non-propositional; as structured conceptually or non-conceptually; as Russellian or Fregean, or in terms of possible worlds; and as environment-dependent or independent.

It is important to emphasise that since action-oriented perceptual content is not committed to this content being carried by a particular neural or psychological mechanism, then *a fortiori* it is not committed to the existence of action-oriented perceptual mechanisms. This will be further discussed later in Section 7.

5 The phenomenal character of perception as action-oriented

Philosophers draw a distinction between the content of perception and its phenomenal character. To talk of perceptual content is to talk of how perception presents the world; whereas to talk of the phenomenal character of perception is to talk of the qualitative properties of the perceptual state, or what it feels like to undergo the perceptual experience. If we assume that perceptual states need not be consciously experienced, then is possible for a perceptual state with the sort of action-oriented content outlined above to lack phenomenal character altogether. And where a perceptual state has phenomenal character and action-oriented content, that action-oriented content might be "non-soliciting" (Siegel, 2014): one could consciously perceive something as affording action without experiencing the motivation to perform that action. Similarly, Prosser (2011) proposes that the phenomenal character of a perceptual state correlates with its action-oriented perceptual content, but doesn't claim that the phenomenal character itself is action-oriented.

But it looks like there could be cases where the phenomenal character of a perceptual state is itself action-oriented. Siegel (2014), for example, claims that perceiving what the environment affords can sometimes be accompanied by a felt quality of solicitation, where the perceptual experience has a qualitative feel of inviting or prompting us to act. If we think of this as action-oriented phenomenal character, then it looks like phenomenal character might be action-oriented to differing degrees: some perceptual experiences might not just feel like invitations or prompts, but more like urges that motivate us to act. Siegel (2014) proposes that this particular subset of perceptual states with action-oriented phenomenal character have a "feeling of answerability" and are experienced as mandates for action.

Notice that how one understands the relation between action-oriented content and action-oriented phenomenal character will depend on one's attitude more generally to the relation between perceptual content and phenomenal character. A proponent of strong intentionalism about phenomenal character, for example, would presumably claim that any action-oriented aspects of phenomenal character reduce to the action-oriented properties of the perceptual content. At the other end of the spectrum, one might think that phenomenal character is entirely independent of perceptual content, and thus that it is possible to have two perceptual experiences with the same content but distinct characters.

6 The nature of perception as action-oriented

I have suggested that claims about perception being action-oriented can be understood as making a variety of different claims about the evolution, mechanisms, content, or character of perception. But there remains a further question about the modal strength of the claims being made. Should we understand any of these claims as proposing that perception is *necessarily* action-oriented in any of these senses? And if so, what kind of necessity is involved?

I take it that most of those making action-oriented claims concerning the evolution of perception or the mechanisms of perception are making largely empirical claims about features of perception in the actual world. But at least some proponents of action-oriented perception seem to be making modal claims, which can be interpreted in at least two ways. On one hand, the necessity at play might be nomological necessity. To say that perception is necessarily action-oriented in the nomological sense is not to say that perception is action-oriented in all possible worlds, but rather to say that the laws of nature in the actual world make it physically impossible for perception not to be action-oriented in the actual world. Alternatively, the necessity in question might be metaphysical necessity, concerning the nature of perception across possible worlds. To posit a metaphysically necessary connection between perception and action is to claim that there is no possible world in which perception is not action-oriented in some appropriate sense.

The claims made by some proponents of sensorimotor theories of perception, such as Alva Noë (2004) and Susan Hurley (1998), seem to suggest that there is a constitutive dependence of perception on our capacity for action: that to be a perceiver is necessarily to be an agent. It is not clear what kind of necessity is involved, but their reliance on empirical evidence suggests that they are primarily concerned with making a case for nomological necessity. The case for action being metaphysically necessary for perception is perhaps found in Schellenberg's (2010) exploration of the relationship between perception, self-location, and spatial know-how. Her key claim, that perception requires the capacity to know what it would be to act in relation to objects, seems to rely on our intuitions about how perception could be in possible worlds rather than how it is in the actual world.

Notice that even where claims of nomological necessity are made and supported by empirical evidence about the mechanisms of perception, there is no direct entailment from action-oriented perceptual mechanisms to the claim that perception is necessarily action-oriented. We cannot simply read off claims of nomological necessity from empirical facts (Callender, 2011). And claiming that perception is necessarily action-oriented in the sense that Schellenberg intends does not entail that either perceptual content or phenomenal character is action-oriented. In fact, Schellenberg proposes that perception's dependence on action capacities is what allows us to perceive more than just the relational or perspectival properties associated with affordances: she claims that the action-oriented nature of perception can account for our access to the intrinsic, observer-independent properties of the objects perceived.

7 What about action-oriented representation?

I've differentiated several different approaches to perception that fall under the umbrella-term 'action-oriented', and demonstrated that each approach makes claims that are independent from the others. There is no entailment from one variety of action-oriented perception to another without the addition of further premises.

While I've talked about the mechanisms of perception and the content of perception, I have said little about the matter of perceptual *representations*: content-bearing internal states of the perceiver. In fact, many people working in action-oriented perception want to play down the role of representations. This is because in cognitive science, representational mechanisms are traditionally posited to account for the flexibility of intelligent human behaviour: our capacity to respond differently to similar stimuli, and similarly to different stimuli, in a way that can't be easily accounted for in terms of reflex-like responses. Positing internal representations allows cognitive science to say that the same stimulus can be represented in different ways: the same extensional content can be carried by different representational 'vehicles' which play distinct causal roles within the cognitive mechanism.[1] These vehicles are just internal representations individuated by the non-semantic properties in virtue of which they are causally efficacious: the formal, functional, or physical properties to which the cognitive mechanisms are sensitive. Thus two representations are tokens of the same vehicle type when they are treated similarly by the cognitive processes at play (Shea, 2007).

Traditionally, cognitive science suggests that these representational mechanisms are

---

[1] For naturalistic reasons, cognitive science is generally committed to extensional content.

genuinely cognitive in the sense of being non-perceptual: they are distinct from the mechanisms of perception, and they are what allow the same perceptual input to result in different action outputs. Proponents of embodied cognitive science, however, deny that explanations of flexible and context-specific behaviour require this sort of mediating representational mechanism. They argue instead that if perceptual mechanisms are action-oriented in the sense discussed in Section 3, then these sorts of coupled mechanisms can account for the behaviour.

But not everyone agrees that the insights from action-oriented approaches to perception should lead to the rejection of internal representational states. There are many within embodied cognitive science who allow that there are internal representations, but deny that these are the sort of amodal cognitive representations in a central cognitive architecture. Andy Clark (1997), for example, argues that behaviour is not mediated by "action-independent inner states; ones which require additional cognitive operations to drive appropriate behavior", but rather by action-oriented representations which are "poised between pure control structures and passive representations of external reality" (Clark, 1997, p. 49). A similar approach is put forward by Michael Wheeler, who understands action-oriented perceptual representations as "special-purpose adaptive couplings" that are "tailored to a particular behavior" (Wheeler, 2005, p. 196). Both Clark and Wheeler propose that the sorts of action-oriented mechanisms of perception discussed in Section 3 are best understood as representational mechanisms: as containing perceptual states which are both causally efficacious and semantically evaluable. They argue that these perceptual representations themselves, and not just the mechanisms in which they occur, should be understood as action-oriented in virtue of the way that they are causally coupled to certain motor processes.

To claim that a mechanism is representational is to be committed to vehicles of content. But this does not commit one to any particular views about the nature of that content: what it is or how it is determined, evaluated, or structured. (We might agree that a list of items is a representation, while disagreeing over whether it is a description or an instruction, for example; we might believe a drawing to be a map without knowing what it is of; we might know that someone's utterances are linguistic without knowing what they mean.) To understand an internal state as a representational vehicle does not, therefore, entail a particular view of its content. *A fortiori*, it does not entail that its content is action-oriented in the sense of presenting the world in action-relevant ways. The reverse is also true, as already discussed in Section 4: action-oriented perceptual content is neutral with regard to the mechanisms of perception, and therefore compatible with a range of cognitive architectures. Action-oriented perceptual content does not entail action-oriented perceptual mechanisms, and *a fortiori* does not entail that there are action-oriented vehicles of representation bearing the action-oriented perceptual content.

Interestingly, however, both Clark and Wheeler seem to assume that their action-oriented vehicles of perceptual representation have action-oriented perceptual content. Clark claims that action-oriented perceptual representations "simultaneously describe aspects of the world and prescribe possible actions", with the result that "to thus know the world is at once to know what possibilities it affords for action and intervention" (Clark, 1997, p. 49), while Wheeler claims that action-oriented perceptual states "represent the world in terms of specifications for possible actions" with "bearer-relative content" (Wheeler, 2005, p. 196). Their action-oriented representations, in other words, seem to be examples of action-oriented perceptual mechanisms with action-oriented perceptual contents.

Neither Clark nor Wheeler explain why they think that action-oriented vehicles will have action-oriented contents. Any attempt to reconstruct their arguments will, I suggest, encounter the following problem: reasons to think of perceptual vehicles as action-oriented seem to undermine our reasons to think of perceptual contents as action oriented, and vice-versa.

To see this, consider the role played by representational vehicles. Recall that vehicles are just internal representations individuated by the non-semantic properties in virtue of which they are causally efficacious within the representational mechanisms. Vehicles provide a naturalistic way of understanding how the same extensional content can play distinct cognitive roles, such as when the same heavenly body is thought about in different ways according to the time of day it is observed. The extensional content is the same, but the representational vehicles have distinct causal properties in each case and thus interact differently with other states of the mechanism. If we claim, as Clark and Wheeler do, that representational vehicles should be individuated according to their action-guiding properties, then this enables us to show how the same extensional content (e.g. a mountain) can be represented as climbable in one context and unclimbable in another context: distinct vehicles, with different causal relations to action, can carry the same extensional content. But if this is the role of representational vehicles, to provide a naturalistic means of accounting for modes of presentation, then it removes the need to think of the content itself as action-oriented. The perceptual content can be action-neutral because the relevant action-guiding information is supplied by the perceptual mechanisms.

In the account above, the mountain is presented to us in perceptual experience, but to say that we perceive it as climbable to is to say that the vehicle of the perceptual content plays a certain causal role. But what if one thinks that the climbability of the mountain is presented to us in experience? On such a view, the perceptual content of the experience would involve a climbable mountain in one case, and an unclimbable mountain in another case. The two states would have different action-oriented contents. In this case, we wouldn't need to explain how the same content (the mountain) could function as climbable in one context and not climbable in another. And without this explanatory need, it's not clear what motivation we'd have for positing vehicles of representation, rather than non-representational mechanisms. This, I take it, is the essence of Gibson's (1979) view of perception. And even if we had reason to posit representational mechanisms, we'd then need a further argument for individuating the representational vehicles according to their action-guiding properties.

In summary, the arguments for taking perceptual vehicles to be action-oriented seem to undermine the arguments for taking perceptual contents to be action-oriented, and the arguments for taking perceptual contents to be action-oriented seem to undermine the arguments for taking perceptual vehicles to be action-oriented. None of this rules out that there may be arguments from the action-oriented status of either vehicles or contents to the other, but it suggests that the onus is on proponents of such views to provide them.

Bibliography

Akins, K. (1996). Of sensory systems and the "aboutness" of mental states. *Journal of Philosophy, 93*(7), 337–372.

Aloimonos, Y. (Ed.) (2013), In *Visual navigation: from biological systems to unmanned ground vehicles*. Psychology Press.

Briscoe, R. (2014). Spatial content and motoric significance. *Avant: Trends in Interdisciplinary Studies, 2*, 199–216.

Brooks, R. (1991). Intelligence without representation. *Artificial Intelligence, 47*, 139–159.

Callender, C. (2011). Philosophy of science and metaphysics. In S. French, & J. Saatsi (Eds.), *Continuum companion to the philosophy of science* (pp. 33–54). Continuum.

Cañal-Bruland, R., & van der Kamp, J. (2009). Action goals influence action-specific perception. *Psychonomic Bulletin & Review, 16*(6), 1100–1105.

Chemero, A. (2009). *Radical embodied cognitive science*. MIT Press.

Churchland, P. S. (1987). Epistemology in the age of neuroscience. *Journal of Philosophy, 84*(10), 546–553.

Clark, A. (1997). *Being there: Putting brain, body, and world together again*. MIT Press.

Fajen, B. R. (2005). Perceiving possibilities for action: On the necessity of calibration and perceptual learning for the visual guidance of action. *Perception, 34*(6), 717–740.

Froese, T., & Spiers, A. (2007). Toward a phenomenological pragmatics of enactive perception.

Gallagher, S., & Zahavi, D. (2014). Primal impression and enactive perception. *Subjective Time: The Philosophy, Psychology, and Neuroscience of Temporality*, 83–99.

Gibson, J. J. (1979). *The ecological approach to visual perception*. Houghton Mifflin.

Gora, P., & Wasilewski, P. (2014). Adaptive system for intelligent traffic management in smart cities. In *International conference on active media technology* (pp. 525–536). Springer International Publishing.

Heyes, C. (2009). Where do mirror neurons come from? *Neuroscience and Biobehavioral Reviews*.

Hoffman, D. D., Singh, M., & Mark, J. (2013). Does evolution favor true perceptions? In *Proc. SPIE 8651, Human vision and electronic imaging XVIII* (Vol. 865104).

Hurley, S. (1998). *Consciousness in action*. Harvard University Press.

Milner, A. D., & Goodale, M. A. (1995). *The visual brain in action*. Oxford University Press.

Morse, A., & Ziemke, T. (2007). Cognitive robotics, enactive perception, and learning in the real world. *Cognitive Science*.

Nanay, B. (2012). Action-oriented perception. *European Journal of Philosophy, 20*, 430–446.

Noë, A. (2004). *Action in perception*. MIT Press.

Pijpers, J. R., Oudejans, R. R., & Bakker, F. C. (2007). Changes in the perception of action possibilities while climbing to fatigue on a climbing wall. *Journal of sports sciences*, *25*(1), 97–110.

Prosser, S. (2011). Affordances and phenomenal character in spatial perception. *Philosophical Review*, *120*(4), 475–513.

Richardson, M. J., Shockley, K., Fajen, B. R., Riley, M. A., & Turvey, M. T. (2008). Ecological psychology: Six principles for an embodied-embedded approach to behavior. In *Handbook of cognitive science: An embodied approach* (pp. 161–187).

Ridderinkhof, K. R. (2014). Neurocognitive mechanisms of perception–action coordination: A review and theoretical integration. *Neuroscience & Biobehavioral Reviews*, *46*, 3–29.

Schellenberg, S. (2010). Perceptual experience and the capacity to act. In M. M. Gangopadhay, & F. Spicer (Eds.), *Perception, action, and consciousness* (pp. 145–159). Oxford University Press.

Shea, N. (2007). Content and its vehicles in connectionist systems. *Mind and Language*, *22*(3), 246–269.

Siegel, S. (2010). *The contents of visual experience*. Oxford University Press.

Siegel, S. (2014). Affordances and the contents of perception. In B. Brogaard (Ed.), *Does perception have content?* (pp. 39–76). Oxford University Press.

Thompson, E. (2007). *Mind in life: Biology, phenomenology, and the sciences of mind*. Harvard University Press.

Weyns, D., Steegmans, E., & Holvoet, T. (2004). Towards active perception in situated multi-agent systems. *Applied Artificial Intelligence*, *18*(9–10), 867–883.

Wheeler, M. (2005). *Reconstructing the cognitive world: the next step*. MIT Press.

Whitehead, S. D., & Ballard, D. H. (1990). Active perception and reinforcement learning. *Neural Computation*, *2*(4), 409–419.

Witt, J. K. (2011). Action's effect on perception. *Current Directions in Psychological Science*, *20*(3), 201–206.

Author biography. Zoe Drayson is Assistant Professor of Philosophy at the University of California, Davis. She previously held positions at the Australian National University and the University of Stirling, and gained her PhD from the University of Bristol. She works in philosophy of mind and cognitive science, with a focus on the nature of psychological explanations across philosophy and the mind sciences. Her published work explores the distinction between personal and subpersonal approaches to the mind; the commitments of embodied and extended cognitive science; the relation between perception, cognition, and action; the metaphysics and epistemology of explanation; and various issues surrounding the nature of mental representation.

16 Philosophy of science in practice: A proposal for epistemological constructivism

MIEKE BOON [*]

Abstract. *Philosophy of science in practice* (PoSiP), besides other things, aims at an epistemology of scientific practices that addresses questions such as: How is the construction of knowledge for epistemic uses possible? Epistemology that goes beyond *formal epistemology* has been banned from mainstream philosophy of science since the Wiener Kreis (Vienna Circle) in order to avoid drifting into the realm of metaphysics. However, formal epistemology cannot account for the possibility of *epistemic uses* of scientific knowledge (e.g. Cartwright, 1983). For, this account would involve as an epistemological presupposition that knowledge about new, previously unexamined systems can be derived from logical or mathematical structures that are ultimately grounded on experiential and experimental data. The plausibility of this presupposition draws on the belief that there *exist* such structures. Hence, metaphysics re-enters through the backdoor. When acknowledging that any claim to the *certainty* of knowledge involves metaphysics, the entire idea of *certain* knowledge may be abandoned (as in Van Fraassen's anti-realism). At this point, it becomes apparent that there is a need for an epistemology that suits scientific practices, especially those which aim at knowledge for practical uses. Indeed, a reconsideration of some of the presup-

---

[*]Department of Philosophy, University of Twente, PO Box 217, 7500 AE Enschede, The Netherlands. Email: m.boon@utwente.nl.

positions that vindicated formal epistemology seems in order, in particular: on the aim of science, on how to avoid metaphysics, and on the task of the philosophy of science. In this article Kant's epistemology is taken as a starting point for such reconsideration, since his work aimed to explain the *possibility* of knowledge. Based on contemporary interpretations of Kant's epistemology (e.g. Neiman, 1994), *epistemological constructivism* is proposed as a view in which the aim of science is to construct knowledge for epistemic uses. It involves the idea that scientific knowledge (patterns in data, and scientific laws, models and concepts) is constructed to enable and guide epistemic uses, which also entails that scientific practices develop *epistemic strategies* for the production of knowledge that meets this purpose. Accordingly, one of the tasks of PoSiP is to reconstruct, investigate, and evaluate epistemic strategies by means of which knowledge is constructed.

Keywords: philosophy of science in practice, Wiener Kreis, Vienna Circle, Kantian epistemology, synthetic a priori, constructive empiricism, representation, pragmatism, model construction, non-representationalism, epistemic tool, epistemic use, engineering sciences.

1 Philosophy of science in practice

The conference *Philosophical Perspectives on Scientific Understanding*, in August 2005 (De Regt, Leonelli, & Eigner, 2009), at the Free University in Amsterdam, was the prelude to a movement called *Philosophy of Science in Practice*.[1] In the invitation to the introductory meeting held at PSA 2006[2], the founders (Boon, Chang, Ankeny, Boumans and De Regt) wrote: "We have begun organizing a network of scholars under the working name of "Society for Philosophy of Science in Practice" (SPSP). We aim to promote a philosophy of science that engages more closely with scientific practice, and with the practical uses of scientific knowledge." Different from other significant movements that originated from the philosophy of science, such as the sociology of scientific knowledge (SSK), and science, technology and society (STS), philosophy of science in practice (PoSiP) aims to maintain close ties to mainstream philosophy of science (also see Ankeny, Chang, Boumans, & Boon, 2011; Soler, Zwart, Lynch, & Israel-Jost, 2014).

One of the reasons why PoSiP crystalized at this conference on scientific understanding may have been that focus was turned to the scientists – the user and producer of knowledge – whereas traditional philosophy of science has frenetically aimed at an account of scientific knowledge in terms of a mere two-way relationship between world and knowledge only. Conversely, philosophers within this new movement aim at an understanding of science that avoids the belief that the objectivity of knowledge

---

[1] http://philosophy-science-practice.org/en/mission-statement/
[2] http://philosophy-science-practice.org/en/events/introductory-spsp-meeting-psa-2006/

can be warranted by an account of knowledge-justification that eliminates the role of scientists, but that also avoids a mere psychological and sociological interpretation of scientists' subjectivity. Instead, the philosophy of science in practice aims to include the scientist by focusing on an *epistemology of scientific practices*.

The movement of philosophy of science in practice shows similarities to founding movements of traditional philosophy of science, such as the Wiener Kreis and in its wake Logical Empiricism. It is well known that those movements had pertinent societal concerns,[3] which motivated their members to aim at a philosophical account of knowledge that warrants objectivity (Hahn, Neurath, and Carnap, 1929/1973, and also see Uebel, 1996, 2008, 2016). Almost one century later, we live in a society that is thoroughly dependent on science in diagnosing problems, forecasting complex natural and societal processes, and in the development of (socio-)technology for welfare, well-being, and societal justice. The *philosophy of science in practice* is similar to its historical predecessors in the sense of being motivated by societal concerns. In this case, our concern is science's ability to play this role well. This implies a philosophy of science that, on the one hand, supports the production of knowledge that meets society's expectations of science, and on the other hand, explains the character of knowledge in such manner that possible misuses of the authority of science can at least be made transparent.

Concerning the societal role of science, our grandfathers in the philosophy of science wagered on the objectivity of science. In a rigorous, dedicated, and insistent manner, they strived for a philosophy of science that explains how the objectivity of science can be justified. Such objectivity essentially requires that the role of the scientist can be eliminated. However, critics within the analytical tradition of the philosophy of science have convincingly shown that objectivity cannot possibly be achieved by reducing knowledge to atomic sentences that are then used to describe observed (atomic) facts independent of whatever context, nor does reducing knowledge to formal logical structures void of meaning result in the desired objectivity (e.g. Quine, 1951; Feyerabend, 1962). A problem of warranting the objectivity of knowledge is that atomic facts cannot be generated by reading out (atomic) facts, implying that knowledge always involves one or another kind of *construction*, be it the ways in which facts are "pointed out" (i.e., discerned and given meaning), or the ways in which facts are connected to each other, and also, how they are related to theories. Moreover, the

---

[3] In the 1930s, Hitler, in his speeches on his political ideology, explicitly expressed that, since science is not objective, it must be supervised by the state: "Der Gedanke einer freien voraussetzungslosen Wissenschaft konnte nur im Zeitalter des Liberalismus auftauchen. Er ist absurd. Die Wissenschaft ist ein Soziales Phänomen, und wie ein jedes solches begrentzt durch den Nutzen oder Schaden, den es für die Allgemeinheit stiftet. Mit dem Schlagwort von der objectiven Wissenschaft hat sich die Professorenschaft nur von der sehr nötigen Beaufsichtigung durch die staatlichen Macht befreien wollen" (quote from Klages, Mohr, and Sontheimer, 1970, and Feix, 1978). "The idea of a free unconditional science could emerge only in the age of liberalism. It is absurd. Science is a social phenomenon, and as such limited by the benefits or harm that it bestows upon the community. With the slogan of objective science the caste of professors just wanted to get rid of the very necessary supervision by the state power" (my translation).

objectivity of knowledge is no warrant for its *predictive* certainty – it requires skillful scientists to figure out how to apply scientific knowledge such as to produce reliable predictions about real-world problems (Cartwright, 1983; Boon, 2006). Therefore, in PoSiP we argue for the abandonment of the belief that societally desired qualities of science are warranted by objectivity of knowledge in terms of a two-way relationship between world and knowledge, and instead, adopt the idea that these qualities depend on a three-way relationship between world, knowledge, and scientist. Unavoidably, meeting societal goals through science must also be safe-guarded by moral and intellectual abilities of scientist, and these *abilities* are partially covered by notions such as objectivity and rationality. Philosophy of science in practice, therefore, cannot ignore these notions but rather than hunting down the objectivity and rationality of knowledge, it focuses on how scientific practices achieve these goals. Therefore, PoSiP aims at an *epistemology* of these practices that elucidates how knowledge is generated, which may also involve re-framing traditional notions such as truth, certainty, observation, explanation, experiment, and justification.

In this article, I take epistemological issues encountered in scientific research practices of the engineering sciences as paradigm example, and raise the question: How is construction of knowledge that allows for epistemic uses (in practical applications such as problem-solving) possible? Compared with traditional philosophy of science, focus is not on scientific knowledge as such (considering issues such as its truth and certainty), but on real-world problems in need of scientific knowledge that aids resolution. In addressing this question, my approach will be to investigate (in Section 2) assumptions and beliefs in mainstream philosophy of science that may hamper the development of an appropriate epistemology, including: the assumption that the construction of knowledge belongs to the *context of discovery* and therefore is a no-go-area; the idea that knowledge *represents* reality; and the belief that Kant's idea of *synthetic a priori judgements* has proven false. The connection between these assumptions is that they corroborate each other and entail an empiricism that is still close to Humean empiricism in the sense of assuming that objective knowledge consists of "basic empirical facts" (and plain data) that are to be acquired through *passive* observation (or measurement), and that any other knowledge should be built on (e.g., be reducible to, or verified by) these facts. In Section 3, I aim to present an outline of a revised interpretation of Kant's epistemology by contemporary Kant scholars (amongst others, Neiman, 1994, and Allison, 2004), which in my view is a fertile starting point for developing the desired epistemology of scientific practices. In Section 4, I will pick up on the claim that a representational view of knowledge is not well suited to an epistemology that accounts for the possibility of constructing knowledge for epistemic uses, and I will summarize why the alternative of *knowledge as epistemic tool* is a better fit for this purpose. Finally, in Section 5, I propose *epistemological constructivism* as a position that takes "the production of knowledge for epistemic uses" as the aim of science, and that adopts (the proposed version of) Kantian epistemology to explain how the construction of knowledge for epistemic uses is possible. Epistemological constructivism steers away from the realism/anti-realism debate – and from refined positions in this debate such as Van Fraassen's (1980) constructive empiricism and Giere's (1999) constructive realism – as this debate focuses on arguments for the justification or acceptance of scientific knowledge in accordance with one or another representational view of knowledge. Conversely, epistemological constructivism takes "the construction of knowledge for epistemic uses" as its guiding issue, including epistemic strategies for producing and using knowledge. One of the tasks of the philosophy of science is to investigate and evaluate these epistemic strategies.

2   How is construction of knowledge for epistemic uses possible?

What is scientific knowledge? And what purposes does it serve? In the philosophy of science it is commonly assumed that scientific knowledge, such as theories and scientific models, laws and concepts *represent* reality. The question that has led to the idea of scientific knowledge as *epistemic tool* as an alternative to this representational view (Boon & Knuuttila, 2009; Knuuttila & Boon, 2011; Boon, 2012), and to the idea of *epistemological constructivism* as a philosophical position that circumvents the realism/anti-realism debate (Boon, 2015), is: How is the construction of knowledge for epistemic uses possible? This philosophical question and proposed solutions emerged from epistemological and methodological problems experienced in concrete scientific research practices of the *engineering sciences*. Since most current scientific research is performed in one or another problem- or application context, I take it that this question and the proposed solutions are relevant to the philosophy of science in general.

Clearly, there is some friction between raising this question and several principles (norms, beliefs and presuppositions) in main-stream philosophy of science that have been accepted since, say, the establishment of the Wiener Kreis (Hahn et al., 1929/1973; Uebel, 2008, 2016). Most prominently, it is still assumed – as once clearly phrased by Carnap (1934) in the first issue of *Philosophy of Science* – that philosophy of science should only deal with formal aspects of science:

> Philosophy deals with science only from the logical viewpoint. Philosophy is the logic of science, i.e., the logical analysis of the concepts, propositions, proofs, theories of science, as well as of those which we select in available science as common to the possible methods of constructing concepts, proofs, hypotheses, theories. [What one used to call epistemology or theory of knowledge is a mixture of applied logic and psychology (and at times even metaphysics); insofar as this theory is logic it is included in what we call logic of science; insofar, however, as it is psychology, it does not belong to philosophy, but to empirical science.] (Carnap, 1934, 6, his brackets.)

Hence, in traditional philosophy of science, epistemology as a theory of knowledge should focus on the logic of knowledge and avoid psychology, which underpins the dogma that the *construction* of knowledge belongs to the *context of discovery* – a prohibited area for mainstream philosophy of science.[4] Furthermore, the philosophical question, summarized as "how ... possible," touches on basic assumptions in the philosophy of science, concerning what is: "science," "scientific knowledge," "the aim of science," "the task and territory of the philosophy of science," and also "the justification of *philosophical* claims." In the cited quote, Carnap (1934) is very clear on what philosophy of science is. Therefore, we may first ask whether the very question of "how the construction of knowledge for epistemic uses is possible" *should* be addressed in the philosophy of science. If so, productively addressing it brings with it questions about assumptions of the philosophy of science itself.

An almost implicit assumption in the philosophy of science, but also in much science education, is that scientific knowledge can be applied in solving concrete problems by means of deductive reasoning upon laws and theories. This idea draws on the assumption that there is symmetry between prediction and explanation. Indeed, Hempel (1965) believed that laws discovered in the laboratory can be applied in practical applications:

> The formulation of laws and theories that permit the prediction of future occurrences are among the proudest achievements of empirical science; and the extent to which they answer man's quest for foresight and control is indicated by the vast scope of their practical applications. (Hempel, 1965, p. 333.)

Cartwright (1974, 1983) is one of the first authors criticizing this belief (see also Boon, 2006). Her fundamental critique is that scientific laws are only true – i.e., give true predictions – for circumstances close enough to the "nomological machine" that produced the regular behavior represented in the law:

> [Usually] it takes what I call a *nomological machine* to get a law of nature. ... A nomological machine is a fixed arrangement of components, or factors, with stable capacities that in the right sort of stable environment will, with repeated operation, give rise to the kind of regular behaviour that we represent in our scientific laws. (Cartwright, 1999, pp. 49-50.)

The *semantic view* suggests a similar assumption about how scientific knowledge is applied for solving concrete problems. In this view, mathematical models are derived from abstract theories (called "models of the theory")[5], and thus involve the idea that

---

[4] The terms "context of discovery" and "context of justification" are often associated with Hans Reichenbach's work, but see Schickore (2014).

[5] One of the philosophical problems inherited from logical empiricism and the so-called *syntactic view* that Suppes (1960a) and succeeding defenders of the *semantic view* of theories aimed to solve, is that the relationship between the theory and the object "referred to" by the theory cannot be understood in terms of a *correspondence* relationship. Therefore, Suppes (1960a) argued that the concept of model used by mathematical logicians should be the basis for a fundamental concept of model in any branch of, what he called the empirical sciences, which includes both the natural and the social sciences. Suppes adopted the

the application of theory is through "models of the theory" that are expected to correctly represent the real-world target system. However, this idea can be criticized for similar reasons as Cartwright (1983, 1999) has put forward concerning laws: usually these models are not correct about real-world systems – they only produce correct predictions for carefully designed experimental systems, and even then, before we can compare models derived from theory with experimentally derived data, the latter need to be corrected by means of "theories of data" and "theories of experiment" (Suppes, 1960a, 1960b)[5] A theorist who takes a semantic view may admit that the primary function of "models of the theory" is testing theories, but still assume symmetry between the role of models in testing theories and in predicting the behavior of real-world target-systems. Yet, this idea is criticized by Cartwright (1983) and also easily rebutted by concrete scientific practices. The outlined view does, however, reveal another assumption of the philosophy of science: that the aim of science is theories (not practical application), and that the task of philosophy is to account for the justification of theories (but not the justification of how knowledge is produced and epistemically used in solving practical problems).

In accordance with Cartwright's critique, concrete scientific research practices show that philosophical assumptions on the applicability of scientific knowledge by mere deductive reasoning are inadequate, which is why the question "how ... possible" should be addressed in the philosophy of science. The need for an epistemology that addresses this question is therefore evident. But some of the assumptions and beliefs of mainstream philosophy of science seem to discourage these kinds of philosophical investigation. The examples discussed so far include the idea that the construction of knowledge, if it cannot be formalized, belongs to the *context of discovery*, which is beyond the domain of the philosophy of science; the idea of *symmetry* between explanation and prediction, which justifies the assumption that scientific knowledge

---

concept of *satisfaction* to specify the semantic relationship between theories and models. Hence, *models* are the objects that *satisfy* the axioms of the theory (i.e., the model provides a realization in which the theory is satisfied). He calls these models "models of the theory." Additionally, he introduced the notion "models of data," which are the data-structures that, according to the theories of data and of the experiment, would be generated by the experiment but restricted to those aspects of the experiment which have a parametric analogy in the theory (Suppes, 1960b). Suppes' formal approach in set-theoretical terms makes his approach to models somewhat tedious from the perspective of scientific practices. Van Fraassen (1980, p. 64) puts the relationship between models of the theory and experiments in more accessible terms: "To present a theory is to specify a family of structures, its models; and secondly, to specify certain parts of those models (the empirical substructures) as candidates for the direct representation of observable phenomena. The structures which can be described in experimental and measurements reports we can call appearances." Van Fraassen uses the concept of *isomorphism* to specify the semantic relationship between "models of the theory" and what he calls "appearances." It can then be specified how these "appearances" or "measurement reports" play a role in testing the theory: "The theory is empirically adequate if it has some model such that all appearances are isomorphic to empirical substructures of that model." In the language of a scientist, this means for instance that abstract theories are tested by comparison of models that are derived from the theory for relevant conditions of the experiment (i.e., mathematical structures), and data-models produced in experiments (i.e., "empirical substructures") – where the theoretically generated structure must be isomorphic to the experimentally produced structure (also see Suppe, 1989). Still closer to the language of scientists is Giere's (1988) account of the semantic view, which he summarizes in Giere (2010).

allows for correct predictions in new situations; the idea that knowledge such as models *represents* aspects of the real-world, and that its epistemic use is by virtue of this representational relationship; and, the idea that the *aim of science* is theories, while the *task of the philosophy of science* is their justification by formal methods.

Here, I will focus on the *representational* view of knowledge as one of the deeply rooted ideas in current philosophy of science that obscures an epistemology of "how the construction of knowledge for epistemic uses is possible." More precisely, I aim to provide an alternative to the interconnected beliefs that scientific knowledge (theories, scientific models, laws and concepts) represents aspects of the world, and that its epistemic usefulness is due to this representational capacity. First, I will expand on earlier work, which proposes the idea of "knowledge as epistemic tool" as a viable alternative to "knowledge as representation." Considering knowledge as epistemic tool stresses the fact that humans construct knowledge for epistemic uses of all kinds. Therefore, this notion is one element of the epistemology aimed at. The further development of this epistemology requires an account of how epistemic tools are constructed, justified, and actually used. Important in my argument is that the *construction* of knowledge should not be conflated with discovery, nor should the *application* of knowledge be conflated with representation or algorithmic reasoning. Rather, in scientific research practices, construction and use are less far apart than we tend to believe. *Constructing* an epistemic tool involves its justification, not only in view of established knowledge and data, but also in view of intended applications. *Using* it (in new, real-world situations) involves the assessment and justification of whether it suits this particular application, but also constructive activities by means of the epistemic tool (rather than merely using it as a basis for description or algorithmic reasoning). These two aspects of using knowledge require an understanding of its original construction, such as the idealizations and simplifications used in the construction of the epistemic tool.

Second, next to the claim that "knowledge as representation" hampers an epistemology of "how constructing knowledge for epistemic uses is possible," I will argue that it is also hampered by the dogma that the construction of knowledge is beyond the scope of the philosophy of science. Although I do not aim to make a claim on the history of philosophy, I will start from the assumption that this dogma reflects the rejection of Kant's notion of synthetic a priori judgments. This rejection happened already at the establishment of the philosophy of science by members of the Wiener Kreis, and was motivated by their urge to ensure that science did not become tainted by metaphysics and psychology. In the Manifesto they state:

> In such a way logical analysis overcomes not only metaphysics in the proper, classical sense of the word ..., but also the hidden metaphysics of Kantian and modern *apriorism*. The scientific world-conception knows no unconditionally valid knowledge derived from pure reason, no 'synthetic judgments a priori' of the kind that lie at the basis of Kantian epistemology and even more of all pre- and post-Kantian ontology and metaphysics. The judgments of arithmetic, geometry, and *certain fundamental principles of physics, that Kant took as examples of a priori*

knowledge will be discussed later. *It is precisely in the rejection of the possibility of synthetic knowledge a priori that the basic thesis of modern empiricism lies. The scientific world-conception knows only empirical statements about things of all kinds, and analytic statements of logic and mathematics."* (Hahn et al., 1929/1973, 7, my emphasis.)

Epistemological analysis of the leading concepts, of natural science has freed them more and more from *metaphysical admixtures* which had clung to them from ancient time. In particular, Helmholtz, Mach, Einstein, and others have cleansed the concepts of *space, time, substance, causality, and probability*. The doctrines of absolute space and time have been overcome by the theory of relativity; space and time are no longer absolute containers but only ordering manifolds for elementary processes. Material substance has been dissolved by atomic theory and field theory. Causality was divested of the anthropomorphic character of 'influence' or 'necessary connection' and reduced to a relation among conditions, a functional coordination. (Hahn et al., 1929/1973, 10, their emphasis.)

However, current Kant scholars, such as Allison (2004) and Neiman (1994) suggest that the above reading of Kant is flawed.[6] Put very briefly, the *synthetic a priori* is not about fundamental principles of physics as suggested in these quotes – it is not about concepts in the sense of empirically revisable knowledge about reality. Instead, it concerns "the conditions for the possibility" of such knowledge, which Kant attributes to human cognition. Due to this flawed understanding of Kant's epistemology, philosophers of science were trapped in a position that is closer to Humean empiricism (although extended with modern logic). Instead of adopting Kant's basic idea that empirical knowledge of the world must already be understood as *actively constructed* by humans (which means accepting that knowledge-production results from a three-partite relationship between world, knowledge, and the human cognitive system), these philosophers retreated into the idea that empirical knowledge of the world *emerges passively* through observation (perception), and that (in order to warrant objectivity and avoid metaphysics) epistemology must ban the idea that some kind of synthetic a priori knowledge plays a role.[7]

---

[6] Nevertheless, these scholars admit that Kant himself gave reasons for this flawed understanding, especially as a result of his *Metaphysische Anfangsgründe der Naturwissenschaft* (1786/2004) (Metaphysical Foundations of Natural Science) which he wrote as an application of his principles on human cognition in *Kritik der Reinen Vernunft* (1787/1998) (Critique of Pure Reason).

[7] Also see Massimi (2008, 2011), who argues that the empiricist idea that "literally true description of the way things are" is possible, is highly problematic even for the observable part of the world. She argues that a Kantian stance on (observable) phenomena studied in science can offer a genuinely new perspective on the issue of how we infer phenomena from data, by distancing both from a metaphysics of ready-made phenomena and from conventionalist ideas of how phenomena are established (e.g., as sometimes proposed in Logical Positivism and Logical Empiricism). Accordingly, Massimi (2011) aims at solving the controversy between Woodward's (1989, 1998) realist and McAllister's (1997) anti-realist account of how phenomena are inferred from data.

Although, mainstream philosophy of science has criticized, abandoned, and revised many of the initial ideas of members of the Wiener Kreis, their claim that Kant's notion of synthetic a priori judgments has proven false seems still widely accepted. As is probably already clear, I actually consider Kantian rather than Humean empiricism a better starting point for contemporary philosophy of science, and it is from this standpoint that I suggest the use of modern interpretations of Kantian epistemology in developing an epistemology that explains how the construction of knowledge for epistemic uses is possible.

3 Kantian epistemology[8]

In the last chapter of her monograph, which addresses "The task of philosophy," Neiman (1994) argues that Kant's work reflects a tension between two wholly diverse conceptions of philosophy and hence of his own procedure: a *regulative* conception and a *constitutive* conception. The latter "is reflected in Kant's determination to 'put metaphysics on the sure path of a science' and to complete a necessary edifice that will never need to be revised" (Neiman, 1994, p. 185), which agrees to how Kant is usually understood. In contrast, the regulative conception connects with an anthropological interpretation of the task of philosophy (see last part of note 8), which is the one I propose to endorse.

Kantian epistemology as it is proposed here, promotes the idea that scientific knowledge is not the product of algorithmic (deductive and inductive) reasoning upon experiences, but results from the capacity to demand explanations of experience, which in turn, "requires the capacity to go beyond experience, for we cannot investigate the given until we refuse to take it as given. To ask a question about some aspect of experience, we must be able to think the thought that it could have been otherwise. Without this thought, we cannot even formulate the vaguest why" (Neiman, 1994,

---

[8] Although I do not consider myself a Kant scholar, I have been studying Kant's work repeatedly (both in German and in English translations) over the past 30 years. It provided me with clues for better understanding of scientific reasoning in research practices, which I did not find in mainstream philosophy of science. When still being a practicing scientist, in 1986, I read Putnam's (1981) *Reason, Truth and History*, which has significantly affected my appreciation of Kant's ideas, and my conviction that his ideas are crucial to an understanding of how we produce knowledge about the world (also see note 12). Even more so, Kant and Putnam's work made me understand what constitutes a Copernican revolution in our own naïve presuppositions: it cured me of basic intuitions (scientism) that in many respects were close to ideas that motivated members of the Wiener Kreis. Allison's (2004) detailed study of the *Critique of Pure Reason* was a feast of recognition, as, except for Putnam, I had not read anything so near to my own interpretation of Kant's work. But most of all was I touched by Neiman's (1994) work on Kant. She shows that instead of the common caricature of Kant as a shriveled scholar who is suffering of systematism, his work circles around the question of what it means to be human – i.e., Kant's fourth question: "What is the human being?" on which Kant concludes provocatively, that the whole field of philosophy could be reckoned to anthropology, whose business it is to tell us what is human (CPR IX, 25; Neiman, 1994, p. 185). Kant offers an extremely rich, deep and quite coherent philosophy on that very issue.

p. 59). Hence, the demand for explanation of experience is crucial to the *construction* of scientific knowledge. Importantly, in a Kantian epistemology an explanation is not regarded as a *representation* of the mind-independent world behind the phenomena (i.e., the picture or animation as seen by a God's eye; also see Putnam, 1981). Instead, an explanation of experiences is constructed, enabled, and guided by our ways of structuring and conceptualizing even at the level of "observable" phenomena (also see Massimi, 2008, and Boon, 2012).

According to Kant, in seeking the explanation of experience, reason seeks its systematization, which involves the employment of *regulative principles* of reason (such as, the principle of induction; the principle that every event has a cause; the principle that nature forms a system according to laws; and the principle of the unity of nature). Importantly, regulative principles are crucial to the possibility of science – they are not, as in a constitutive conception of Kant's epistemology metaphysical (or even scientific) claims about the world.[9] Explanation also involves surpassing the confines of experience, which allows theory to be extended to the realm of the unobservable. In this way, reason both constructs and anticipates entities that could not otherwise be incorporated into science (Neiman, 1994, p. 71). "[Reason] seeks for the unity of this knowledge in accordance with ideas which go far beyond all possible experience (CPR, A662/B690)."[10]

The following quote of Kant may illustrate the difference of view and appreciation of science in Kantian and Humean epistemology:

> When Galileo rolled balls of a weight chosen by himself down an inclined plane, or when Torricelli made the air bear a weight that he had previously thought to be equal to that of a known column of water, or, when in more recent times Stahl changed metals into calx and then changed the latter back into metals by first removing something and then putting it back

---

[9] Very similar to Kant's notion of regulative principles as principles that make science possible, Hasok Chang (2009) introduces the notion of *ontological principles*, which, similar to Kant, is not knowledge empirically derived from the world, but concerns principles that allow for the intelligibility of epistemic activities – for example, the epistemic activity of *counting* involves the ontological principle of *discreteness*. Similarly, but closer to Kant's vocabulary, I have argued that the notion of "same conditions, same effects" is a regulative principle that cannot be proven or disproven, but that we must adopt, as without this principle, scientific experimentation for testing theories would not be possible (Boon, 2015).

[10] Kant's idea that "[Reason] seeks for the unity of this knowledge in accordance with ideas which go far beyond all possible experience" also concerns the formation of scientific concepts. On this aspect of the formation of scientific concepts Rouse (2011) has challenged the assumption of the Vienna Circle and Logical Empiricism that the aim of science is true (or empirically adequate) theories. He argues in favor of an image of science that would place *conceptual articulation* at the heart of the scientific enterprise: "Conceptual articulation enables us to entertain and express previously unthinkable thoughts, and to understand and talk about previously unarticulated aspects of the world" (quote in Rouse's paper at the San Francisco State Workshop, March 2009, on *The Role of Experiment in Modeling*, but not in Rouse, 2011). To this idea, I add that aiming at strict certainty – that is, avoiding any content that goes beyond what is empirically given – would drastically reduce our ability to develop concepts that enable us to think about interventions with the world; interventions that are 'unthinkable' without the added content of those concepts (Boon, 2012).

again, a light dawned on all those who study nature. *They comprehended that reason has insight only into what it itself produces according to its own design*; that it must take the lead with principles for its judgments according to constant laws and compel nature to answer its questions, rather than letting nature guide its movements by keeping reason, as it were, in leading-strings; for otherwise accidental observations, made according to no previously designed plan, can never connect up into necessary law, which is yet what reason seeks and requires. Reason, in order to be taught by nature, must approach nature with its principles in one hand, according to which alone the agreement among appearances can count as laws, and, in the other hand, the experiments thought out in accordance with these principles – yet in order to be instructed by nature not like a pupil, who has recited to him whatever the teacher wants to say, but like an appointed judge who compels witnesses to answer the questions he puts to them. Thus even physics owes the advantageous revolution in its way of thinking to the inspiration that what reason would not be able to know of itself and has to learn from nature, it has to seek in the latter (though not merely ascribe to it) in accordance with what reason itself puts into nature. This is how natural science was first brought to the secure course of a science after groping about for so many centuries. (CPR, Bxii-xiv.)

But if the business of science is that of constructing our own questions and answers, is there still a point in scientific knowledge being true or false? In Kantian epistemology, empirical truth can no longer be grounded in alleged insight into the nature of things-in-themselves. Kant proposes, instead, a version of what is known today as a *coherence theory of truth.* Reason's search for systematic unity is the search for this coherence. While isolated observation-sentences may be confirmed by sense experience, any statement of empirical law can only be regarded as such by virtue of its incorporation in a system of beliefs as a whole (Neiman, 1994, p. 75; and note that Quine, 1951 is close to Kant's view).

Kantian epistemology does not entail a *representational* image of scientific knowledge, that is, the idea that scientific knowledge represents the independently existing order, or structure, or "furniture" of the real world. Moreover, it aims to avoid the idea that pre-given objects or structures "out there" are somehow presented in our observations (or, in the data taken from measurements) and form the solid ground of scientific knowledge (as in Humean empiricism). Instead, in Kantian epistemology, synthetic a priori categories or concepts are crucial to the possibility of knowledge, and are therefore called *regulative principles.*

4 A non-representational view of knowledge

For understanding how "the construction of knowledge for epistemic uses is possible," I propose to adopt the presented interpretation of Kant's epistemology. Crucially,

Kant changed the problem of knowledge. He abandoned some of the ideals and assumptions maintained before and after his days: the ideal of truth as correspondence between world and knowledge; the belief that *certain* knowledge is possible; the basic idea of Humean empiricism that knowledge of basic facts can and should come about by *passive* observation; and, the philosophical question itself, of how truth is justified.[11] Central to Kantian epistemology is that the human mind plays an *active* role at different levels of structuring and conceptualizing the "sensible manifold." Nature imposes the limits, but the mind produces conscious perceptions and judgments, and adds (morphological, logical, mathematical, causal, mechanistic, functional, etc.) structure by the way it asks questions and constructs answers. Without these functions of the mind there would be no conscious perception, nor empirical or scientific knowledge of any physical phenomenon. Furthermore, by these activities of the mind, empirical and scientific knowledge is constructed that goes far beyond experience. As a consequence, as Kant admits, knowledge about nature is never certain – it may prove incorrect when confronted with new experiences.

To the questions raised in Section 2 – What is scientific knowledge? And what purposes does it serve? – another one can be added: If knowledge is "only" a representation of (aspects of) the world, what would it be good for? It is commonly assumed that knowledge can be *used* (for instance in explanation and prediction) by virtue of this representational relationship. However, the philosophy of science has been unsuccessful in answering the question how scientific knowledge (e.g. scientific models) allows for epistemic uses. Knuuttila and Boon (2011) deny that models give us knowledge (i.e., allow for epistemic uses) *because* (i.e., *by virtue*) of this representational relationship, but instead, because they are representations in the sense of being conceptually meaningful constructs that allow for epistemic uses by the model-user.[12] We call these constructs *epistemic tools*, that is, conceptually meaningful tools that guide and

---

[11] Putnam (1981) – who was a former student of Reichenbach – came to reject the anti-metaphysical project of the Wiener Kreis. Affected by Kant's epistemology, he concluded that Logical Positivism & Empiricism is mistaken in holding on to the distinction between primary and secondary properties, which is fundamental to the idea that objective observational sentences are possible and that knowledge can be reconstructed in terms of elementary statements and logical or mathematical structures (also see Suppe's 1974/1979 outline of the so-called Received View). According to Putnam (1981, pp. 60-61): "What Kant did say has precisely the effect of giving up the similitude theory of reference. ... I suggest that (as a first approximation) the way to read Kant is as saying that what Locke said about secondary qualities is true of all qualities ... *If all properties are secondary*, what follows? It follows that *everything* we say about an object is of the form: it is such as to affect us in such-and-such a way. *Nothing at all* we say about any object describes the object as it is 'in itself', independently of its effect on us, on beings with our rational natures and our biological constitutions. It also follows that we cannot assume any similarity ('similitude', in Locke's English) between our idea of an object and whatever mind-independent reality may be ultimately responsible for our experience of that object. Our ideas of objects are not copies of mind-independent things."

[12] Both Giere (2010) and Suárez (2003, 2004) adopt an account of the representational relationship between model and world in which the model-user plays a key role. However, they intentionally developed a *deflationary* notion of representation that only minimally characterizes it. As a consequence, their accounts are not very informative as to the epistemic functioning of models. They shift the problem of how it is possible that models are used to the competent and informed agent.

enable epistemic uses.[13] By epistemic uses we mean scientific reasoning in the broad sense, which not only involves explanation and prediction, but also, asking new questions, crafting new hypotheses, generating ideas on possible interventions with the target system, and even, reasoning towards models of non-existing physical phenomena. Furthermore, we argue that the construction of models involves prudently putting together *heterogeneous* aspects relevant to the problem at hand, in order to form a coherent whole that allows for specific kinds of reasoning. We show, for instance that the model of the *ideal heat-engine* (which is the precursor of thermodynamics) was constructed by Sadi Carnot, not to describe or explain an existing phenomenon, but to represent and investigate an imagined phenomenon (the cycle of conversion of heat into motive power and reverse). Carnot's modelling is a clear example of how the construction of a model involves putting together *heterogeneous* aspects, including empirical knowledge of his days on the behavior of gasses and steam-engines, experiential principles such as "the restoration of equilibrium at difference in temperature," theoretical principles, such as "heat (caloric) will always flow from a hot body to a cold body until the two bodies have the same temperature, by which equilibrium is restored," and the introduction of new theoretical concepts such as "reversibility of the process." This model allowed Carnot to theoretically predict the maximum efficiency ("as set by nature") of real heat-engines. Significantly, Carnot did not test the model by means of experiments. Knuuttila and Boon (2011) follow Boumans (1999) by showing that in scientific practices much of the *justification* of a model is in how it is constructed instead of how it is derived from theory (also see Hacking, 1992). This reconstruction of Carnot's modelling shows several things: firstly, that models are not always derived from theories,[14] but instead, constructed so as to meet *epistemic* criteria (e.g., empirical adequacy in view of accepted empirical knowledge, relevancy, and reliability of empirical and theoretical content, epistemic acceptability of simplifications, and coherency between all elements of the model) as well as *pragmatic* criteria (e.g., the intended epistemic uses, and simplicity and intelligibility so as to allow for specific types of scientific reasoning upon the model); secondly, that a sharp distinction between discovery and justification of scientific knowledge does not smoothly apply to knowledge constructed for epistemic uses; and thirdly, that the model "gives knowledge" (i.e., is epistemically useful), but *not* by virtue of a two-way representational relationship between the model and its real-world target.

---

[13] The notion *epistemic tool* is different from Rheinberger's (1997) notion of *epistemic things*. An epistemic tool entails a *conception* of the target system (represented by means of representational means such as text, analogies, pictures, graphs, diagrams, mathematical formula, but also 3D material entities), and the epistemic tool is constructed such that we can use it in performing epistemic tasks (e.g., reasoning about possible interventions with the target system). Differently, an epistemic thing in Rheinberger's writing is the research object or the scientific object. Epistemic things are material entities or processes – physical structures, chemical reactions, biological functions – that constitute the objects of inquiry.

[14] Also see Morrison and Morgan (1999), who have criticized the semantic view, which assumes that models are derived from theories. They argue that, if attention is paid to the actual construction of models and their uses in science, it becomes evident that models are partially autonomous both of theory and of reality, playing different kinds of epistemic roles.

In a similar fashion, I propose that *scientific concepts* should be considered as *epistemic tools* (concepts may function as definitions, but this is just one way in which they function as epistemic tools; see Boon, 2012, also see Nersessian, 2009, and Andersen, 2012). My argument expands on Feest (2008, 2010), who harks back to Logical Empiricism by explaining scientific concepts in terms of operational definitions that are cast in terms of a paradigmatic experiment that generates the (purported) phenomenon indicated by the concept.[15] I agree with Feest that scientific concepts are tools for investigating the purported phenomenon. Yet, in accordance with Kantian epistemology, I propose that the formation of the scientific concept involves the scientists subsuming the initial operational definition under more abstract concepts (such as "object," "property," "process") or theoretical concepts (e.g., force, field, etc.) or even analogies (also see Hesse, 1966; Hanson, 1969). In a Kantian epistemology, structuring and interpreting experimental data by means of concepts is not fundamentally different from bringing structure to them by means of using logic and mathematics. In other words, mathematical structuring (which involves employing "mathematical templates," e.g., Humphreys, 2004) is similar to the use of "more abstract and theoretical concepts" in structuring and interpreting experimental data. Although formal epistemology only accepts the use of logic and mathematics in structuring data, there is no good reason why this should be preferred to the use of abstract and theoretical concepts. Again similar to structuring by means of logic and mathematics, the crux of subsuming experimental data under concepts is that abstract and theoretical concepts add hypothetical content to the preliminary concept (e.g., the operational definition). This additional content goes beyond the empirically given, and allows scientists to reason upon those concepts, and so derive new predictions and ask new (testable) questions about the purported object or property indicated by the concept.

In short, understanding the construction of scientific models and the formation of scientific concepts in terms of a Kantian epistemology allows for a non-representational understanding of scientific knowledge (knowledge as "epistemic tool" rather than as "representation" in the case of models and "definition" in the case of scientific concepts). In this manner, Kantian epistemology explains how the construction of knowledge for epistemic uses is possible.[16]

---

[15] Important to the question "how the construction of knowledge for use is possible," is: how we bring an observation under existing concepts, and how we generate new concepts for describing "apparently" similar phenomena. Kant uses the notions *determinative* judgment and *reflective* judgement as powers of the mind involved in these activities of applying and constructing knowledge. See Procee (2006) for a clarifying explanation of the two notions.

[16] Peirce's pragmaticism is deeply indebted to Kant's epistemology. Peirce defended an account of scientific reasoning in terms of inductive, deductive, and retroductive reasoning (which are defined more broadly than their narrow logical meanings) that includes how knowledge is constructed. A concise account of these ideas can be found in Chapter III of *A neglected argument for the reality of God* (Peirce, 1908).

5 Conclusions – Epistemological constructivism

Since its foundation by movements such as the Wiener Kreis (Hahn et al., 1929/1973), mainstream philosophy of science has banned metaphysics, and therefore, rejects epistemology that entails metaphysical notions such as Kant's synthetic a priori concepts. With the exception of a few outspoken dissidents (e.g. Hesse, 1966; Hanson, 1969) most of philosophy of science has accepted that the construction of scientific knowledge belongs to the context of discovery, which is beyond the liable domain of the philosophy of science. This belief has had at least two societal disadvantages. Firstly, after Kuhn (1970) for example, this domain has been hijacked by other movements such as the sociology of scientific knowledge (SSK), which have emphasized the subjectivity and irrationality of science to the point of skepticism, leading to a loss of societal trust in science. Secondly, this narrow understanding of its own tasks has resulted in a philosophy of science that is of limited value to scientific practices.

Philosophy of science in practice, however, is a movement that encompasses epistemological issues of scientific practices in its remit, one example being an epistemology for the appropriate construction of knowledge. A focal point of this article, for instance, is an epistemology that facilitates an explanation of how the construction of knowledge for practical uses is possible. It has been argued that a revised, contemporary understanding of Kantian epistemology allows for avoiding well-known problems of (modern versions of) Humean empiricism. The core of the argument in favor of Kantian epistemology is, firstly, that the distinction between primary and secondary properties, needed to justify formal approaches in both the Received View and the Semantic View, cannot be maintained; and secondly, that according to new interpretations Kant's epistemology must be understood as *regulative*, rather that *constitutive*: regulative principles such as synthetic a priory concepts must be pre-supposed as "conditions for the possibility" of generating knowledge. Synthetic a priori in the sense of regulative principles, therefore, does not mean that these principles are *a priori* truths about how the world really is – but instead, that humans, in order to generate knowledge about the world, need to adopt these kinds of principles while knowing that proving them is not possible for fundamental reasons. Therefore, Kantian contrary to Humean empiricism, entails that even the most elementary bits of knowledge result from activities of the mind, not from passive observation.

Next, it has been argued that a Kantian epistemology means that it is not necessary to adhere to a representational view of scientific knowledge according to which knowledge "gives" us knowledge by virtue of the (two-way) representational relationship between knowledge and world.[17] Instead, it is proposed to view knowledge as epistemic tools, which are epistemic entities – such as descriptions of physical phenomena,

---

[17] In several respects Giere's (2006) and Van Fraassen's (2008) notions of representation agree with the 'non-representational' view I aim to pursue, but there are also significant differences (see Boon, 2012, 2015).

and scientific laws, concepts, and models – that must be considered as representations of our conceptions, rather than representations in the sense of a two-way relationship. Indeed, in this manner, the strict separation between primary and secondary properties, that is, between knowledge in terms of formal (logical and mathematical) structures, and knowledge in terms of conceptual content that goes beyond direct experience, disappears.

All this results in a position that entails a view on the character of scientific knowledge and the aim of science, as well as on the tasks of the philosophy of science, which I propose to call *epistemological constructivism*. In this view, the aim of science is to construct knowledge for epistemic uses, and this can range from the highly abstract to the very concrete, as long as this knowledge can be comprehended by human cognition and fits with our experiences of the world. Epistemological constructivism adopts and stresses the idea that science cannot give certainty, as certainty would involve a metaphysical belief in the existence of a structure to the world that can be discovered by science, and from which relevant and true knowledge about concrete target systems can be deduced. Nevertheless, science (should) aim to meet relevant epistemic and pragmatic criteria. Although, epistemological constructivism takes Kantian epistemology as its starting point, it expands on it by assuming that in the history of science, humans have *constructed* more general scientific concepts (such as force, energy, field, evolution, cyclic, symmetry, reversible, operator, amplifier) and epistemic strategies (such as generalization, abstraction, mathematization, reduction, idealization, categorization, explanation, conceptualization, drawing analogies, and crafting graphs, diagrams and models) that are re-used and applied in the construction of knowledge in distinct scientific disciplines (also see Humphreys, 2004). In accordance with Kantian epistemology, these concepts and strategies suit human cognition and human experience of the world, rather than being determined by "how the world really is," independent of humans. This is why their application and re-use in the construction of knowledge is possible. Epistemological constructivism, therefore, does not only put emphasis on the idea that knowledge is constructed, but also entails that scientific practices invent concepts and epistemic strategies for the construction of knowledge.[18] In conclusion, epistemological constructivism provides a preliminary answer to the main question of this paper: how the construction of knowledge for epistemic uses is possible, and also illuminates one of the aims of the philosophy of science in practice, which is to investigate and evaluate general epistemic strategies in science that enable the construction of knowledge for use.

---

[18] Several authors have suggested that Kuhn defends a historicized (Neo-)Kantian position. Friedman (2008), for instance cites Kuhn, who in an Afterword (first published in 1984) appended to his book on Planck and black-body radiation, states: "The concept of historical reconstruction that underlies [this book] has from the start been fundamental to both my historical and my philosophical work. It is by no means original: I owe it primarily to Alexandre Koyré; its ultimate sources lie in Neo-Kantian philosophy." Thus, according to Friedman, Kuhn, not only characterized his philosophical conception as a dynamical and historicized version of Kantianism, but also explicitly acknowledged the background to his own historiography in Neo-Kantian philosophy.

Acknowledgements. This paper has been presented as an invited talk at the CLMPS conference in Helsinki 2015, for which I would like to thank Hanne Andersen. This work is financed by an Aspasia grant (2012-2017) of the Dutch National Science Foundation (NWO) for the project "Philosophy of Science for the Engineering Sciences." Professional text editing with valuable suggestions is by Claire Neesham http://www.rops.org/cen/experience.html.

Bibliography

Allison, H. E. (2004). *Kant's transcendental idealism. An interpretation and defense*. New Haven: Yale University Press.

Andersen, H. (2012). Concepts and conceptual change. In V. Kindi, & T. Arabatzis (Eds.), *Kuhn's The structure of scientific revolutions revisited* (pp. 179–204). Routledge.

Ankeny, R., Chang, H., Boumans, M., & Boon, M. (2011). Introduction: Philosophy of science in practice. *European Journal for Philosophy of Science, 1*(3), 303–307.

Boon, M. (2006). How science is applied in technology. *International Studies in the Philosophy of Science, 20*(1), 27–47.

Boon, M. (2012). Scientific concepts in the engineering sciences: Epistemic tools for creating and intervening with phenomena. In *Scientific concepts and investigative practice* (pp. 219–243). Berlin: De Gruyter.

Boon, M. (2015). Contingency and inevitability in science - Instruments, interfaces and the independent world. In L. Soler, E. Trizio, & A. Pickering (Eds.), *Science as it could have been: Discussing the contingent/inevitable aspects of scientific practices* (pp. 151–174). Pittsburgh: University of Pittsburgh Press.

Boon, M., & Knuuttila, T. (2009). Models as epistemic tools in engineering sciences: A pragmatic approach. In A. Meijers (Ed.), *Philosophy of technology and engineering sciences. Handbook of the philosophy of science* (pp. 687–720). Elsevier/North-Holland.

Boumans, M. (1999). Built-in justification. In M. S. Morgan, & M. Morrison (Eds.), *Models as mediators - Perspectives on natural and social science* (pp. 66–96). Cambridge: Cambridge University Press.

Carnap, R. (1934). On the character of philosophic problems. *Philosophy of Science, 1*(1), 5–19.

Cartwright, N. (1974). How do we apply science? In *PSA: Proceedings of the biennial meeting of the Philosophy of Science Association* (Vol. 1974, pp. 713–719). Reidel Publishing Company.

Cartwright, N. (1983). *How the laws of physics lie*. Cambridge University Press.

Cartwright, N. (1999). *The dappled world. A study of the boundaries of science*. Cambridge University Press.

Chang, H. (2009). Ontological principles and the intelligibility of epistemic activities. In H. De Regt, S. Leonelli, & K. Eigner (Eds.), *Scientific understanding: Philosophical perspectives* (pp. 64–82). Pittsburgh: Pittsburgh University Press.

De Regt, H. W., Leonelli, S., & Eigner, K. (2009). *Scientific understanding: Philosophical perspectives*. Pittsburgh: Pittsburgh University Press.

Feest, U. (2008). Concepts as tools in the experimental generation of knowledge in psychology. In U. Feest, G. Hon, H. J. Rheinberger, J. Schickore, & F. Steinle (Eds.), *Generating experimental knowledge* (pp. 19–26). Berlin: Max Planck Institute for the History of Science: MPI-Preprint 340.

Feest, U. (2010). Concepts as tools in the experimental generation of knowledge in cognitive neuropsychology. *Spontaneous Generations: A Journal for the History and Philosophy of Science, 4*(1), 173–190.

Feix, N. (1978). *Werturteil, Politik und Wirtschaft: Werturteilsstreit U. Wissenschaftstransfer Bei Max Weber*. Vandehoeck & Ruprecht.

Feyerabend, P. (1962). Explanation, reduction, and empiricism. In H. Feigl, & G. Maxwell (Eds.), *Minnesota studies in the philosophy of science* (Vol. 3, pp. 29–97). Minneapolis: University of Minneapolis Press.

Friedman, M. (2008). Ernst Cassirer and Thomas Kuhn: The neo-Kantian tradition in history and philosophy of science. In *The Philosophical Forum* (Vol. 39, 2, pp. 239–252).

Giere, R. N. (1988). *Explaining science: A cognitive approach*. Chicago: University of Chicago Press.

Giere, R. N. (1999). *Science without laws*. Chicago: University of Chicago Press.

Giere, R. N. (2006). *Scientific perspectivism*. Chicago: University of Chicago Press.

Giere, R. N. (2010). An agent-based conception of models and scientific representation. *Synthese, 172*(2), 269–281.

Hacking, I. (1992). The self-vindication of the laboratory sciences. In A. Pickering (Ed.), *Science as Practice and Culture* (pp. 29–64). Chicago: University of Chicago Press.

Hahn, H., Neurath, O., & Carnap, R. (1973). The Vienna Circle of the scientific conception of the world // Wissenschaftliche Weltauffassung: Der Wiener Kreis. In M. Neurath, & R. S. Cohen (Eds.), *Empiricism and sociology* (pp. 299–318). Dordrecht: Springer Netherlands. (Original work published 1929)

Hanson, N. R. (1969). Seeing and seeing as. In N. R. Hanson (Ed.), *Perception and discovery* (pp. 91–110). San Francisco: Freeman, Cooper.

Hempel, C. G. (1965). *Aspects of scientific explanation*. New York: Free Press.

Hesse, M. B. (1966). *Models and analogies in science*. Notre Dame: University of Notre Dame Press.

Humphreys, P. (2004). *Extending ourselves: Computational science, empiricism, and scientific method*. New York, NY: Oxford University Press.

Kant, I. (1998). *Critique of pure reason* (P. Guyer, & A. W. Wood, Trans.). Cambridge: Cambridge University Press. (Original work published 1787)

Kant, I. (2004). *Metaphysical foundations of natural science*. Cambridge Texts in the History of Philosophy. Cambridge University Press. (Original work published 1786)

Klages, H., Mohr, H., Sontheimer, K., et al. (1970). *Weltgespräch - Möglichkeiten und Grenzen der Zukunftsforschung*. Herder, Arbeitsgemeinschaft Weltgespräch.

Knuuttila, T., & Boon, M. (2011). How do models give us knowledge? The case of Carnot's ideal heat engine. *European Journal for Philosophy of Science, 1*(3), 309–334.

Kuhn, T. S. (1970). *The structure of scientific revolutions* (2nd ed.). Chicago: University of Chicago press.

Massimi, M. (2008). Why there are no ready-made phenomena: What philosophers of science should learn from Kant. *Royal Institute of Philosophy Supplement, 83,* 1–35.

Massimi, M. (2011). From data to phenomena: a Kantian stance. *Synthese, 182*(1), 101–116.

McAllister, J. W. (1997). Phenomena and patterns in data sets. *Erkenntnis, 47*(2), 217–228.

Morrison, M., & Morgan, M. S. (1999). Models as mediating instruments. In M. Morrison, & M. Morgan (Eds.), (pp. 10–37). Cambridge: Cambridge University Press.

Neiman, S. (1994). *The unity of reason: Rereading Kant.* Oxford: Oxford University Press.

Nersessian, N. J. (2009). *Creating scientific concepts.* Cambridge, MA: MIT press.

Peirce, C. S. (1908). A aeglected argument for the reality of God. Retrieved from http://en.wikisource.org/wiki/A_Neglected_Argument_for_the_Reality_of_God

Procee, H. (2006). Reflection in education: A Kantian epistemology. *Educational Theory, 56*(3), 237–253.

Putnam, H. (1981). *Reason, truth and history.* Cambridge University Press.

Quine, W. V. O. (1951). From a logical point of view. *The Philosophical Review, 60,* 20–43.

Rheinberger, H.-J. (1997). *Toward a history of epistemic things.* Stanford: Stanford University Press.

Rouse, J. (2011). Articulating the world: Experimental systems and conceptual understanding. *International Studies in the Philosophy of Science, 25*(3), 243–254.

Schickore, J. (2014). Scientific discovery. *The Stanford Encyclopedia of Philosophy (Spring 2016 Edition).*

Soler, L., Zwart, S., Lynch, M., & Israel-Jost, V. (2014). *Science after the practice turn in the philosophy, history, and social studies of science.* Routledge.

Suárez, M. (2003). Scientific representation: Against similarity and isomorphism. *International Studies in the Philosophy of Science, 17*(3), 225–244.

Suárez, M. (2004). An inferential conception of scientific representation. *Philosophy of Science, 71*(5), 767–779.

Suppe, F. (1979). *The structure of scientific sheories.* Urbana: University of Illinois Press. (Original work published 1974)

Suppe, F. (1989). *The semantic conception of theories and scientific realism.* Urbana: University of Illinois Press.

Suppes, P. (1960a). A comparison of the meaning and uses of models in mathematics and the empirical sciences. *Synthese, 12,* 287–301.

Suppes, P. (1960b). Models of data. *Logic, Methodology and Philosophy of Science,* 252–261.

Uebel, T. (1996). The enlightenment ambition of epistemic utopianism: Otto Neurath's theory of science in historical perspective. In W. R. N. R. A. Giere (Ed.), *Origins of Logical Empricism* (Vol. 16, pp. 91–112). Minneapolis: University of Minnesota Press.

Uebel, T. (2008). Writing a revolution: On the production and early reception of the Vienna Circle's Manifesto. *Perspectives on Science, 16*(1), 70–102.
Uebel, T. (2016). Vienna Circle. *The Stanford Encyclopedia of Philosophy*.
Van Fraassen, B. C. (1980). *The scientific image*. Oxford University Press.
Van Fraassen, B. C. (2008). *Scientific representation: Paradoxes of perspective*. Oxford: Oxford University Press.
Woodward, J. (1989). Data and phenomena. *Synthese, 79*(3), 393–472.
Woodward, J. (1998). Data, phenomena, and reliability. *Philosophy of Science, 67*, S163–S179.

Author biography. Mieke Boon is full professor in Philosophy of Science in Practice at the University of Twente (The Netherlands), where she is a core teacher in the programs *Philosophy of Science, Technology and Society* (PSTS) and in the University College ATLAS. In 2005 she initiated the *Society for Philosophy of Science in Practice* (SPSP) and was member of the SPSP organizing committee until 2016. Since 2015 she is member of the *European Philosophy of Science Association* (EPSA) steering committee, and chair of the chamber theoretical philosophy of *National Dutch Research School in Philosophy* (OZSW). She studied philosophy at the University of Leiden. Her scientific background is in chemical engineering in which she received an MSc and PhD, both with honor. In 2003-2008 and 2012-2017 she worked on personal research grants of the Dutch National Science Foundation (NWO-Vidi and NWO-Vici/Aspasia), aiming to develop a *philosophy of science for the engineering sciences*.

17 Patchwork narratives for tumour heterogeneity

GIOVANNI BONIOLO [*,†]

Abstract. Molecular biology, in particular post-genomics, is determining the proximal and the distal future of both biomedical research and clinical practice. Thanks to the advancements in understanding the molecular bases of diseases, we are drastically changing our approach to prevention, diagnosis and therapy. In particular, through recently available sequencing techniques, we have realised that each instance of cancer affecting a particular individual is actually composed of a set of different cancer subpopulations. This phenomenon, called *tumour heterogeneity*, is posing a tremendous challenge to biomedicine but also to philosophy. In this paper the notion of *patchwork narrative* is introduced in order to offer a philosophical framework for the description and the explanation of the many different events and processes ascribable to tumour heterogeneity and to its complexity.

Keywords: tumour heterogeneity, patchwork narrative, complexity, explanation, description.

1 Introduction

Molecular biology, in particular post-genomics, is an important game-changer that is determining the proximal and the distal future of both biomedical research and clinical

---

[*] Dipartimento di Scienze Biomediche e Chirurgico Specialistich, Università of Ferrara, Via Fossato di Mortara, 64A, 44121-Ferrara - Italy.
[†] Institute for Advanced Study, Technische Universität München - Germany.

practice.[1] Thanks to the advances in understanding the molecular bases of diseases –which makes individuals' genetic makeup, their life styles and the environments in which they live, amenable to systematic scrutiny– we have the promise to drastically change our approach to prevention, diagnosis and therapy.

These achievements, especially in the oncological field on which this paper is focused, have suggested to some that we are in the age of *personalized medicine*.[2] Nevertheless, despite its importance and novelty, personalized medicine seems to attract more researchers and clinicians, rather than patients. Indeed, understandably, the latter have shown much more interest in the "care" side of the matter, starting from the consideration that healthcare delivery should not be limited to disease treatment in a strict technical sense, but in the act of taking care of a human being as a whole (van Heist, 2011; Cornetta & Brown, 2013; Boniolo & Sanchini, 2016).

However, the passage from *personalized medicine* to *personalized care*, especially in the oncological field, is made more intricate by an aspect, that is, *tumour heterogeneity*, which so far has not been discussed sufficiently from an epistemological perspective (see Bertolaso, 2011; Germain, 2012; Plutynski, 2017; Fagan, 2017).

Tumour heterogeneity means not only that each cancer has to be individualised in a specific patient, but, more importantly, that each individual cancer affecting an individual is actually composed of a set of different cancer subpopulations. That is, cells belonging to the same cancer show distinct genetic and phenotypic characteristics (such as gene expression, metabolism, motility, and angiogenic, proliferative, immunogenic, and metastatic potential).

This fact, which emerged from recently available sequencing technologies and which is at the centre of an intensive biomedical research program[3], is posing a huge challenge to personalized medicine and thus to personalized care.

Needless to say, if we were able to provide a good epistemological analysis of this issue, we would be able to understand the core of contemporary biomedicine and its epistemic role in the fabric of knowledge. It is a task that we should undertake, if we want to comprehend the current status of biomedical research and clinical practice and

---

[1] Of course, these new ways of seeing disease and health should be at the core of the philosophers' attention (see Boniolo & Nathan, 2017). By *post-genomics* we usually refer to a research field that extends the domain of genomics (characterised, in particular, by genome sequencing, functional genomics, gene architecture) to include transcriptomics (the study of patterns of gene transcription), proteomics (the study of patterns of protein expression), metabolomics (the study of patterns of chemicals influencing our cellular biochemistry and metabolism), and, above all, epigenomics (the study of gene expression due to micro and macro-environmental inputs particularly affecting DNA methylation and histone modifications).

[2] It can be broadly defined as the tailoring of medical treatment to the individual patient's (especially genetic and epigenetic) characteristics, needs and preferences during all stages of care, including prevention, diagnosis, treatment and follow-up. See the *Report on Personalized Medicine* by European Science Foundation and the report *Paving the way for personalized medicine* by Food and Drug Administration and Golubnitschaja et al. (2016, 23).

[3] See the recent issue of *Nature* devoted to it (Vv.Aa., 2013).

if we want to be more aware of the vagaries that life could present whenever we are in the unfortunate role of patient or patient's relative.

In order to philosophically set the question, after a first survey on what tumour heterogeneity is and on what its clinical consequences are, I introduce the notion of *patchwork narrative*. This concept allows me to indicate how molecular oncologists describe and explain the extremely complex set of events and processes ascribable to tumour heterogeneity by resorting to conceptual terms belonging to other biological fields (in particular, evolutionary biology, developmental biology, ecology and stem cells biology). Then I present the main accounts of tumour heterogeneity in order to make clear their patchwork narrative structure. I conclude by discussing the reasons why speaking in terms of evolutionary, developmental, or ecological models is not satisfactory and why a patchwork account could be ideal to cope with the epistemological complexity of this biomedical field.

2 Tumour heterogeneity

Any multicellular organism shows cellular heterogeneity in its several compounding tissues. For example, in humans there are about 200 different types of cells. Nevertheless, such heterogeneity is guided by an ordered developmental program starting from the zygote's single initial genome and governed by a proper coordination of dynamic signal transduction and then by long-term maintenance of gene expression patterns due to epigenetic mechanisms. These processes ensure a balance between cells capable of continuous self-renewal or remaining in stem-cell-like states and their progenies committed to tissue lineages and differentiation. The same processes occur, more or less, in a neoplastic tissue but in that case, we do not have the initial developmental program starting from the genome and resulting in ordered structures. Instead, the neoplastic process results in a complex situation—*tumour heterogeneity*—that is difficult to describe and to explain and even more difficult to clinically treat.

It is important to recall that since the advent of cancer nosology, an organ based classification system has been used with good success in clinical practice and research. Thus, for a long time we have spoken about breast cancer, lung cancer, prostate cancer, liver cancer, etc. With the development of molecular technologies, both genetic (e.g., KRAS mutation for colon and lung cancer; EGFR mutation for lung cancer; BRAF mutation for colon cancer and melanoma) and epigenetic (e.g., DNA methylation alterations) biomarkers have began playing a major role both in the diagnostic phase and in the drug discovery process that could lead to new therapies (see Febbo et al., 2011; Baylin & Jones, 2011; Boem, Pavelka, & Boniolo, 2015). But the introduction of extensive sequencing technology has revealed the genetic and epigenetic heterogeneity among tumours and, thus, not only that there are no two identical tumours, but that not even two samples from the same patient's cancer are exactly the same. So, step-by-step, we have understood that cancer is not a single disease, rather that any cancer is a "different disease", and that in the same cancer actually we have "many cancers", each with its own histopathological and biological features.

In a slogan, as any individual human being is unique (Boniolo & Testa, 2012; Boniolo, 2013a), so any tumour is unique, or, rather, any tumour cell population in an individual tumour is unique.

Such uniqueness has two facets: the *intertumour heterogeneity* (the variability occurring between tumours arising in the same organ) and the *intratumour heterogeneity* (the variability occurring in the same individual tumour) (see Burrell, McGranahan, Bartek, & Swanton, 2013).

In what follows, I will focus my analysis particularly on intratumour heterogeneity, since it is the more relevant from a research and clinical point of view and the one which necessitates a more accurate epistemological investigation.

Before beginning, it is worth recalling two pairs of notions that will help understanding what follows. The first pair is the *cell-of-origin* and the *cancer stem cell* (CSC). By the former we intend a normal cell that acquires the first cancer driven mutation. The latter is a cell that possesses the characteristics usually associated with normal stem cells, that is, self-renewal (which, in our case, produces daughter cancer stem cells) and potentiality (which, in our case, begets sub-type cancer cells).

The second pair has to do with *spatial heterogeneity*, indicating that different regions of a tumour present different series of genetic aberrations, and *temporal heterogeneity*, referring to the course of disease progression (see Geyer et al., 2010; Torres et al., 2006; Martelotto, Ng, Piscuoglio, Weigelt, & Reis-Filho, 2014). Concerning this latter aspect, one should note that heterogeneity within primary tumours is only one of the many aspects of cancer heterogeneity. Cancer is a systemic disease: over time, malignant cancers shed a large number of cells into the blood stream and lymph vessels; some of these cells find a place in distant sites and develop into metastases. Therefore, to have a really complete understanding of cancer heterogeneity we should also understand metastatic tumours, which, as is known, are the most fearsome aspects, since they are responsible for the majority of cancer-related deaths.

3 Clinical relevance

Needless to say, spatial and temporal heterogeneity –that is, inter and intratumour heterogeneity and heterogeneity between primary and metastatic lesions– have profound implications for patient care (Bedard, Hansen, Ratain, & Siu, 2013, see).

Concerning the diagnostic aspect, heterogeneity poses a challenge to personalized cancer medicine because a single needle biopsy, or a single surgical excision, almost never accurately captures the complete genomic landscape of a patient's cancer. One might think that serial tumour sampling, at crucial times, may help to monitor the temporal heterogeneity. In many cases, however, this multiple sampling is not clinically feasible. On the other hand, heterogeneity also involves clinically relevant biomarkers. This means that, due to genetic differences within and between tumours, biomarkers that may predict treatment responses or prognoses vary from situation to situation.

Concerning therapies, the extent of heterogeneity observed in various cancer types suggests we need a new paradigm for thinking about how they should be delivered. Heterogenic tumours may exhibit different sensitivities to cytotoxic drugs among different clonal populations (see Yap, Gerlinger, Futreal, Pusztai, & Swanton, 2012). This is attributed to clonal interactions that can inhibit or alter therapeutic efficacy. Thus, drug administration in heterogenic tumours, even if it results in initial tumour shrinkage, seldom kills all tumour cells. The initial tumour population can bottleneck, such that few drug resistant cells can survive.[4] This allows resistant tumour populations to replicate and grow a new tumour, which is resistant to the initial drug therapy and sometimes even more aggressive (see Sequist et al., 2011; Turner & Reis-Filho, 2012; Landau et al., 2013) (see **Figure 1**, from Kleppe and Levine, 2014).

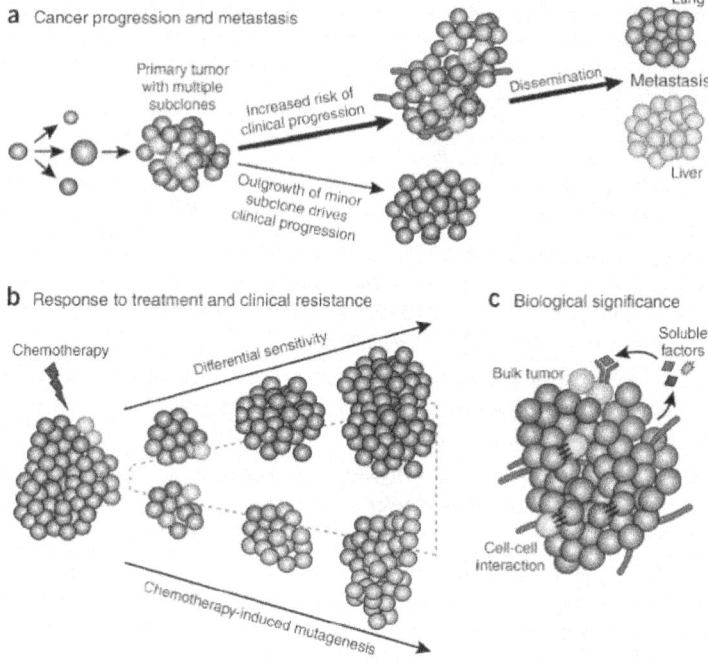

Figure 1.

---

[4]For example, *temozolamide*, the standard first-line therapy for glioblastoma multiforme, induces mutations in tumour DNA. Some are deleterious for the cells and result in death, others neutral and act as passenger mutations, but others such as mutations in mismatch repair genes are potentially advantageous for tumour cells (see Johnson et al., 2014). *Imatinib mesylate* is the gold standard of treatment of chronic myeloid leukaemia. Although response rates to imatinib are high at 5 years, approximately 6% of patients progress to the accelerated phase and 3% have a haematological relapse (see Druker et al., 2006).

4 The patchwork narrative

Given this brief introduction on tumour heterogeneity, let us move on to the philosophical tool I wish to introduce.

In order to describe and explain scientific events and processes, we need a particular kind of *narrative*, which, depending on the specific field, takes different forms. Here by 'narrative' I mean a form of human discourse that, more or less coherently, communicates something about something. Thus, a 'scientific narrative' communicates something, which should be epistemologically valid, about a given piece of nature, also by offering a representation of it. Different scientific fields allow for different kinds of scientific narratives. Just as there is the research field of gravitation, there is also the research field of tumour heterogeneity. But while a narrative concerning gravitation has the form of a theory (i.e., the theory of gravitation), it seems that this is not true of tumour heterogeneity: we do not have any theory of tumor heterogeneity. Maybe here 'theory' could be used in a very loose and non-technical manner, but without any strong epistemological commitment concerning its logical structure and its predictive power. On the other hand, we know that the question of whether there are theories in biology and medicine has a long and controversial history, recently revamped (Pigliucci, Sterelny, & Callebaut, 2013).

Nevertheless, even if we do not have a theory of tumor heterogeneity, we need some kind of narrative in this field, since we have to describe and explain why there is such a genotypic and phenotypic diversity among the neoplastic cells of a single tumour, or among those of a primary tumour and those of its metastatic products. In order to do this we have to consider a matter of great great complexity, both from a spatial and a temporal point of view, involving both differences among cellular populations and intricate intersecting probabilistic causal pathways where genes, proteins, chemicals, intrinsic and extrinsic cellular properties[5], etc. have a role. It is not important here to address the problem of whether this extreme ontological complexity has to be considered in a metaphysical way –that is, as something pertaining to the real nature of tumors, or just epistemically –that is, as something concerning our level of knowledge of what tumors are and of how they develop. What is at stake is the fact that we have to describe and explain such a complexity in the most accurate and reliable possible manner. This is not merely of philosophical interest, and not even a sheer matter of scientific curiosity, but a practical need since we want to cure actual, and to take care of potential, patients affected by cancer pathologies.

---

[5] By *cell intrinsic properties* we mean the properties related both to the genetic mutations driving the primary tumour formation (for example, HER2, or BRCA1 and 2 mutations for breast cancer), and to the alterations of the epigenetic landscape (disruption of DNA methylation, histone modification, and chromatin compartments). Instead the *cell extrinsic properties* concern the usual interactions between cells (in this case cancer cells) and the surrounding microenvironment, connected with the recruitment of both endothelial cells to generate the blood supply for the developing tumour and other stromal cells (inflammatory cells, fibroblasts, pluripotent mesenchymal stem cells, etc.).

We could think pluralistically that in the field of tumour heterogeneity we have a narrative realised by a set of different phenomenological models, each of which is able to grasp a particular aspect.[6] This may resemble what happens in population genetics or nuclear physics. We know that there is no theory of population genetics as a whole and (for totally different reasons) that there is no theory of nuclear physics as a whole. Instead, we have a set of (more or less mathematized) phenomenological models able to describe and explain events and processes belonging to the field of population genetics (see Hartl & Clark, 2007; Ewens, 2004) and, similarly, we have a set of (mathematized) models able to describe and explain events and processes belonging to the field of nuclear physics (see Boniolo, Petrovich, & Pisent, 2002). Nevertheless, the narratives concerning both population genetics and nuclear physics have their own language, with their own conceptual terms on (or by means of) which the (more or less mathematized) phenomenological models are constructed. But this does not happen with tumour heterogeneity. As I will show in the next sections, in order to describe and explain events and processes, each researcher constructs a peculiar narrative borrowing concepts from other biological fields, mainly evolutionary biology, developmental biology, ecology, and stem cell biology.

Certainly, as could be easily observed by reading the literature, there are molecular oncologists constructing narratives in terms of an 'evolutionary/Darwinian model of tumour heterogeneity', 'cancer stem cells model for tumour heterogeneity', 'ecological model for tumour heterogeneity', etc. Actually, they are using the term 'model' in a very loose way. As far as I know, in none of their papers is there an analysis of what a model is, or an indication on what they mean by 'model'. On the other hand, the fact that they use a philosophically (epistemologically) loaded term like 'model' implies neither that they use it in a technical sense, nor that they recognize the need to clarify in which sense they are adopting it. They simply use it in an "epistemologically relaxed" way, as has happened with terms like 'information', in the field of coding DNA, or 'mechanism', especially in molecular biology (see Boniolo, 2003, 2013b). And, I think, nothing should hinder this way of proceeding. They are not philosophers of science and they are not publishing in philosophical journals.

The same problem attaches to their narratives: molecular oncologists claim that the latter are evolutionary/Darwinian, or developmental, or ecological. Certainly, what they are providing could have a certain *Familienähnlichkeit* with an evolutionary/Darwinian account, or with a stem cell account, or with an ecological account. But, as we will see, it is just a *Familienähnlichkeit*.

This points have induced me to introduce the idea of *patchwork narrative*. By this

---

[6]There is an extremely long (and well-known) story about what a model is and about what are (if there are) the differences between a theory and a model. There is also a long story about different taxonomies of models in different scientific fields. Of course, it is not worth entering now such a topic (see Boniolo, 2007). For the sake of what I am discussing here, it suffices to speak in terms of 'phenomenological models', intended as those models constructed by the researchers in order to *save the phenomena* in a given domain in which there is no theory available.

concept I mean a scientific narrative which is not a theory (in any technical sense), not even a model (in any technical sense) or a set of models, but a descriptive and explanatory discourse which, by borrowing conceptual terms (the "pieces of the patchwork") belonging to evolutionary biology, developmental biology, stem cells biology, ecology, etc., offers a logically coherent and epistemically valid view of a given aspect of tumour heterogeneity.

Let us see if and how this philosophical idea really works. In order to accomplish this, I will review the main narratives adopted to describe and explain events and processes concerning (intra)tumour heterogeneity. Then, I will return to philosophy.

5 Describing and explaining intratumour heterogeneity

As mentioned, by *intratumour heterogeneity* we mean the complex coexistence of subpopulations of cancer cells that differ in their genetic and phenotypic characteristics within a given primary tumour, and between a given primary tumour and its metastases. In order to describe and explain it, four main narratives have been offered: (i) *the clonal evolution narrative*; (ii) *the cancer stem cell narrative*; (iii) *the cell plasticity narrative*; (iv) *the ecological narrative*.

5.1 The clonal evolution narrative

The clonal evolution narrative was brought to wide attention in the Seventies (see Cairns, 1975; Nowell, 1976; Attolini & Michor, 2009; Polyak, 2014). It is based on the idea that cancer cells evolve progressively during multistep tumourigenesis and heterogeneity is caused by heritable genetic and epigenetic changes, which form the material for the selection and the clonal development of novel cell populations.

Tumours arise from a single mutated cell (the cell-of-origin) and accumulate additional mutations as they progress. Such mutations give rise to additional subpopulations, each of which has a different ability to divide and mutate further. As a consequence, there could be subclones possessing an evolutionary advantage over the others within the tumour microenvironment, and these subclones may also become dominant in the tumour over time. It is important to note, however, that the generation of variants can occur more rapidly than the elimination of less-fit clones, resulting in an increase of heterogeneity.

Already from this, it is evident that such a narrative borrows its jargon mainly from evolutionary biology: the evolution concerns cancer cells and the selection pressure is given by the tumour microenvironment.

For the sake of simplicity, tumour evolution is often described as a *linear succession of clonal expansion rounds*, where every new step is driven by the acquisition of an additional mutational event, which leads to a new selective situation (**Figure 2a**, from Marusyk and Polyak, 2010).

However, this linear representation does not reflect the dynamics of any tumour evolution, even if it has been observed in multiple myeloma and in acute myeloid leukaemia. More common appears to be a so-called *branched-evolution*, found in many tumours (breast, ovarian, prostate, pancreatic, and bladder cancers, as well as in chronic lymphocytic leukaemia, multiple myeloma, acute myeloid leukaemia, glioma and clear cell renal cell carcinoma) (**Figure 2b**, from Marusyk and Polyak, 2010). Here, random mutations are constantly produced as a result of both proliferation and increased genomic instability and then tested by selection. Concerning these mutations, only a minority of them is selectively advantageous, while the majority is discarded by selection. Nevertheless, there are many neutral or even slightly disadvantageous mutations that can be retained and even undergo some expansion due to genetic drift. These are the so-called *passenger mutations*, which are acquired as incidental by-products of the cancer cells' high mutability (they are the *hitchhiker mutations* of the evolutionary account; see Greaves and Maley, 2012).

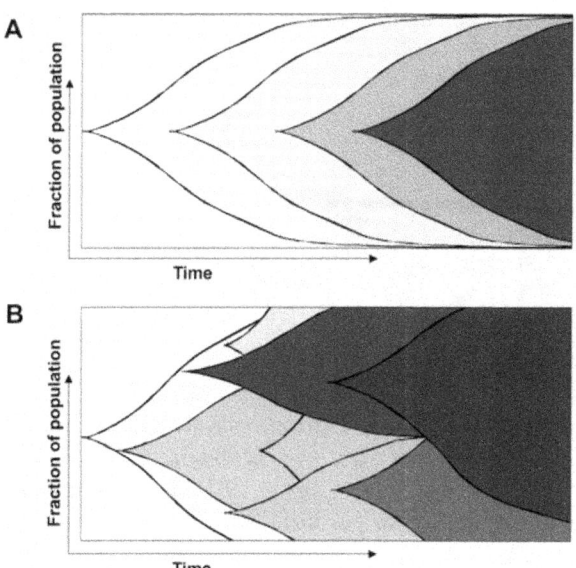

Figure 2.

As the selective pressure changes, for instance, between the microenvironment of the primary and metastatic site or due to the administration of systemic therapies, a different set of mutations may become advantageous. The presence of differential selective pressures can, therefore, cause subclones to diverge, generating additional intratumour heterogeneity. Moreover, since selection is context-specific and blind to the future, some of the mutations that are selectively advantageous at certain stages of tumour progression may lead to evolutionary dead ends and, therefore, cannot be present at a successive time.

It should also be noted that in addition to genetic alterations, epigenetic events could be heritable and subject to selection. But the complexity of tumour evolution is further influenced by the continuous variation of the tumour microenvironment, which alters the selective pressures on tumour cells.[7]

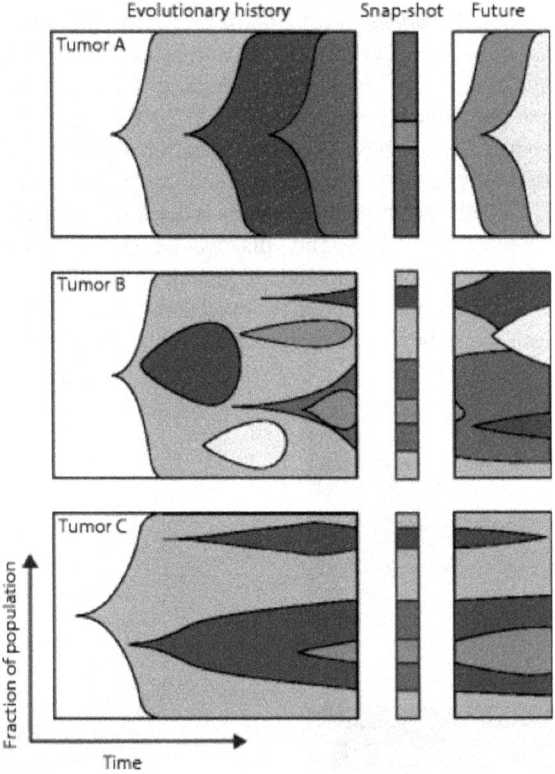

Figure 3. *The left panel shows the evolutionary history of a tumour; the central panel represents a snapshot of the tumour at a given time; the right panel shows the potential future development. Tumour A shows a linear evolution pattern; tumours B and C show a branched pattern. Single snapshots of tumours B and C could suggest that they have identical evolutionary processes, but their past and future evolution actually follow different patterns.*

Note that the subclonal diversity within a tumour if viewed as a snapshot, rather than longitudinally, provides little information about the future evolution paths that the tumor and its cellular populations might take. On the other hand, as already noted above, the acquisition of mutations is a stochastic process, and it is blind to the future.

---

[7] In addition, there could a change of tumour cell populations over time due to the *epistatic relationships* that there are among the effects of driver mutations (see Weigelt & Reis-Filho, 2014). Generally speaking, in genetics by *epistasis* we mean that the effect of more genes together is different from the effect of each one considered individually (see Cordell, 2002; Phillips, 2008).

All of this implies a lot of difficulties for clinicians in prognosticating what will happen to that particular patient in the near and more distant future. (**Figure 3**, from Hiley, de Bruin, McGranahan, and Swanton, 2014).

5.2 The cancer stem cell narrative

The concept of *cancer stem cell* (CSC) was proposed to describe and explain heterogeneity of cancer cells more than three decades ago (see Dick, 2008). However, it emerged as a mainstream idea only recently, initially for hematopoietic neoplastic phenomena (Lapidot et al., 1994). Then, it was expanded to solid tumours, such as breast cancer (see Al-Hajj, Wicha, Benito-Hernandez, Morrison, & Clarke, 2003; Visvader & Lindeman, 2008).

The CSCs narrative proposes that, within a given tumour, a phenotypic hierarchy exists, with a minor subset of the CSCs at the apex and highly proliferating, lineage-committed progenitors and terminally differentiated cells at the base. This means that tumour growth, disease progression and heterogeneity generation are driven by a small population of tumor cells, while the vast majority do not contribute and would be unable to regenerate a tumour after a successful therapeutic intervention. In principle, CSCs could self-renew indefinitely, drive growth and differentiate into virtually all cancer cell types, thereby producing heterogeneity. On the other hand, progenitors and terminally differentiated cells should be highly proliferative, display lineage commitment, have limited proliferative potential[8] and very poor capacity to contribute to disease progression.

However, the existence of CSCs is under debate. One reason is that is not easy to reproduce markers for CSCs across multiple tumours. Moreover, in order to determine tumourigenic potential, we utilize xenograft models. Unfortunately, these methods suffer from limitations such as the need to control immune response in the transplant animal, and the significant difference in environmental conditions between the primary tumour site and the xenograft site (see Quintana et al., 2008; Anderson et al., 2011; Meacham & Morrison, 2013). This has raised some doubts about the accuracy and the relevance of the results concerning CSCs. Further, the CSCs hypothesis is disputed given the evidence showing the existence of a dynamic equilibrium between differentiated cells and CSCs (Gupta et al., 2011). It seems not only that CSCs can differentiate into terminally differentiated cells, but terminally differentiated cells can also de-differentiate into a CSCs state. This means that, in some contexts, the CSCs phenotype may represent a state that cancer cells within a tumour can acquire rather than an isolated population of cancer cells that constantly exhibit those properties. Moreover, the difficulty in replicating solid-CSCs markers, the variability from patient to patient, and the dissimilarity in results from different xenograft models have

---

[8]The cellular division is linked to the so-called *Hayflick's limit* (see Hayflick, 1965).

made it even more unclear which cancers can be positively grasped with this approach (Magee, Piskounova, & Morrison, 2012).

5.3 The cell plasticity narrative

The two approaches above might not be mutually exclusive. For example, a driver mutation could occur in a cell with stem cell properties. Or a mutation might promote cancer stemness by reprogramming a cell that lacked this potential before (see Visvader, 2011). The *cell plasticity narrative* develops this idea.

It is an approach, mixing the evolutionary jargon and the stem cells jargon, according to which cancer heterogeneity is due to a series of processes concerning different cell populations in which committed, non-stem cells and non-CSCs can undergo a de-differentiation program and re-enter the cancer stem state (**Figure 4**, from Marjanovic, Weinberg, and Chaffer, 2013).

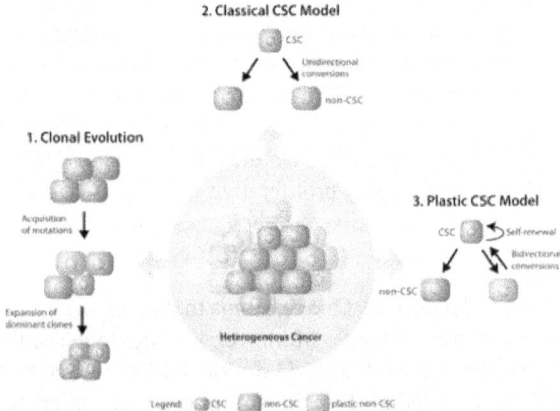

Figure 4.

Note that epigenetic variation might be one of the changes contributing to cellular plasticity. On the other hand, this would be an additional factor in countering the strictly hierarchical hypothesis of the existence of CSCs as a stable and isolated population (see Easwaran, Tsai, & Baylin, 2014).

5.4 The ecological narrative

We know that the tumour microenvironment is a highly heterogeneous mix of cellular and non-cellular components, consisting of the extracellular matrix, vasculature, fibroblasts, smooth muscle cells, immune cells, nerves, and proteins in the immediate extracellular environment. As briefly described, it plays a major role in heterogeneity, since it is the selective agent.

There is not only a one-way direction: from the microenvironment to the tumour; but the reverse pathway also occurs: from the tumour to the microenvironment. Any tumour creates its own particular microenvironment, and since any tumour is unique, due to its heterogeneity, each microenvironment is unique. On the other hand, any microenvironment selectively acts differently on tumour cells, and since any microenvironment is unique, any selective action is unique (see **Figure 5**, from Ungefroren, Sebens, Seidl, Lehnert, and Hass, 2011, 1). Moreover, the tumor microenvironment is modified by drug therapies, but different microenvironments allow for a different efficacy of drug therapies (Junttila & de Sauvage, 2013). Starting from these considerations, there are many molecular oncologists who have elaborated narratives, by borrowing concepts from ecology and by emphasising the idea of *niche construction*, based on a strong mutual causal relation between tumour heterogeneity and tumour microenvironment (see Odling-Smee, Laland, & Feldman, 2003; Merlo, Pepper, Reid, & Maley, 2006; van Dijk, Göransson, & Strömblad, 2013; Barcellos-Hoff, Lyden, & Wang, 2013; Yang et al., 2014; Chen et al., 2015; Kareva, 2015; Amend & Pienta, 2015).

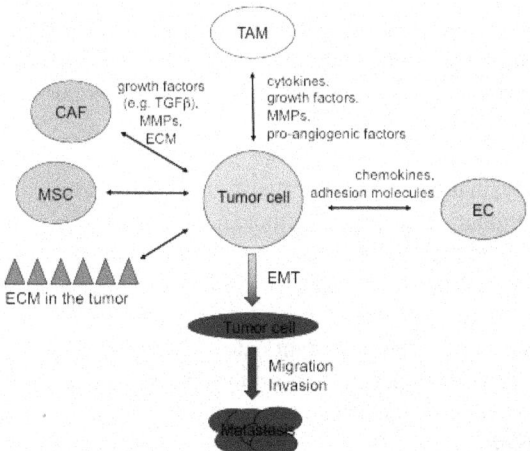

Figure 5. *Mutual interactions between tumour cells and the extracellular matrix (ECM), tumour-associated macrophages (TAM), carcinoma-associated fibroblasts (CAF), mesenchymal stem cells (MSC), endothelial cells (EC).*

6 The patchwork narrative and its competitors

As seen, many scientists working on tumour heterogeneity speak of an evolutionary, or developmental, or ecological model. Unfortunately, there is no indication of the criteria to be satisfied in order that those accounts could be really considered as such. This mission has been taken by some (not so many) philosophers who have exercised their analytical expertise exactly on this point.

One of the first (if not the first) works facing tumour heterogeneity from a philosophical perspective was a paper in which the author argued that the 'Darwinian (or evolutionary) model' is not Darwinian (or evolutionary) in a paradigmatic sense (Germain, 2012).

Germain first recalled the characteristics which should be satisfied in order for a population to be considered "Darwinian". Then he showed that among the events and processes ascribable to tumour heterogeneity, there were some that could not exactly satisfy those criteria. Note that Germain had many possibilities of identifying the set of characteristics considered as individuating a real Darwinian population, and therefore a real Darwinian model. Instead of choosing among the evolutionary biologists' proposals (for example, that one suggested by Mayr, 2002 or by Ridley, 2007), he decided for a philosopher's account, that is, the one indicated by Godfrey-Smith (2009). According to Godfrey-Smith, a population in order to be Darwinian should be characterised by (i) fidelity of heredity; (ii) abundance of variation; (iii) continuity, or smoothness, of the fitness landscape; (iv) dependence of reproductive differences on intrinsic characters; (v) reproductive specialization; (vi) integration, or the extent of mutual dependence. Germain showed, by resorting to the scientific literature, that although cancer population meets these requirements to some degree, and can therefore be considered minimal Darwinian populations, they do not quite qualify as paradigmatic Darwinian populations:

> Let us now summarize our evaluation of cancer cells. Cancer cells have sufficient variation for evolutionary change, and although their relatively low fidelity of heredity might threaten the possibility of complex adaptation, it is generally in the range compatible with minimal Darwinian processes. However, cancer cells display a level of integration reminiscent of organs, and share reproductive fate to an important extent. Much of the fitness differences between cells does not depend on their intrinsic features, which strongly suggests that these differences cannot be the basis for paradigmatic Darwinian processes. Perhaps the most crucial dimension, the degree of reproductive specialization will vary from case to case, depending on the extent to which a given cancer follows a CSC model. When it does, long-term proliferation will be restricted to a tiny subpopulation of the tumour, rendering most cancer cells unable to accumulate evolutionary changes [... Therefore c]ancer cells are at least minimal Darwinian populations [that is, there are heritable variations in fitness], but not paradigmatic ones. (Germain, 2012.)

This perspective has recently been challenged by Lean and Plutynski (2016). They claim that Germain is not wrong but that his view is too narrow since he does not take into account a "multilevel evolutionary perspective" (see Sober & Wilson, 1998; Okasha, 2006), especially in the formulation offered by Damuth and Heisler (1988). Here there are some points to be noted. First, this perspective is not totally shared both among the scientists and among the philosophers, even if, honestly, this fact is not so important for what is here on the stage.[9] Second, Lean and Plutynski do not show that each of Germain's counter-examples to the evolutionary model of tumour heterogeneity is really graspable by the multilevel evolutionary perspective. Third, they assert that "this multi-level perspective provides a representation of cancer progression that is both predictive and explanatory. The accumulation of mutations (and epigenetic changes) in neoplastic progression involves heritable variation in fitness of cancer cells and lineages, e.g., timing and acquisition of particular mutation types, chromosomal instability, hypoxia, acid resistance, transition to mesenchymal phenotype. Cancer progression involves several transitions in individuality and levels of selection, and, is a unique case of multi-level selection: it evolves at the level of populations of cells, competition among cell lineages, via cooption, a kind of cross-level exaptation, and as a group, or higher level individual" (Lean & Plutynski, 2016, p. 52). In reality, they show that some events and processes ascribable to tumour heterogeneity can be coopted by a multilevel evolutionary account. Nevertheless, as we know, this does not mean that such an account is epistemologically useful in general, but only that it is epistemologically useful in grasping those particular items. Moreover, I have some doubts about "predictability". If they were right, the difficulties indicated towards the end of section 5.1 above (and depicted in Fig. 3) would not exist and the clinical prognosis of any tumour would be easy and sure. Unfortunately for the patients, this is not so.

Certainly, and I could agree with Plutynski (2017), there are no "devastating objections to the evolutionary perspective on cancer", even if she herself concedes that cancer cells do not have the same heritable variation in fitness and the same level of adaptation of other living beings. Nevertheless these objections exist and they weaken the claim that we have an evolutionary model for cancer heterogeneity.

There is another point that is worth mentioning. Speaking in favour of an evolutionary model, it has been said that even if the cancer cells have a short life (due to the death of the host, or to their –fortunate for the host— total elimination due to a working therapy), there is the case of the canine transmissible venereal tumour and of the *in vitro* cancer cells, like the famous HELA cells.[10] These should be examples of immortal, or at least long living, cancer populations. Actually the canine transmissible venereal tumour is one of the very few examples of non-viral transmissible cancer, that is, cancer transmissible but not by oncoviruses or cancer bacteria (others are the Tasmanian devils' facial tumour disease; the Syrian hamsters' contagious reticulum cell sarcoma; the soft-shell clams' neoplasm of the hemolymphatic system). It is important to note that they are transmitted from non-humans to non-humans and that they happen very rarely in humans and in a totally different way.[11] Given their rarity and particularity, instead of being considered as a good proof of the long life of cancer cells and thus

---

[9] For a first hint on the debate, see http://edge.org/conversation/the-false-allure-of-group-selection.

[10] These are immortalised cancer cells derived from a cervical cancer of a woman, Henrietta Lacks, who died of her tumor in 1951 (see Skloot, 2010).

[11] In the literature, very few cases between humans are reported, for example, Gärtner et al. (1996).

as an argument in favour of an evolutionary model, I would consider them as one of many aspects of the complexity of the phenomenon 'cancer', especially considering the fact that species with different genotypes do have or do not have (as it happens in naked mole rats) peculiar cancerogenous manifestations. Concerning the immortalized *in vitro* cancer cells, there is a misunderstanding. The *in vitro* cancer cells (and we have dozens of different varieties of these cancer lines, which are now commercialised for the benefits of the researchers) have almost nothing to do with the real *in situ* cancer cells or with the primary tumour cancer cells, as I discussed elsewhere (Boniolo, 2017).

As seen above, beyond the evolutionary account, there is also the CSCs account. This could be a problem for those who think that the evolutionary model is enough. Nevertheless, Plutynski (2017) claims that there is no reason to consider the CSCs as a counterexample, since instead of considering one single evolving lineage starting from a cell-of-origin, one should think in terms of many evolving cell lineages. She grounds this observation on a paper by Sprouffske and colleagues (2012). Actually, as we know from the Popperian debate on this topic in the Seventies, every counterexample can be inserted in any theory by *ad hoc* modifications. But we know, from that old classical debate, that such a modified account should be empirically more powerful than the unmodified account and it seems to me that neither Sprouffske and colleagues nor Plutynski have provided such a proof.

This mention to stemness and tumours allows me to move to the CTCs "model". As far as I know, among philosophers only Laplane (2014) and Fagan (2017) have offered a view on this issue. I focus on Fagan, since, on the one hand, her conclusions are more or less similar to Laplane's and, on the other hand, her account is more interesting for my aim to show how weak an account can be in terms of CSCs and 'model'.

First of all, while the approach above interpreted tumour progression as an evolutionary process, this reads it as a developmental process starting from an initial cell characterized by stemness. Fagan, honestly, admits that it is not so clear what a 'CSCs model' is. She is not thinking about the 'model' part of that locution (as said, no one tells us in which sense he/she is using the term 'model'). Instead she wants to show the ambiguity of that locution as a whole. She says that, in reality, there is a "minimal model" characterised by two features: "(i) cells comprising a tumour are heterogeneous in phenotype and function; and (ii) these patterns of variation map onto a hierarchical lineage structure, with more tumourigenic cells giving rise to less tumourigenic, more differentiated, progeny". Actually, these two features work well also for a possible evolutionary model, as seen above, since they describe two of the main aspects of tumour heterogeneity in general. Moreover, surprisingly enough, the two usual features of stemness (self-renewal and potency) are not mentioned for this minimal model! But let us move on. Fagan proceeds to show that different authors add different features to the minimal set above. In particular, it could be added that CSCs (iii) are rare within tumours; (iv) are more likely to survive cancer therapy than other tumour cells; (v) either acquire or inherited that molecular characteristic of stemness that protect them from anti-cancer drugs; (vi) divide at a low rate; (vii) exhibit gene expression patterns associated with pluripotency and long-term self-renewal; (viii) derive from a cell-of-origin which is a normal stem cell.

Of course, this means that we do not have just one "model" but many "models", each one characterised by a different set of assumptions added to the "minimal" two. This implies that each one should be articulated differently and therefore be capable of describing and explaining different ensembles of events and processes ascribable to tumour heterogeneity.

There are, however, two other aspects highlighted by Fagan that deserve attention. The first is that some of the assumptions just mentioned do not have adequate empirical support. The second is that any CSCs "model" underestimates events or processes which can be grasped only evolutionarily or by emphasising the role of the tumour microenvironment. That is, none seems to be completely empirically adequate and epistemologically powerful.

With reference to tumour microenvironment and its correlated concepts, while we have (even if very few) philosophical analyses of the evolutionary approach and of the CSCs approach, we do not have, as far as I know, any philosophical analysis of the ecological approach and of its use of the niche construction concept.

Summing up, we have, on the one hand, scientists who speak loosely about evolutionary models, CSCs models, ecological models and, on the other hand, philosophers who argue for the epistemological relevance of the evolutionary "model", or for the CSCs "model" (but not for the ecological "model"), without telling us why it should be a model and without being able to argue for their epistemological capability to exhaustively describe and explain all the facts belonging to tumour heterogeneity. But, more importantly, we have a field, i.e. tumour heterogeneity, whose complexity, maybe due to lack of knowledge maybe due to intrinsic features, is so great that no account alone seems to be able to describe and explain comprehensively. Instead we have molecular oncologists who use terms like heritable genetic and epigenetic changes, selective pressure, evolutionary advantage, less-fit clone, driver mutation, passenger mutation, random mutation, genomic instability, neutral mutation, disadvantageous mutation, genetic drift, developmental program, linear and branched evolution, cellular phenotypic hierarchy, cellular differentiation and cellular de-differentiation, lineage commitment, niche construction, etc. By means of these concepts, borrowed from evolutionary biology, developmental biology, CSCs biology and ecology, they construct what I have called *patchwork narratives*, which allow them to describe and explain a certain set of events and processes in a logically coherent and epistemologically valid way.

More precisely, each paper contains a phenomenological patchwork narrative on a particular aspect of tumour heterogeneity. Such a patchwork narrative can be more evolutionary, or more developmental, or more ecological, depending on the particular situation to be coped with and depending on the particular researcher's perspective. But none is exhaustive; none is able to cover all the aspects of the complexity of tumour heterogeneity. On the other hand, if we really wanted something fully comprehensive, we should collect all the narratives contained in all the papers. In such a way we would have a sort of big patchwork narrative which is able to grasp all the known aspects of tumour heterogeneity and where the compounding narratives frequently intersect or superpose, sometimes using the same jargon sometimes using a jargon borrowed from different biological fields. Nevertheless, while each single patchwork narrative contained in a paper is logically coherent and epistemically valid, this big patchwork narrative as a whole is epistemically complex, superabundant and sometimes also logically incoherent. However, it is surely phenomenologically extremely useful, if only because it is what we have now!

Acknowledgments. Previous versions of this paper have been read and commented on by some friends and colleagues of mine that I would like to thank: P.-L. Germain, M. Nathan, E. Ratti, E. Sober and D. Teira.

Bibliography

Al-Hajj, M., Wicha, M. S., Benito-Hernandez, A., Morrison, S. J., & Clarke, M. F. (2003). Prospective identification of tumorigenic breast cancer cells. *Proceedings of the National Academy of Sciences*, *100*(7), 3983–3988.

Amend, S. R., & Pienta, K. J. (2015). Ecology meets cancer biology: The cancer swamp promotes the lethal cancer phenotype. *Oncotarget*, *6*(12), 9669.

Anderson, K., Lutz, C., Van Delft, F. W., Bateman, C. M., Guo, Y., Colman, S. M., ... Swansbury, J., et al. (2011). Genetic variegation of clonal architecture and propagating cells in leukaemia. *Nature*, *469*(7330), 356–361.

Attolini, C. S.-O., & Michor, F. (2009). Evolutionary theory of cancer. *Annals of the New York Academy of Sciences*, *1168*(1), 23–51.

Barcellos-Hoff, M. H., Lyden, D., & Wang, T. C. (2013). The evolution of the cancer niche during multistage carcinogenesis. *Nature Reviews Cancer*, *13*(7), 511–518.

Baylin, S. B., & Jones, P. A. (2011). A decade of exploring the cancer epigenome—biological and translational implications. *Nature Reviews Cancer*, *11*(10), 726–734.

Bedard, P. L., Hansen, A. R., Ratain, M. J., & Siu, L. L. (2013). Tumour heterogeneity in the clinic. *Nature*, *501*(7467), 355–364.

Bertolaso, M. (2011). Hierarchies and causal relationships in interpretative models of the neoplastic process. *History and Philosophy of the Life Sciences*, 515–535.

Boem, F., Pavelka, Z., & Boniolo, G. (2015). Stratification and biomedicine. How philosophy stems from medicine and biotechnology. In M. Bertolaso (Ed.), *The Future of Scientific Practice*. London: Pickering & Chatto.

Boniolo, G. (2003). Biology without information. *History and Philosophy of the Life Sciences*, *25*(2), 255–273.

Boniolo, G. (2007). *On scientific representations: From Kant to a new philosophy of science*. Houndmills: Palgrave Macmillan.

Boniolo, G. (2013a). Is an account of identity necessary for bioethics? What postgenomic biomedicine can teach us. *Studies in History and Philosophy of Science Part C: Studies in History and Philosophy of Biological and Biomedical Sciences*, *44*(3), 401–411.

Boniolo, G. (2013b). On molecular mechanisms and contexts of physical explanation. *Biological Theory*, *7*(3), 256–265.

Boniolo, G. (2017). Molecular medicine: The clinical method enters the lab. What tumor heterogeneity and primary tumor culture teach us. In G. Boniolo, & M. Nathan (Eds.), *Philosophy of molecular medicine: Foundational issues in research and aractice*. New York: Routledge.

Boniolo, G., & Nathan, M. J. (2017). *Philosophy of molecular medicine: Foundational issues in research and practice*. New York: Routledge.

Boniolo, G., Petrovich, C., & Pisent, G. (2002). Notes on the philosophical status of nuclear physics. *Foundations of Science*, *7*(4), 425–452.

Boniolo, G., & Sanchini, V. (Eds.). (2016). *Ethical counselling and medical decision-making in the age of personalised medicine.* Heidelberg: Springer.

Boniolo, G., & Testa, G. (2012). The identity of living beings, epigenetics, and the modesty of philosophy. *Erkenntnis, 76*(2), 279–298.

Burrell, R. A., McGranahan, N., Bartek, J., & Swanton, C. (2013). The causes and consequences of genetic heterogeneity in cancer evolution. *Nature, 501*(7467), 338–345.

Cairns, J. (1975). Mutation selection and the natural history of cancer. *Nature, 255*(5505), 197–200.

Chen, F., Zhuang, X., Lin, L., Yu, P., Wang, Y., Shi, Y., ... Sun, Y. (2015). New horizons in tumor microenvironment biology: Challenges and opportunities. *BMC Medicine, 13*(1), 1.

Cordell, H. J. (2002). Epistasis: What it means, what it doesn't mean, and statistical methods to detect it in humans. *Human Molecular Genetics, 11*(20), 2463–2468.

Cornetta, K., & Brown, C. G. (2013). Perspective: Balancing personalized medicine and personalized care. *Academic Medicine: Journal of the Association of American Medical Colleges, 88*(3), 309.

Damuth, J., & Heisler, I. L. (1988). Alternative formulations of multilevel selection. *Biology and Philosophy, 3*(4), 407–430.

Dick, J. E. (2008). Stem cell concepts renew cancer research. *Blood, 112*(13), 4793–4807.

Druker, B. J., Guilhot, F., O'Brien, S. G., Gathmann, I., Kantarjian, H., Gattermann, N., ... Stone, R. M., et al. (2006). Five-year follow-up of patients receiving imatinib for chronic myeloid leukemia. *New England Journal of Medicine, 355*(23), 2408–2417.

Easwaran, H., Tsai, H.-C., & Baylin, S. B. (2014). Cancer epigenetics: Tumor heterogeneity, plasticity of stem-like states, and drug resistance. *Molecular Cell, 54*(5), 716–727.

Ewens, W. J. (2004). *Mathematical population genetics.* New York: Springer.

Fagan, M. (2017). Pathways to the clinic: Cancer stem cells and challenges for translational re-search. In G. Boniolo, & M. Nathan (Eds.), *Philosophy of molecular medicine: Foundational issues in research and practice.* New York: Routledge.

Febbo, P. G., Ladanyi, M., Aldape, K. D., De Marzo, A. M., Hammond, M. E., Hayes, D. F., ... Ogino, S., et al. (2011). NCCN Task Force report: Evaluating the clinical utility of tumor markers in oncology. *Journal of the National Comprehensive Cancer Network, 9*(Suppl 5), S1–S32.

Gärtner, H.-V., Seidl, C., Luckenbach, C., Schumm, G., Seifried, E., Ritter, H., & Bültmann, B. (1996). Genetic analysis of a sarcoma accidentally transplanted from a patient to a surgeon. *New England Journal of Medicine, 335*(20), 1494–1497.

Germain, P.-L. (2012). Cancer cells and adaptive explanations. *Biology & Philosophy, 27*(6), 785–810.

Geyer, F. C., Weigelt, B., Natrajan, R., Lambros, M. B., de Biase, D., Vatcheva, R., ... Reis-Filho, J. S. (2010). Molecular analysis reveals a genetic basis for the phenotypic diversity of metaplastic breast carcinomas. *The Journal of Pathology, 220*(5), 562–573.

Godfrey-Smith, P. (2009). *Darwinian populations and natural selection*. Oxford University Press.

Golubnitschaja, O., Baban, B., Boniolo, G., Wang, W., Bubnov, R., Kapalla, M., ... Costigliola, V. (2016). Medicine in the early twenty-first century: Paradigm and anticipation. *The EPMA Journal, 7*. DOI: 10.1186/s13167-016-0072-4.

Greaves, M., & Maley, C. C. (2012). Clonal evolution in cancer. *Nature, 481*(7381), 306–313.

Gupta, P. B., Fillmore, C. M., Jiang, G., Shapira, S. D., Tao, K., Kuperwasser, C., & Lander, E. S. (2011). Stochastic state transitions give rise to phenotypic equilibrium in populations of cancer cells. *Cell, 146*(4), 633–644.

Hartl, D. L., & Clark, A. G. (2007). *Principles of population genetics*. Sunderland: Sinauer.

Hayflick, L. (1965). The limited in vitro lifetime of human diploid cell strains. *Experimental Cell Research, 37*(3), 614–636.

Hiley, C., de Bruin, E. C., McGranahan, N., & Swanton, C. (2014). Deciphering intratumor heterogeneity and temporal acquisition of driver events to refine precision medicine. *Genome Biology, 15*(8), 1.

Johnson, B. E., Mazor, T., Hong, C., Barnes, M., Aihara, K., McLean, C. Y., ... Tatsuno, K., et al. (2014). Mutational analysis reveals the origin and therapy-driven evolution of recurrent glioma. *Science, 343*(6167), 189–193.

Junttila, M. R., & de Sauvage, F. J. (2013). Influence of tumour micro-environment heterogeneity on therapeutic response. *Nature, 501*(7467), 346–354.

Kareva, I. (2015). Cancer ecology: Niche construction, keystone species, ecological succession, and ergodic theory. *Biological Theory, 10*(4), 283–288.

Kleppe, M., & Levine, R. L. (2014). Tumor heterogeneity confounds and illuminates: assessing the implications. *Nature Medicine, 20*(4), 342–344.

Landau, D. A., Carter, S. L., Stojanov, P., McKenna, A., Stevenson, K., Lawrence, M. S., ... Wang, L., et al. (2013). Evolution and impact of subclonal mutations in chronic lymphocytic leukemia. *Cell, 152*(4), 714–726.

Lapidot, T., Sirard, C., Vormoor, J., Murdoch, B., Hoang, T., Caceres-Cortes, J., ... Dick, J. (1994). A cell initiating human acute myeloid leukaemia after transplantation into SCID mice. *Nature, 367*(6464), 645–648.

Laplane, L. (2014). Identifying some theories in developmental biology: The case of the cancer stem cell theory. In A. Minelli, & T. Pradeu (Eds.), *Toward a theory of development* (pp. 246–259). Oxford: Oxford University Press.

Lean, C., & Plutynski, A. (2016). The evolution of failure: Explaining cancer as an evolutionary process. *Biology & Philosophy, 31*(1), 39–57.

Magee, J. A., Piskounova, E., & Morrison, S. J. (2012). Cancer stem cells: Impact, heterogeneity, and uncertainty. *Cancer Cell, 21*(3), 283–296.

Marjanovic, N. D., Weinberg, R. A., & Chaffer, C. L. (2013). Cell plasticity and heterogeneity in cancer. *Clinical Chemistry, 59*(1), 168–179.

Martelotto, L. G., Ng, C. K., Piscuoglio, S., Weigelt, B., & Reis-Filho, J. S. (2014). Breast cancer intra-tumor heterogeneity. *Breast Cancer Research, 16*(3), 1.

Marusyk, A., & Polyak, K. (2010). Tumor heterogeneity: Causes and consequences. *Biochimica et Biophysica Acta (BBA)-Reviews on Cancer, 1805*(1), 105–117.

Mayr, E. (2002). *What evolution is*. New York: Basic books.

Meacham, C. E., & Morrison, S. J. (2013). Tumour heterogeneity and cancer cell plasticity. *Nature, 501*(7467), 328–337.

Merlo, L. M., Pepper, J. W., Reid, B. J., & Maley, C. C. (2006). Cancer as an evolutionary and ecological process. *Nature Reviews Cancer, 6*(12), 924–935.

Nowell, P. C. (1976). The clonal evolution of tumor cell populations. *Science, 194*(4260), 23–28.

Odling-Smee, F. J., Laland, K. N., & Feldman, M. W. (2003). *Niche construction: The neglected process in evolution*. Princeton: Princeton University Press.

Okasha, S. (2006). *Evolution and the levels of selection*. Oxford: Oxford University Press.

Phillips, P. C. (2008). Epistasis—The essential role of gene interactions in the structure and evolution of genetic systems. *Nature Reviews Genetics, 9*(11), 855–867.

Pigliucci, M., Sterelny, K., & Callebaut, W. (2013). The meaning of "theory" in biology. *Biological Theory, 7*(4), 285–286.

Plutynski, A. (2017). Evolutionary perspectives on molecular medicine. In G. Boniolo, & M. Nathan (Eds.), *Philosophy of molecular medicine: Foundational issues in research and practice*. New York: Routledge.

Polyak, K. (2014). Tumor heterogeneity confounds and illuminates: a case for Darwinian tumor evolution. *Nature medicine, 20*(4), 344–346.

Quintana, E., Shackleton, M., Sabel, M. S., Fullen, D. R., Johnson, T. M., & Morrison, S. J. (2008). Efficient tumour formation by single human melanoma cells. *Nature, 456*(7222), 593–598.

Ridley, M. (2007). *Evolution*. Oxford: Blackwell.

Sequist, L. V., Waltman, B. A., Dias-Santagata, D., Digumarthy, S., Turke, A. B., Fidias, P., ... Cosper, A. K., et al. (2011). Genotypic and histological evolution of lung cancers acquiring resistance to EGFR inhibitors. *Science Translational Medicine, 3*(75), 75ra26–75ra26.

Skloot, R. L. (2010). *The Immortal Life of Henrietta Lacks*. New York: Crown Publishing Group.

Sober, E., & Wilson, D. S. (1998). *Unto others: The evolution and psychology of unselfish behavior*. Cambridge: Harvard University Press.

Sprouffske, K., Merlo, L. M., Gerrish, P. J., Maley, C. C., & Sniegowski, P. D. (2012). Cancer in light of experimental evolution. *Current Biology, 22*(17), R762–R771.

Torres, L., Ribeiro, F. R., Pandis, N., Andersen, J. A., Heim, S., & Teixeira, M. R. (2006). Intratumor genomic heterogeneity in breast cancer with clonal divergence between primary carcinomas and lymph node metastases. *Breast Cancer Research and Treatment, 102*(2), 143–155.

Turner, N. C., & Reis-Filho, J. S. (2012). Genetic heterogeneity and cancer drug resistance. *The Lancet Oncology, 13*(4), e178–e185.

Ungefroren, H., Sebens, S., Seidl, D., Lehnert, H., & Hass, R. (2011). Interaction of tumor cells with the microenvironment. *Cell Communication and Signaling, 9*.

van Dijk, M., Göransson, S. A., & Strömblad, S. (2013). Cell to extracellular matrix interactions and their reciprocal nature in cancer. *Experimental cell research*, *319*(11), 1663–1670.
van Heist, A. (2011). *Professional loving care: An ethical view of the healthcare sector*. Leuven: Peeters.
Visvader, J. E. (2011). Cells of origin in cancer. *Nature*, *469*(7330), 314–322.
Visvader, J. E., & Lindeman, G. J. (2008). Cancer stem cells in solid tumours: Accumulating evidence and unresolved questions. *Nature Reviews Cancer*, *8*(10), 755–768.
Vv.Aa. (2013). Tumour heterogeneity. *Nature*, *501*.
Weigelt, B., & Reis-Filho, J. S. (2014). Epistatic interactions and drug response. *The Journal of Pathology*, *232*(2), 255–263.
Yang, K. R., Mooney, S. M., Zarif, J. C., Coffey, D. S., Taichman, R. S., & Pienta, K. J. (2014). Niche inheritance: A cooperative pathway to enhance cancer cell fitness through ecosystem engineering. *Journal of Cellular Biochemistry*, *115*(9), 1478–1485.
Yap, T. A., Gerlinger, M., Futreal, P. A., Pusztai, L., & Swanton, C. (2012). Intratumor heterogeneity: Seeing the wood for the trees. *Science Translational Medicine*, *4*(127), 127ps10.

Author biography.
Giovanni Boniolo has a doctoral degree in Physics and in Philosophy.

Full Professor of Philosophy of Science and Medical Humanities (University of Ferrara).

Anna Boyksen Fellowship (Institute for Advanced Study, Technische Universität München).

Honorary Ambassador of the Technische Universität München.

Former Director of the PhD program in "Foundations of the Life Sciences and Their Ethical Consequences", European School of Molecular Medicine (Milano) and Former Director of the Biomedical Humanities Unit, Istituto Europeo di Oncologia (Milano).

Member or chair of several ethics committees both at Italian and European level.

Research interests: (i) public and individual decision-making, whenever ethical issues are involved; (ii) trust based consent for patients; (iii) foundations of contemporary biomedical research and clinical practice.

Two new books:
1. G. Boniolo & V. Sanchini (eds) (2016). *Ethical counselling and medical decision-making in the age of personalised medicine*. Heidelberg, Springer.

2. G. Boniolo & M. Nathan (eds) (2016). *Philosophy of Molecular Medicine: Foundational Issues in Research and Practice*. New York, Routledge.

Page: http://docente.unife.it/giovanni.boniolo.

Part D

CLMPS 2015 conference theme: Models and Modelling

18 Models and modeling in formal epistemology: Some thoughts on probability aggregation

ELEONORA CRESTO [*]

Abstract. In this paper I discuss the role of normativity in model building, particularly within formal epistemology. I begin by making some distinctions and clarifications, and then I focus on the problem of testing normative models. I suggest a novel way to think about putting normative models to the test, which consists in building meta-models for first order normative settings. I argue that a successful meta-modeling strategy should enable us to illuminate the mechanism that underlies a given normative structure, and in this sense it can further test or refine our intuitions concerning what ought to be the case.

Next I propose a model for probability aggregation that seeks to illustrate our prior discussion on the relevance and purpose of meta-model building for normative modeling in general. I suggest that, under certain circumstances, it can be rewarding to look at probability aggregation as a type of cooperative bargaining. Individual agents are assumed to hold utilities over possible probability assignments to propositions. Given such utilities, I show how to build an appropriate (pseudo)bargaining situation, such that points inside the bargaining set are correlated with sets of probability assignments (on a given proposition) by the individual agents. Solving the bargaining problem helps us figure out the probability that can be credited to the group as a whole. We then obtain a unified perspective on two seemingly disparate phenomena – probability aggregation and cooperative bargaining. The proposal illustrates how normative

---

[*]CONICET, Buenos Aires.

meta-models are meant to work: Bargaining models here act as meta-models that can help us elicit intuitions regarding probability aggregation (the first-order phenomenon) in an indirect way.

Keywords: normative models, probability aggregation, cooperative game theory, bargaining solutions, utilities.

1 Introduction

My aim in this paper is twofold. In the first part (Sections 2 to 4) I will offer some reflections about the role of normativity in model building, and in particular about normativity within formal epistemology. This topic remains largely unexplored, so I would like to establish some ground for future discussion. I will begin by making some distinctions and clarifications, and then I will focus on the problem of testing normative models. I will suggest a novel way to think about putting normative models to the test, which consists in building meta-models for first order normative settings. A successful meta-modeling strategy should enable us to illuminate the mechanism that underlies a given normative structure, and in this sense it can further test or refine our intuitions concerning what ought to be the case.

In the second part (Sections 5 and 6) I will develop one particular model for probability aggregation (a subject that squarely belongs to formal epistemology), which seeks to illustrate some of the general traits of satisfactory normative modeling, including model testing, discussed in the first part of the paper. More precisely, the proposal attempts to illustrate our prior discussion on the relevance and purpose of meta-model building for normative modeling in general. The proposed model might have some interest in itself, regardless of our prior discussion on modeling strategies. I will suggest that, under certain circumstances, it is rewarding to look at probability aggregation as a type of cooperative bargaining. Individual agents can be interpreted as holding utilities over possible probability assignments to propositions, such that, for a given proposition p, each agent gives maximum utility to the probability of p that each one takes to be 'correct' (i.e., to his or her actual credence on p); utility functions are assumed to decrease continuously from there. Given such utilities, I show how to build an appropriate (pseudo)bargaining situation (for proposition p), such that points inside the bargaining set are correlated with sets of probability assignments by the individual agents. I will argue that solving the bargaining problem helps us figure out the probability of p that can be credited to the group as a whole; traditional discussions on the adequacy and correctness of different bargaining solutions become relevant for our current setting as well. We then obtain a unified perspective on two seemingly disparate phenomena, probability aggregation and cooperative bargaining. The proposal illustrates how normative meta-models are meant to work: Bargaining models here act as meta-models that can help us elicit intuitions regarding probability aggregation (the first-order phenomenon) in an indirect way; as with most normative meta-models, the legitimacy of Bargaining models springs from their role in a different context altogether.

2 Models in formal epistemology: The role of normativity

Formal Epistemology is an umbrella term, which includes many different disciplines and research lines. Correspondingly, we should expect to find many different types of models that can be said to belong to the formal epistemology realm, or to formal epistemology modeling. Consider, for example, the idealized setting that we take as a starting point to study the properties of certain voting procedure, within the quarters of social choice theory. Or consider how a system of dynamic epistemic logic explores the ways in which the knowledge of a group of agents should (ideally) evolve in response to new information; or how rational agents are supposed to accept hypotheses in agreement with a particular brand of cognitive decision theory, among many other problems. It is customary to use the word 'model' to refer to these and similar examples. Thus we typically read that a $S5$ system *models* knowledge in a way that allows for negative and positive introspection. Or we say that while AGM proposal in Alchourrón, Gärdenfors, and Makinson (1985) aims at *modeling* the way agents revise their beliefs when they learn new facts, Katsuno and Mendelzon (1991) is *a model* of belief update that is better understood as aiming to capture doxastic modifications that spring from environmental changes. Or we read that Bayesian conditionalization provides a *model* for (partial) belief change that applies to agents in particular contexts of uncertainty. One salient property of all these examples, compared to typical models in the empirical sciences, is that they exhibit strong normative features, in one way or another. We can even wonder whether the term 'model' does not have a different meaning once normativity enters so heavily into the picture. Indeed, the nature and features of normative models appear elusive, and not as well studied as their counterparts within empirical models. In what follows I will try to set some background for further discussion.

We can begin by considering a very broad distinction between normative and descriptive models, where "normative models attempt to fit normative facts" (Titelbaum, forthcoming; more on this later). Other authors would rather commit to a tripartite setting that classifies models into normative, prescriptive and descriptive. So-called prescriptive models are conceived of as tools to try to improve the extent to which practice differs from what it should be (according to some account of normativity); we find a paradigmatic example of this distinction in the work of Jonathan Baron:

> One task of our field is to compare judgments to normative models. We look for systematic deviations from the models. These are called biases. If no biases are found, we may try to explain why not. If biases are found, we try to understand and explain them by making descriptive models or

theories. With normative and descriptive models in hand, we can try to find ways to correct the biases, that is, to improve judgments according to the normative standards. The prescriptions for such correction are called prescriptive models. (Baron, 2004, 19, in the context of a discussion about judgment and decision making.)

Not everybody understands the relation between the normative and descriptive realms in the same way. Some authors tend to conceive of whole areas of inquire normally thought of as normative as ultimately descriptive. Yap (2014) for example, argues that epistemic logic is in fact descriptive: it contains idealizations, in the same way scientific models do; thus "criticizing models of epistemic logic in which agents know all propositional tautologies as being unrealistic would be like criticizing frictionless planes in physics for being unrealistic." (Yap, 2014). More generally, other authors have argued that normative models should be really close to descriptive ones, to actual practice. Gabbay and Woods (2003) are a good example of this standpoint:

how cognitive agents actually do behave is a (substantial approximation to) how they should behave. (Gabbay & Woods, 2003, 605; they make this remark regarding inference, rather than belief.)

And also:

...normativity inheres in how we act and behave... normativity is descriptively immanent, rather than transcendent. (Gabbay & Woods, 2003, p. 605.)

As it is apparent, this position defies a long and venerable tradition that would insist that one cannot derive 'ought' from 'is'. Regardless of the details, even if we embrace a 'descriptively immanent normativity', the case remains that idealized behavior is not actual behavior, and there will always be many different, incompatible ways to settle on what counts as ideal for each particular situation. Notice that, as opposed to other context in which we may find normative constraints, when applied to behavior, 'ideal' amounts to 'desirable' (for certain purposes and goals).

In any case, it might well be that no model is *purely* descriptive, prescriptive, or normative. To accommodate for this idea, let me try here a somewhat different taxonomy – one that seeks to capture the extent to which, and the sense in which, normativity can play a role within a particular model. In this sense the labels I will mention below are not meant to single out disjunctive types of models, but to identify non-exclusive features that might be present in any model. There are (at least) three ways in which normativity can enter into the picture – three ways in which models can be said to have normative traits:

(a) Normativity in the context of a (mostly) descriptive enterprise

It is important to acknowledge that normativity definitively plays a role within models from the empirical sciences. Empirical models can exhibit normative features, for example, at the time of reasoning *about* the model, or at the time of relying on various kinds of idealizations:

(i) No matter which discipline we are working on, eventually we will have to assess whether it is reasonable to assert certain claims within the model; for example, we will have to assess whether we have deductive or inductive reasons to draw certain inferences. This is just part of the business of carrying out what is customarily referred to as "investigation or research *about* the model" – as opposed to: from the model, to the target system.[1] At least in this minimal sense most models share some normative traits: if we accept a given starting point, other claims *should* be asserted too.

(ii) On the other hand, empirical models can be quite removed from reality, as we all know; many models involve what the literature dubbed either Galilean or Aristotelian idealizations (or both); many models are also said to be *caricatures*.[2] Idealizations sometimes seem to implement a sort of 'peeling off' process; at other times they may work by way of *adding* certain features, rather than just removing them.[3] As I see it, many idealizations incorporate normative features, to the extent that ideal entities of various kinds are conceptually required to possess certain traits. Normativity in this sense is not related to what someone should strive to do or achieve, nor is it related to considerations of better or worse performance. As Titelbaum (forthcoming), has rightly pointed out, "there's no sense in which a frictionless plane is *better* than a real plane". This is of course true; however, a frictionless plane *ought* to have certain properties – although this is not a moral 'ought'.

Focusing on idealization mechanisms can establish a link between empirical models and models within the formal epistemology realm (more on this later). However, typical models in formal epistemology include normative features in other senses as well, as we will see below.

(b) Constitutive Normativity

Some models seem to have the explicit goal of helping us acquire a better understanding of rationality, or personhood, among other possibilities – or, to be more precise: some models *can be used* with the explicit aim of bringing about such understanding, regardless of the original motivation of the authors who first developed them. We might conceive of them as aiming to represent normative facts (as Titelbaum would put it), or perhaps as aiming to represent our *intuitions* regarding the features of various types of non-empirical phenomena (fairness, rationality, correct language usage, or many others). Under the last conception, a big question at this juncture is whose in-

---

[1] Cf. also Williamson (forthcoming); models (including normative models) can give us both "vague unconditional knowledge" that compares one model to another, as well as precise conditional knowledge of the form "if a given case satisfies the model description, then it satisfies this other description too".

[2] For an overall perspective on idealized models cf. Frigg and Hartmann (2012).

[3] It is not always obvious which of the two procedures is actually in place in a given example. Cf. for example Mäki (2009): "In assuming perfect information on the part of economic agents, a model appears to add a feature that does not obtain in the real world: an excessively powerful mental capacity seems to be attributed to the model agents. However, it seems to me that the correct reading of the function of this idealizing assumption is that it is used to remove certain real-world features from the model world: the search, acquisition and processing of information."

tuitions we are seeking to model: those held by the layperson? By the expert? Which expert, anyway? The resulting model can be very different in each case.

(c) Normativity within a prescriptive setting

Some models depict ideal features of different phenomena with the explicit purpose of providing a guide to action; even if the full-fledged picture remains ultimately unattainable, it can function as a regulative ideal. One way of putting it could be to say that such models seek to capture 'prescriptive intuitions': intuitions *on what we ought to do*. In any case, we should avoid thinking of prescriptive and normative models as mutually exclusive types. Consider again an $S5$ system of epistemic logic with Common Knowledge. What is the target here? The model can be taken to represent certain aspects of the behavior of rational agents, and hence constitutive features of ideal reasoners. But at the same time, by way of modeling what perfect reasoners will do we may also prescribe what real agents *should* try to do, even if they will ultimately fail to do so, if they find it desirable to think of themselves as rational creatures. Consider now various voting models, within social choice theory. We could say that the goal here is mostly prescriptive, in the sense that it is possible to take conscious implementations seriously; at the same time, however, we can seek to use such models to have a better understanding of rational features of fair choice, or impartiality, among other things.

The present reflections point to the fact that normativity is not an all-or-nothing affair, which speaks in favor of treating empirical models and models in formal epistemology (and, in general, in philosophy) along similar lines. Indeed, several accounts try to assimilate normative to empirical models, by claiming that the most characteristic features of normative models are in fact idealizations akin to the ones we find in the empirical sciences (cf. Baron, 2004, p. 24; Williamson, forthcoming; or Colyvan, 2013). This has led some authors to claim, for example, that rationality models are not 'fully' normative.[4] I find this move curious, as I tend to think that the situation is exactly the opposite: the fact that idealizations are widespread shows that there are normative elements within empirical models, rather than non-normative elements in models from logic or formal epistemology. In any case, regardless of whether we take idealizations to inject normativity (to descriptive models) or to take it away (from models of rationality), it is not clear to me whether what I have called 'constitutive normativity' can be fully reduced to the presence of various idealizations of actual behavior – even though idealizations can of course play a role at the time of building such models.

---

[4] Cf. Colyvan (2013), or Yap (2014). Colyvan contends that theories of rationality within formal epistemology are normative, but "they are not normative through and through" (2013, p. 1338); he wants to separate the normative elements from other kinds of idealizations, which he takes to be non-normative. On the other hand, Yap (2014) follows Weisberg (2007) in identifying three kinds of idealizations in science (Galilean, minimalist and multiple-models); all three of them are present in models within epistemic logic. Yap suggests that, to the extent that we can show that the unrealistic features are in fact idealizations of various types, a given model is not truly normative, but descriptive.

3 A problem

In light of the above, we could be tempted to conclude that the distinction between models in the empirical sciences and models in formal epistemology is really a matter of degree. However, things are slightly more complicate. Even authors sympathetic to pragmatist conceptions of models, like Suárez (for whom the reference to agents and purposes is essential, and models do not 'mirror' the world in any interesting sense) are fast to point out that "scientific representations have cognitive value because they aim to provide us with specific information regarding their targets" (Suárez, 2004). As I see it, one of the main problems we face here is that, when constitutive normativity enters into the picture, the target system becomes elusive. What type of information are we aiming at, exactly? It is true that gathering data can always be a difficult business, regardless of the discipline. However, if our data are normative facts the problem is compounded. How can we ever ensure we are getting such facts right? We could try to overcome this problem, to some extent, by claiming that normative models target our intuitions on various normative matters. But this answer does not solve our worries.

The reason is the following. An assessment of the correctness of models in the empirical sciences can proceed, among other things, by way of keeping track of their predictions. How does the predictive enterprise work in the case of (mostly) normative models? Suppose we agree that such models seek to capture certain intuitions as a starting point; later on we draw consequences from there, which might force us to commit ourselves to claims we would not have thought of before building the model. Such consequences should be equally intuitive. To put it differently, normative models can be said to make predictions regarding which other claims we will find intuitive. So far so good. The problem is that, as I have already pointed out, intuitions on various phenomena are seldom universally shared. Therefore, even though we can still assess whether certain sets of intuitions have been more or less well captured by a given model, it is not clear how to assess whether the intuitions that lie at the heart of it are misguided in the first place, or perhaps whether the model itself is faulty. Unwanted consequences (i.e., consequences that are no longer intuitive, or which are at odds with the intuitions that constituted our starting point) can mean either than the model was not a good one, or that some of our intuitions should be re-educated: perhaps we should bite the bullet and accept the odd consequences, or perhaps we should reconsider the plausibility of our starting point altogether. In this sense, modeling intuitions fosters endogamy: the quality control mechanisms for adequate model building are not independent of the model we are using.

It should be clear that what we may call 'the endogamy problem' of normative models is not tied to our description of the target system in terms of intuitions. Consider, for example, Titelbaum's analysis (in Titelbaum, forthcoming). He endorses a well known distinction among the model framework, its interpretation, and the individual model that results from so interpreting the framework. In the empirical sciences, we accommodate the particular model (i.e., the particular interpretation) whenever the predictions do not match the data; on the other hand, if there is a significant mismatch, we tend to question the framework itself. According to Titelbaum, normative models are analogous to empirical models in this respect.[5] Take once again the AGM model for belief change: "after the loss of belief, rationality requires the agent to believe such-and-such propositions. Now suppose that this prediction mismatches the data: It's not the case *that rationality requires the specified beliefs* [my emphasis]. This calls the combination of AGM and our interpretation into question." But the problem is how we can ever check whether rationality requires or does not require certain things. The very possibility of putting the model to the test requires our having some mechanism to gather reliable data which is at least partially independent of the modeling process itself. This is sometimes a problem for the empirical sciences as well, but it is not clear how normative models can ever avoid it. It is of course true, as Williamson points out, that "if we started in total ignorance about the target, we could hardly expect to learn much about it by modelling alone" (Williamson, forthcoming), but by itself this observation leaves us in the dark as to how to proceed when we face competing candidates to function as the *real* set of normative facts.

Endogamy, in the sense just explained, might be a trait of normative models we have to learn to live with. Is there anything we can do to make things better? In the next section I will make a suggestion that goes some way towards presenting a possible improvement, at least for some cases.

4 Modeling first-order normativity

The problem described in the previous section can be summarized as the problem of choosing among competing normative models. This is related (though not identical) to the problem of choosing among competing sets of intuitions. My suggestion here is that in order to choose among various normative models it could be helpful to count with a *second*-order model. In other words, sometimes we can profit from developing meta-models for first order normative settings. I do not want to say that every legitimate modeling strategy for normative models should rely on second-order considerations, but only that this is an interesting possibility, which enables us to think about the problem of testing normative models, in particular models within formal epistemology, under a somewhat different light.

Very briefly, this is what I have in mind. We begin the process, at the very first moment, by acknowledging that we have certain intuitions (on rationality, on equity, on grammatical correction, etc); if you prefer, we can take such intuitions to correspond

---

[5]Sometimes, when predictions are not satisfied, we can explain away the inconsistency by better specifying the intended domain of application (on this cf. also Colyvan, 2013); by contrast, *genuine* counterexamples (i.e., those that amount to a real threat for the model) are those that belong to the intended application domain.

in each case to a particular element from a normative space.[6] We are supposed to elicit our normative intuitions, or perhaps our intuitions on what we have to do, through a model encompassing constitutive normativity, or a model with prescriptive features. But very often we will find incompatible sets of intuitions, at level zero, which may give rise to different normative models at level one. For example, sometimes we have a list of criteria, each element of which sounds perfectly reasonable to us, but which cannot be jointly satisfied (think for example of the several desiderata we can use to fix of a voting procedure, or of the many ways to escape from Arrow's impossibility result). How shall we choose among different normative models, then?

A possible way to go is to try to find a model that was originally developed for a (possibly) very different target system (ideally, from the empirical sciences), and use it *to represent features of some of the first order normative models*.[7] The second-order model then represents, indirectly, a particular normative or prescriptive element from our normative space. Such a second-order model can only be said to represent such elements (or our level zero intuitions) indirectly, because the immediate goal is to explain the mechanism behind a given normative model. By doing so it can reveal not so obvious features of the normative model in question. *Then a first order normative model is worth selecting if we find a suitable second-order model that represents it.* In short, the tool to choose can be precisely another model, preferably some other model with roots in empirical phenomena.

Ultimately, this can be seen as a more sophisticated way to test our intuitions regarding what ought to be the case. By proceeding in this way we can notice that the original normative fact or the original prescriptive task (for which we are trying to elicit intuitions) shares structural features with other phenomena, so we can come to see the original problem as part of a more general question. In this sense the second-order model can provide *unification*: it can help us see apparently disparate phenomena in a unified way.

A short clarification: The way I am conceiving of it here, the second-order model is not meant to supersede the first-order model(s) that it represents. Rather, the aim of a particular meta-model is to provide *validation* for a given first-order model. When this happens, we can say that the given first-order model is robust. Sometimes none of the elements from our prior pool of first-order models gets so validated; instead, the second-order model points to the possibility of building an alternative first-order structure, of which the modelers where not aware before. In this case no robustness is revealed, which is in itself a kind of progress, in a sort of Popperian way (though of course I do not mean to say that first-order models which are not found robust are

---

[6]Concomitantly, we can take the target system to be a particular 'normative fact'. For the purposes of this section, it is irrelevant whether we take target systems to be sets of intuitions (on normative facts), or the normative facts themselves.

[7]Several authors acknowledged that models can be reinterpreted; in Titelbaum's terms, the same structure can be used in multiple applications, some of them normative and some of them descriptive (Titelbaum, forthcoming). Cf. also Weisberg (2013).

thereby *disproved*).

Finally, it is clear that not all (first-order) normative scenarios will allow for second-order modeling. Still, the fact that this maneuver is in principle available contributes to ease our worries regarding the testing of normative models.

In what follows I will illustrate this proposal with the problem of probability aggregation. I will suggest that we turn our attention to bargaining models; we will see how a single (meta)model (the bargaining model) can be profited to do more than one task at the same time. In this way we can come to see probability aggregation as a sub-problem of cooperative game theory. For our present purposes, the target system of the meta-model is a first order normative model, which will typically consist in sets of equations; our meta-model could eventually help us decide among an array of first-order models, and choose the "right" set of equations.

5 An illustration: A (meta)model for probability aggregation

Consider a set of probability functions. The individual measures may actually be the measures of the individual members of a given group; alternatively, they can also represent different attempts to capture experimental results, perhaps by a single agent, among many other possibilities. How shall we represent the probability *of the set* as a whole? A rather straightforward answer would be to represent it as a set as well – perhaps a convex set containing the individual measures. Others would argue that what we actually need is a bona fide aggregation method, a method that delivers one single probability function. I will not try to settle this discussion here. Rather, in what follows I will just *assume*, for the sake of the argument, that we are interested in finding *the* single measure that can be said to correspond to the probabilistic attitude of a group.

How shall we combine the individual functions in order to obtain a single measure, then? Notoriously, there are many aggregation rules we could follow here. Among the most common solutions we should include:[8]

- Linear opinion pools (e.g., the arithmetic mean)
  $F(P_1...P_n) = w_1 P_1 + ... + w_n P_n$

- Geometric opinion pools
  $F(P_1...P_n) \propto P_1^{w_1}...P_n^{w_n}$

---

[8]For a classic overview on probability aggregation cf. Genest and Zidek (1986); cf. also Dietrich and List (forthcoming).

- Supra Bayesian approaches
  $F(P_1...P_n)(A) = F(A|P_1...P_n)$
- Multiplicative opinion pools:

$$\pi \propto \frac{\pi_1...\pi_n}{P_1...P} F(P_1...P_n)$$

(where 'F' is our aggregation function; '$P_1...P_n$' are the individual probability functions, for some finite n; '$w_1...w_n$' are weights that add up to 1; and '$\pi_1...\pi_n$' refer to i's posterior probability, for each $i \in (1...n)$.)

How shall we proceed, then? This is a non-exhaustive list, of course, but even if we restrict our attention to these few options, there is no consensus regarding which strategy is best all things considered. The standard answer is to examine which properties are fulfilled in each case, and let such properties guide our choice of an aggregation method. For example, if we want aggregation to commute with conditionalization, we could pick a geometric opinion pool; if we want an analogous of the so-called independence of irrelevant alternatives to hold in this context, the linear opinion pool seems to be a good choice, whereas if we favor an independence preservation property, linear opinion pools no longer look promising.[9] Some sets of criteria may be more desirable than others for particular purposes and in particular contexts (for example, Supra Bayesian accounts seem appropriate for contexts in which the aggregation is performed by a single agent external to the group). But in many cases it may be unclear how to assess the properties, or how to decide which set of criteria is more relevant.

---

[9] A property such as:

$T(P_1,...,P_n)(A) = F[(P_1(A),...,P_n(A)]$, for some arbitrary $F : [0,1]^n \to [0,1]$, and every event A in the algebra

implies that the aggregation is a linear opinion pool. On the other hand, linear opinion pools cannot satisfy the so-called 'independence preservation property':

$T(P_1,...,P_n)(A \cap B) = T(P_1,...,P_n)(A)T(P_1,...,P_n)(B)$, whenever $P_i(A \cap B) = P_i(A)P_i(B)$ for some A and B in the algebra,

unless they are dictatorial or trivial (i.e., unless weights are 0 for all agents except for one). Cf. Genest and Zidek (1986, p. 117).

6 The proposal

Let us look at the problem from a different perspective. Let me clarify at the outset that this proposal will not work for every possible aggregation problem. Still it can model quite well, I think, the attempt to arrive at a single function for a group of agents when there is no third party agent, external to the group, in charge of producing the aggregation. In other words, the present account attempts to capture the mechanism of probability aggregation from the inside, as it were – as experienced from the first person perspective. In addition, the present account is not meant to work for a complete probability distribution; rather, it seeks to determine which probability the group is entitled to have, qua group, on a particular proposition, or on a particular partition of rival propositions or hypotheses. The proposal is still programmatic, but I hope the broad picture is clear enough.

In what follows we assume that each agent thinks of herself as well suited to have a well-founded opinion – at any rate, each agent thinks of herself as being at least as much entitled to having a well-founded opinion as any of her peers. In this context, we will also assume that the members of the group have *utilities* over the possible probability assignments, such that each of them gives maximal utility to her actual credence; utilities are assumed to decrease continuously from there. Utilities here do not measure the desirability of a given proposition, of course, but *the desirability of adopting a particular (not necessarily precise) probability on that proposition*. Each agent would like to impose her view regarding that particular proposition on the group; as far as each individual agent is concerned, the closer the collective probability is to her own credence, the better.

By way of concreteness, let me illustrate with a simple example. For the easiest case, consider a group made of two people, agents A and B, with definite probability assignments on some proposition p. Let $P_A(p) = 0.9$ and $P_B(p) = 0.5$. Consider now each agent's utilities on the possible probability of p. We can take utility scales to be equivalent under positive linear transformation; to simplify we will consider values in the range $[0,1]$. Different utility functions may be appropriate here. In what follows I will illustrate with quadratic functions; this is not the only way to go, of course, but there are some technical and conceptual advantages in proceeding in this way. Agent A gives maximal utility to probability 0.9, and minimum utility to probability 0, whereas agent B gives maximal utility to probability 0.5, and minimum utility to both probabilities 0 and 1. Thus we have:

- For A: $u_A(x) = -100x^2/81 + 20x/9$
- For B: $u_B(x) = -4x^2 + 4x$

We can draw the two curves on the same graph, such that the probability of p is placed on the x-axis:

Models and modeling in formal epistemology

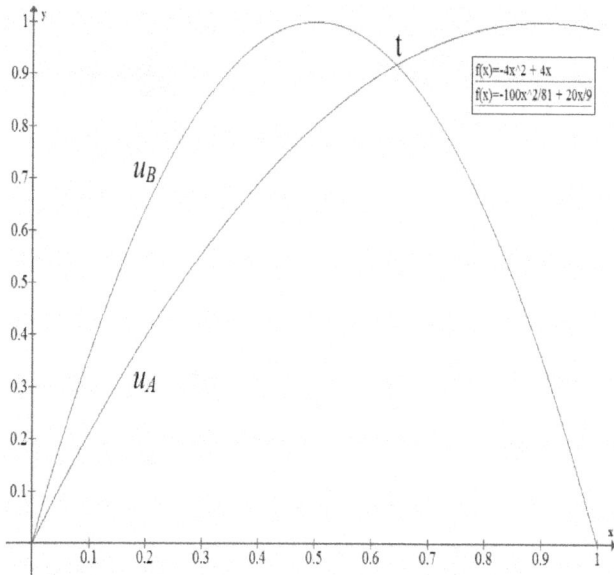

We can see that u_A and u_B intersect at $t = (0.643, 0.918)$, which maximizes the two functions simultaneously. Can we say that t captures the probability of the group? Ultimately, the answer is 'yes', but a proper justification for this claim requires a few more steps. Notice that point s corresponds neither to the arithmetic mean (0.7), nor to the geometric mean (0.671) between P_A and P_B. How shall we interpret t? Let us look at this problem more closely.

To have a better understanding of the task utility functions are performing, we shift to a second graph. Let us place utility scales of agents A and B along the x- and y-axes, respectively. Consider first the set of all points on the square $[0,1] \times [0,1]$. We will interpret each point as *a set of probabilities*: the probabilities that correspond to the utility level of the two agents at that point.[10] Thus, for our example, the point $(1,1)$ corresponds to $\{0.5 \, ; \, 0.9\}$. We then build a (pseudo) bargaining situation $S \subseteq [0,1] \times [0,1]$. We define S as the smallest convex set that includes all points for which the utilities of the two agents agree on the same probability; we can take the defect point to be $(0,0)$.[11] Clearly, points on the boundary of S will represent singletons (unique probabilities); by way of illustration, consider the following points on the boundary of S, up to three decimal places:

---

[10] There is a possible alternative interpretation here, according to which such sets are taken to be convex. In what follows I will not adopt this path, but nothing changes substantially if we do.

[11] The defect point is not playing any substantial role in the present meta-model. This might motivate us to explore in the future the use of bargaining proposals that do not demand a defect point (thanks to Ted Seidenfeld for making this suggestion).

$(0,0)$	represents	$\{0\}$
$(0.802, 1)$	represents	$\{0.5\}$
$(0.889, 0.96)$	represents	$\{0.6\}$
$(0.918, 0.918)$	represents	$\{0.643\}$ (as captured by s)
$(0.935, 0.883)$	represents	$\{0.671\}$ (the geometric mean)
$(0.951, 0.84)$	represents	$\{0.7\}$ (the arithmetic mean)
$(0.988, 0.64)$	represents	$\{0.8\}$
$(0.988, 0)$	represents	$\{1\}$
$(1, 0.36)$	represents	$\{0.9\}$

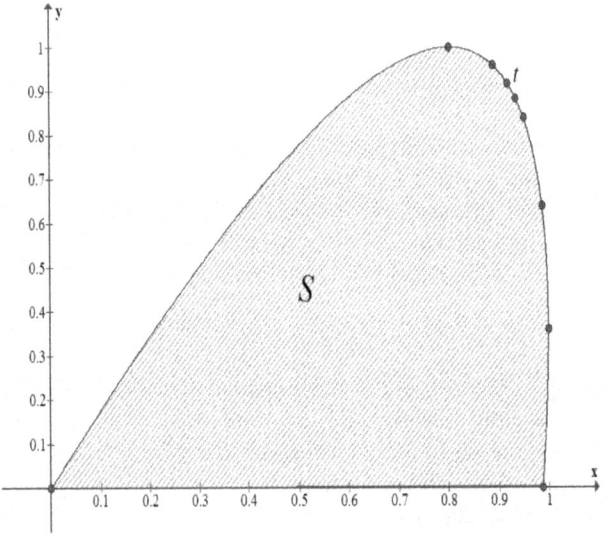

It is easy to see that t will lie in the diagonal; the probability captured by t is represented by the furthest North-Eastern point for which (at least) one probability assignment of agent A coincides with (at least) one probability assignment of agent B, and hence the set is a singleton. This amounts to the Kalai-Smodorinsky and the Egalitarian bargaining solutions.[12] Notice that Nash solution does not coincide with s, and neither do the arithmetic or the geometric mean between $P_A(p)$ and $P_B(p)$, as we have already pointed out. In any case, both the arithmetic and the geometric mean lie within the borders, so neither of them is Pareto inferior. Notice, however, that no stan-

---

[12] For an overview cf. for example Gartner (2009, chapter 8).

dard game-theoretic solution provides justification for picking them, under the current utility functions.

It is also interesting to notice that, within this setting, point $(1,1)$ is not in S. This fact rules out the possibility that the probability of the group be represented by the (possibly convex) set containing the preferred credences of each agent. The decision to proceed in this way can be thought to be the result of a trade-off between maximizing utility and minimizing higher order uncertainty – where a set containing more than one probability measures can be said to carry more uncertain (at least in some sense of this word) than a singleton. The current setting, in this sense, favors uniqueness.

Is the Kalai-Smodorinsky solution the way to go? We need not take a definite stance on this question for the moment. Regardless of the details, the core idea is that we can look at probability aggregation *as a type of cooperative bargaining*. Thus, we can borrow well-known discussions on the adequacy and correctness of different bargaining solutions and apply them here. We then obtain a unified approach to seemingly disparate phenomena.

This approach is still programmatic, but we can think of many different ways to continue it in the future. To begin with, notice that the graphics might look very different if we change the scenario a bit. For example, one of the agents may be truly uncertain about her first order probability assignments, so she might not give utility 1 to any probability value whatsoever (in which case the $(1,1)$ point will not be feasible). Or she might have a (convex) set of probabilities herself, and assign utility 1 to a whole interval. Alternatively, for highly opinionated individuals utility functions may descend abruptly, and agents might feel compelled to give *utility* 0 to a whole interval. In this scenario u_A and u_B might fail to intersect. Once again, at the time of constructing the bargaining set we will have to make a decision as to which points to include; we can again think of a trade-off between the amount of utility captured by each point, and the size of the set of probabilities such points represent in each case (the smaller the better). It would also be nice to work out generalizations to take care of n-person groups.

Related to this, an interesting project would be to look for the precise conditions (such as the specific utility functions) that will enable us to represent traditional methods for probability aggregation as solutions to bargaining problems, such as the arithmetic or the geometric mean. Once we accomplish this, we are left with the empirical problem of finding out whether actual people can be credited with utility functions close to those that would validate traditional proposals from the point of view of the bargaining model. Last, but not least, we can also investigate the extent to which agents can be motivated to adopt the aggregated measure as their own, in which case we will find a natural connection with the peer disagreement literature.

7 Conclusions

In the previous section I have suggested a principled account for probability aggregation that puts the *value* of probabilities at the center of the discussion. The proposal seems particularly well suited to deal with cases in which there is no external agent in charge of producing the aggregation; in that sense it can be said to honor a first person perspective.

Let me go back to our starting point now – to the discussion of normative models in formal epistemology – to see what we have accomplished. The proposal discussed in the previous section constitutes a good illustration of how normative meta-models are supposed to work. Bargaining models as I have used them here can be taken to aim at two different target systems at the same time: (i) they seek to capture intuitions on how groups of agents should negotiate their differences, and (ii) they seek to represent (in the mathematical sense) models that in turn seek to capture our intuitions on probability aggregation in general. Notice that bargaining models can help us elicit intuitions regarding probability aggregation in an indirect way: Strictly speaking, different bargaining models represent intuitions *regarding the way to settle a particular type of group disagreement*. Now, as it happens, some such settlement strategies can *in addition* represent particular sets of equations, which in turn may capture (normative) intuitions on probability aggregation – where the legitimacy of the bargaining models is inherited from their role in a different context altogether. In our particular example, the settlement strategies we selected did not coincide with any of the traditional approaches on probability aggregation (none of the standard first-order models got validated by the meta-model), so in this case they actually *defined* a new aggregation proposal.

As I have anticipated, relying on meta-models can help us unify apparently disparate situations and, at the same time, it helps us offer an illuminating interpretation of the mechanism that underlies the normative phenomenon. This fosters a better understanding of the phenomenon under consideration, and gives us the opportunity to further test our intuitions concerning what ought to be the case.

Bibliography

Alchourrón, C. E., Gärdenfors, P., & Makinson, D. (1985). On the logic of theory change: Partial meet contraction and revision functions. *The Journal of Symbolic Logic, 50*(2), 510–530.
Baron, J. (2004). Normative models of judgment and decision making. In D. J. Koehler (Ed.), *The Blackwell handbook of judgment and decision making* (pp. 19–36). Blackwell Publishing.
Colyvan, M. (2013). Idealisations in normative models. *Synthese, 190*(8), 1337–1350.
Dietrich, F., & List, C. (forthcoming). Probabilistic opinion pooling. In C. Hitchcock, & A. Hajek (Eds.), *Oxford handbook of probability and philosophy*. Oxford: Oxford University Press.
Frigg, R., & Hartmann, S. (2012). Models in science. In E. N. Zalta (Ed.), *The stanford encyclopedia of philosophy (Fall 2012 Edition)*. Retrieved from http://plato.stanford.edu/archives/fall2012/entries/models-science/
Gabbay, D. M., & Woods, J. (2003). Normative models of rational agency: The theoretical disutility of certain approaches. *Logic Journal of IGPL, 11*(6), 597–613.
Gartner, W. (2009). *A primer in social choice theory*. Oxford: Oxford University Press.
Genest, C., & Zidek, J. V. (1986). Combining probability distributions: A critique and an annotated bibliography. *Statistical Science, 1*, 114–148.
Katsuno, H., & Mendelzon, A. O. (1991). On the difference between updating a knowledge base and revising it. In *Principles of knowledge representation and reasoning: Proc. Second international conference (KR '91)* (pp. 387–394). San Francisco, California: Morgan Kaufmann.
Mäki, U. (2009). MISSing the world. Models as isolations and credible surrogate systems. *Erkenntnis, 70*(1), 29–43.
Suárez, M. (2004). An inferential conception of scientific representation. *Philosophy of Science, 71*(5), 767–779.
Titelbaum, M. G. (forthcoming). Normative modeling. In J. Horvath (Ed.), *Methods in analytic philosophy: A contemporary reader*. Bloomsbury Academic Press.
Weisberg, M. (2007). Three kinds of idealization. *The Journal of Philosophy, 104*(12), 639–659.
Weisberg, M. (2013). *Simulation and similarity: Using models to understand the world*. Oxford: Oxford University Press.
Williamson, T. (forthcoming). Model building in philosophy. In R. Blackford, & D. Broderick (Eds.), *Philosophy's future: The problem of philosophical progress*. Oxford: Wiley.
Yap, A. (2014). Idealization, epistemic logic, and epistemology. *Synthese, 191*(14), 3351–3366.

Author biography. Eleonora Cresto (Ph.D. in Philosophy, *Columbia University*, 2006). She is Philosophy Professor at Universidad Torcuato Di Tella and UNTREF

(Buenos Aires), as well as a Permanent Researcher at the CONICET (National Council for Scientific and Technical Research, Argentina). Her main interests lie in both formal and mainstream epistemology; she has also written on philosophy of science, decision theory, and philosophy of logic, among other topics. Currently she is Associate Editor of *Erkenntnis*, and Area Editor of *Ergo*, for Epistemology.

19 Unrealistic models, mechanisms, and the social sciences

HAROLD KINCAID

Abstract. Assessing the status of unrealistic models in economics and the social sciences more generally is an intellectual puzzle and a practical problem of some import. I argue that attempts to provide a general defense of and metric for evaluating unrealistic social models is misguided. Rather, compelling assessments have to look at specific models used for specific purposes and in specific empirical contexts. I outline how arguments can sometimes be made that unrealistic assumptions do not matter and that indeed on occasion they are essential. Across the discussions I look at the claim that mechanisms are needed for unrealistic models and argue that again there is no universal answer–sometimes they are and sometimes not.

Keywords: realism, models, idealization, philosophy of social science.

1 Introduction

Some highly unrealistic economic models were behind various policy pronouncements of economists in the period leading up to the great crash of 2008. Those pronouncements we now know were very far off target, and some blame for the 2008 downturn probably has to go to those models and those who purveyed them. It seems obvious to many in retrospect that these models should not have been trusted because of their quite unrealistic assumptions. Yet simplified models are absolutely essential in much of science and much of social science. So in evaluating unrealistic models it is important to avoid throwing the baby out with the bathwater. Yet how to do that is still very much an open question in the social sciences.

This paper tries to help us decide when we should throw the bathwater out without worry because social science models are so distant from reality and to tell when models tell us something despite their irrealism. My theme is that these problems are unlikely to be solved on general grounds; instead a host of different issues and techniques are involved depending on the kinds of models, their purposes, the empirical domains involved, and the testing techniques used. There is of course a vast philosophy of science literature on models, no doubt some of it of relevance to debates in the social sciences about unrealistic models. In a short paper I cannot pretend to tease out all of the possible connections, though my general approach is that the problem of unrealistic models cannot generally be decided on general conceptual grounds, that it must be attacked by looking for local methods for specific kinds of models, and to show examples of how that can be done. I am doubtful that the problems can be handled by first giving a general account of model representation and using that then to decide which unrealistic models explain and which do not–a general account of model representation seems unlikely (van Fraassen, 2008), and even if we had that, it is 1.) not clear what it would say about unrealistic models and 2.) local detail would likely still be essential to any assessments. My claims are also local in that they are mostly about specific kinds of models and applications in economics and some applications in political science, areas I know best. While some of my claims may generalize to the use of models across other sciences, I am not going to assert that here.

However, in a sense there are some more general morals lurking in my discussion, because evaluating unrealistic models overlaps with general questions about the place of mechanisms. Mechanisms have been offered as a tool for dealing with unrealistic models in the social sciences (Mäki, 2009) and I look at specific kinds of cases where they can help. Still my conclusions will be that evaluating the importance of mechanisms, as in evaluating unrealistic models, depends strongly on the local details involved. As I have argued elsewhere (Kincaid, 2011, 2014), there are multiple independent sense of mechanisms and there is no reason they should all the same import. One of my main and more novel conclusions perhaps is that there are circumstances where providing mechanisms for unrealistic models leads to error; in some cases models need to be unrealistic to be believable.

Throughout I presuppose no general account of what models are (and doubt that there is one to be had). I take the term "model" in the way it is commonly used by social scientists, where the paradigm case comes from economics which overwhelmingly thinks of models as an a equation or set of equations which is accompanied by a verbal narrative linking the model to economic phenomena. I think my conclusions can also be applied to other senses of models such as Schelling's physical checkerboard model of racial segregation.

The paper is organized as follows. Section 2 elaborates on the importance and nature of the issues just sketched above and then looks at some all purpose attempts to evaluate the status of unrealistic models and argues that they are insufficient. Sections 3-5 then look at specific circumstances where parts of economics and political science can successfully ignore the irrealism of their models. My goal is to defend the strong thesis that unrealistic social scientific models sometimes provide us with well-confirmed explanations, where the explanations in question are causal explanations. I make two claims: 1.) that there are methods by which we can show that the irrealism of models does not matter in producing well-confirmed causal explanations and 2.) that sometimes only unrealistic models can provide well-confirmed causal explanations. Claim 1.), I should note, is not the standard and true point that some factors can be left out because they are of negligible importance. The kinds of cases I have in mind concern factors that can be causally *very* important. Rather the idea is that given the right kind of model and method of verification, even causally important factors can be neglected.

Section 3 looks at competitive equilibrium and game theory models, arguing that the typical unrealistic assumptions about the abilities of agents do not necessarily undermine the equilibrium explanations given because some of those explanations do not depend on cited presuppositions about how agents reach equilibrium. Section 4 looks at instrumental variables and related techniques that can ground arguments that models can leave out variables without sacrificing validity, though economists err in thinking that such arguments can be purely statistical. Section 5 demonstrates cases where leaving out causes that have significant effects is essential, not just defensible.

2 Issues and past approaches

Macroeconomic models used in the run up to the 2008 crisis (and still used) illustrate the problem of unrealistic models vividly. These models are called dynamic stochastic general equilibrium (DSGE) models. General equilibrium means they are supposed to consider all market interactions at once. They are dynamic in the obvious sense that they are supposed to describe changes over time and they are stochastic in that they describe probabilistic relations. These models are the dominant trend in economics because they are supposed to be grounded in mechanisms involving constrained optimizing of individuals, unlike traditional Keynesian models whose ties to individual behavior is less determinate.

However, the actual ties to individual behavior of DSGE models are much less clear than the rhetoric advertises. DSGE models depend essentially on "representative agents": for instance, they assume that all consumers can be treated as if they were one agent. In other words they assume that there is only one consumer. They do so for tractability reasons. However, it is widely known that aggregate consumption mirrors individual choice only under stringent assumptions that are seldom met (Kirman, 2010). Aside from this aggregative assumption, DSGE models also assume that we can treat all firms as a single agent and that financial markets are fully efficient, and that relations are linear.

So these models are not making assumptions that certain factors only play a small role and can be ignored nor just transforming complicated relationships into somewhat more tractable simpler ones. They are starkly at odds with reality. Thus the pressing question is how we can explain the real world using them. Models as unrealistic as the DSGE models are not an exception in economics and political science. Two work horses in economic analysis are competitive general equilibrium models and game theory models, with the latter increasingly important in political science as well. Both appear to rest on assumptions about agents that are often highly unrealistic. How can they produce well-confirmed explanations–or do they?

Before I address these questions it is important to note that models may serve multiple functions other than providing well-confirmed explanations. For example, models may:

1. be essential for tractability–to get any way of systematically thinking about the phenomena.

2. be used to discover new possible phenomena.

3. be used to show how certain phenomena might come about—Schelling's models of residential segregation are a favorite example cited in this regard (see Ylikoski & Aydinonat, 2014).

4. be used to get a clearer formulation of a verbal argument. This is certainly quite true of the motivation for models in economics.

5. be used to identify robustness—to show that with weakened or different assumptions results still hold (Kuorikoski, Lehtinen, & Marchionni, 2010).

None of these uses of models are committed to the idea that models have to provide well-confirmed causal explanations of specific real world phenomena. So, given these goals, unrealistic models are not necessarily a problem.

Relatedly, unrealistic social science models may also be unproblematic in that their purpose is just to provide successful forecasts or predictions of the data. There is much science that counts success as the ability to trace variables through state space without a pretense of providing realistic causal mechanisms. Of course, it is a very open question whether some or any social science models provide the ability to predict variables through state space with anything like the success of some natural sciences. Certainly some parts of macroeconomics at least hope to provide prediction models without any claim to have realistic causal explanations. However, if unrealistic models are able to show such predictive success, then it seems that they tell us something about how the world is despite their unrealistic assumptions. This defense of unrealistic models needs to be distinguished from instrumentalism, because it does not claim that all social science is only about predicting observables but only that some models do. A charitable reading of Milton Friedman's (1953) classic paper might see this thesis as one of his more defensible points.

Despite these various functions that unrealistic models can perform, we still of course would like them to provide well-confirmed explanations; social scientists routinely claim that their models can do so, as witnessed by the fact that they are willing to make policy recommendations based on them. Thus while recognizing that the function of models in science is many and that questions about realism may not always be problematic, the status of unrealistic models in the social sciences is still pressing.

There are of course many defenses of unrealistic models in the social sciences, especially in economics, extant in the literature. None of them are entirely convincing. I want to sketch some from social scientists and some from philosophers as a way of setting up the approach that I think is more fruitful and that I apply in the rest of the paper.

A first response frequently found among social scientists is to dismiss the problem outright as unimportant on the grounds that all models misrepresent, as claimed recently for example by political scientists Clarke and Primo (2012). This "be happy, don't worry attitude" will, of course, not do if we want models to give us well-confirmed explanations of how the social world actually works. Since some models with unrealistic assumptions clearly do not tell us how the world works, we need to know how to tell those that do from those that do not. So the very same social scientists who advance this view about models will go on to argue in their actual work that some models, namely their own, are better than others; the question is how that judgment is grounded. They will also use their models to make policy recommendations, implicitly implying that they are close enough to reality to allow causal inferences.

Another common claim from producers and consumers of unrealistic models in the social sciences is that despite their departure from reality they provide "insight." The problem here is how to know when "insight" is anything more than a warm and fuzzy intellectual feeling? What does the insight come to? The idea can be developed into an account of explanatory understanding of possibilities as in Ylikoski and Aydinonat (2014) mentioned above. Yet that still does not tell us how can we use it to ground real world causal knowledge like that needed for policy advice.

Morgan (2012) in a recent interesting book of case studies of economics models defends unrealistic models by claiming that economists provide narratives and use tacit knowledge to tie them to reality just as scientists do in the natural sciences. While this is no doubt true in some sense, the open question is still when and how unrealistic social models do so *successfully*. There are well-developed methods in the natural sciences for tying unrealistic models to reality (see for example Wilson, 2006) but the question is whether the social sciences have any real parallels. Talk of providing narratives and using tacit knowledge is a promisory note that needs to be filled out, something Morgan only hints at. I will make some efforts in that direction below.

Finally, an important trend argues that unrealistic social science models are nonetheless explanatory when they capture or isolate the essence of a causal mechanism. This view has been argued for by Cartwright (1989), at length by Mäki (2009), and very recently by, on my view, one of the more methodologically sophisticated economists, Rodrik (2015). These approaches seem to me fundamentally on the right track. But the question is how and when do we know that unrealistic models actually isolate causal mechanisms? How do we know that a causal process in a model isn't disrupted in reality when other left out factors are involved? These authors by and large do not address that essential question, and it is this question that I am pursuing. What a "causal mechanism" comes to is also often left unspecified.[1]

I want to argue that answering this question about when we really have isolated a real causal process has to be largely a local affair, where the local parameters include the specific structure of the model in question and the specific techniques used to access it. A lack of realism may be no problem for some model structures and a serious obstacle for others; a lack of realism may be no problem for some models using certain techniques and a serious obstacle when using other methods. Mechanisms may be essential in some senses and some cases but not in others. Relatedly, I want to argue that it is unlikely that questions about unrealistic models have general, purely statistical noncausal answers, a view social scientists often harbor.

3 Showing that unrealistic assumptions are not part of the explanation

My first example of unrealistic models that can be nonetheless well-confirmed really builds off the previous point that questions of about the realism of models have to be cognizant of what models are used for. Two paradigm cases of unrealistic models in economics are perfect competition and game theory models, workhorses of the profession. Both kinds of models seem to make incredible assumptions. Perfect competition models often assume among other things:

- complete markets over all goods at all times (e.g. used umbrellas in 2027)
- individuals with complete preferences over all goods (e.g. those umbrellas)
- every buyer and seller is a price taker
- commodities are infinitely divisible
- buyers and sellers have full information

Typical game theory models assume:

- that agents can calculate the relevant best response equilibrium strategy, e.g. the subgame perfect Bayesian Nash equilibrium

---

[1] We could put this approach in terms of the idea of a partial isomorphism, one standard way of talking about model representation. However, we would still need to know which isomorphism and how the isomorphism obtains if the rest of the model is false.

- there is common knowledge of the payoffs, strategies and all the specific players

A considerable amount of ink has been spent on various twists and turns trying to justify such assumptions. However, there is good reason to think that in some applications, these modeling assumptions are no worry (Smith (2008) is the main inspiration for this insight). They are no worry because they need not be part of the explanation that is given. How can that be?

The insight is that assumptions of the perfect competition and game theory models may just be assumptions the *analyst* –the economist or political scientist–uses to identify equilibria. However, in certain empirical applications, the explanations are equilibrium explanations that make no commitment to what process leads individuals to find equilibrium. The analyst uses the unrealistic assumptions to identify an equilibrium. Then experimental studies are used to show that individuals will reach the equilibrium predicted from the models. So Smith shows in multiple experimental setups that subjects reach competitive equilibrium in continuous double bid auctions. Economists identify competitive equilibrium from assumptions such as that every individual is a price taker, there are an indefinite number of individuals in the market, and so on. Obviously these assumptions are not necessary for a competitive equilibrium to exist, since the experimental setup is nothing like this. Nonetheless competitive equilibrium is reached. Thus the unrealistic assumptions are useful to the analyst, but they are not part of the explanation of the experimental results.

Binmore (2007) does something similar for game theory. For example, in a finite, two person zero sum game the Nash equilibrium first proved by von Neuman was the minimax value. Borel first proposed the solution but had no proof; von Neuman proved it was the Nash equilibrium. Binmore notes that his subjects are probably not smarter than Borel, but given repeated play find their way to the minimax equilibrium.

These explanations work by showing that all individuals are making best responses to the actions of others. Equilibrium persists because deviation from it would produce inferiior results for individuals. You may or may not think these are causal explanations (cf Sober, 1983). However, the evidence for them comes from carefully designed experimental setups where the equilibrium behavior predicted by the model is observed (Smith is a Nobel prize winner and Binmore among the most highly respected experimentalists in the field). These explanations are a species of optimality analysis, but they do not have the kind of complications that plague optimality in nonexperimental situations (Orzack & Sober, 2001). For example, the costs and benefits individuals face are well known because they are set by the experimenters, a decided advantage over observational studies.

Notice the role mechanisms play in these cases. A mechanism in the sense of the causal process that leads individuals to find equilibrium is not needed. We don't know how they get there but we have strong evidence that they do. On the other hand, mechanisms are given in the form of appeal to best responses to the actions of others.

This defense of unrealistic competitive equilibrium and game theory models need not generalize to their use in other settings. When we have only observational evidence of behavior outside the lab, we are on much shakier grounds in claiming that observed behaviors are optimal equilibrium behaviors. Then the lack of a realistic mechanism that would lead to an optimum–for example, appeal to individuals calculating subgame perfect equilibrium by backward induction–matters.

4 Causal methods to compensate for unrealistic models

One standard worry about unrealistic models in the social sciences is that confounding variables have been left out. This is a reasonable concern. I want to show in this section that there are ways and circumstances in which such concerns can be shown not to prevent well-confirmed explanations of unrealistic models. I have in mind three approaches: instrumental variables, sensitivity analyses, and the use of structural equation models to establish causal effects rather than causal effect size.

Instrumental variables are a way to defend unrealistic models. The idea of instrumental variables, for those who are unfamiliar, is that there are certain causal variables such that, if we have the right causal structure, we can make reliable estimates of causal influence even if there are confounding variables that our model leaves out. Leaving out important causes makes the model unrealistic. But given the right information, instrumental variables can show that our model nonetheless provides well-confirmed causal explanations. To use instrumental variables, we need to show that there is a variable influencing the alleged cause but not the dependent variable or effect. We can then use the variability in C caused by the instrument I to estimate its influence on E independently of possible confounding variables effecting C and E. When these conditions are met, standard regression methods –two stage least squares–can reliably estimate the influence of C on E despite ignoring the confounder. So this is a case of a good argument for an unrealistic model.

However, note the nature of the argument: it is a causal argument invoking a quite specific causal structure, not a purely formal statistical argument. Econometricians are often confused about this, defining an instrumental variable as one where the instrument is strongly correlated with the cause and not correlated with the effect. But, as usual, causes cannot be gotten from correlations, and as Reiss (2005) has shown, it is easy to imagine situations where an instrumental variable is not correlated with the outcome or the error term and is correlated with the cause being studied, but the instrumental variable is not a cause of the variable in question.

However, economists often just want an instrumental variable to be able to do consistent estimation in the technical statistical sense, since unknown confounders can produce estimates from samples to populations that do not converge on the correct estimate as the sample size increases. Instruments that only meet the statistical criteria do suffice for this goal, which is a rather different project than providing well-confirmed causal explanations. This is an instance of my claim that whether realism matters depends on the purpose of the model and the methods being used: if you want to confirm a causal explanation with left out confounders, you need a realistic causal model; if you only are using the model to make consistent estimates about a population the realism is not necessary.

A related but alternative way of arguing that we can leave out unknown confounders comes from sensitivity analyses. Sensitivity analyses provide arguments that even if there were confounders that are left out, they would have to take on unreasonably high values to completely undermine the causal relations postulated in the model. Suppose that I believe that C causes E but that I also worry that CC is a common cause confounding the relation. I can ask how large must the relation between the common cause CC and the C and E relation be to make the correlation between C and E to be entirely spurious. There are both structural equation and regression based ways of testing this hypothesis.

Note that sensitivity analysis here is much more helpful than what is commonly called robustness analysis. Robustness analysis is common in economics. It involves seeing whether conclusions from models hold up under changed assumptions. Yet such analyses can move from one unrealistic assumption to another and there is no guarantee that a more robust model is likely to be more reliable guide to reality.

Finally, structural equations, using the Pearl causal calculus approach, provide another sense in which we can argue that unrealistic models can nonetheless provide well-confirmed causal explanations despite unrealistic assumptions. Typically, a structural equation model in the social sciences assumes that the relations between variables are linear. But if my goal is to provide evidence for causation rather than the *size* of causal parameters, I can use such a structural equation model even if the linearity assumption is wrong. Causal relations entail dependencies and independencies among the variables but makes no requirements on their functional forms. So long as the requisite dependencies and independencies show up, it does not matter what their functional form is and thus it does not matter that I have estimated an unrealistic linear model.

Note again the nuanced role of mechanisms in these three approaches to dealing with unrealistic models. Instrumental variables allow us to confirm a causal relation without knowing the confounder(s) so in that sense we do not need mechanisms. Yet we need to have evidence that the instrumental variable does not causally influence the effect. In that sense we need to know about mechanisms. Much the same holds for sensitivity analysis: it allows me to find evidence for a causal relation without knowing what the confounders are–to that extent I do not need to know what the mechanism is. Finally in the case of inferring causal effects (but not causal effect sizes), I do not need to know the details of the causal relations–the precise functional forms of their relations–and in that sense need a minimal description of the mechanism. Yet I need more information if inferring effect sizes is my goal and so then mechanisms carry more weight .

5 Showing that irrealism is a necessity

So the moral of these three examples is that with sufficient information we can confirm unrealistic models, but that information is causal and not just formal or statistical and in some ways mechanisms are irrelevant and in other ways needed. I want to now show that sometimes models that *leave things out* –unrealistic models—are better at getting us well-confirmed causal explanations than models that are more realistic in that they do not leave causes out. The context I have in mind is the traditional multiple regression frame work that is the work horse across the social sciences. The kinds of models I have in mind here are the more empirical models involved in statistical testing, not the abstract theoretic models often involved in economics.

Standard practice for social scientists using multiple regression is to include "control variables" or "covariates" when looking at a specific causal relation of interest. The general tendency is to err on the side of including more such variables rather than fewer. The thought is that doing so gives more realistic models that will help eliminate confounding.

However, this "include everything in the model approach"—which seems like a move towards greater realism—can be the enemy of well-confirmed causal explanations. How so? In two common causal situations, including covariates using multiple regression techniques can result in seriously erroneous causal claims. Here are two clear situations:

1. including mediator variables as controls in regressions

2. including collider variables as controls in regressions

We have mediators when we simply have a causal intermediary between a more distal causal and the final effect we want to explain. A collider is a variable that two or more other variables cause. They "collide" on their common effect.

In both cases having a more realistic model that includes mediators or colliders, given that we are using multiple regression techniques to look for causes, leads to error. If I control for an intermediate variable—hold it fixed—then I remove the connection between the distal cause and the final effect. I will conclude that there is no relation between the distal cause and the effect when there is in fact one. If I control for a collider, I create correlations between its two causes that are spurious; holding the collider constant creates a spurious connection. Thus including the collider leads to mistaken causal conclusions. So in both cases, using multiple regression techniques, I should be less realistic in the sense of leaving such variables out of my models when I go to test them.

These kinds of problems have important real world effects. For example, economists have used data sets from most countries in the world to examine the causes of economic growth. Generally what they have done is to treat all possible causes of growth as independent factors and regressed measures of growth on these variables. However, this is an extremely simple model. Most likely there are complex causal relations between the independent variables causing growth. That means it is likely that the regressions in cross country studies are controlling for mediating causes and colliders. So these studies are quite likely to eliminate connections that are there by controlling for intermediate causes and to create spurious correlations by controlling for colliders. In my previous work (Kincaid, 2014) I have shown this to be the case: the simple additive model where every possible cause is controlled for fits the data very badly; models with more complex causal structure do much better. I show that the standard naïve models which typically find that education has no influence on growth are not well supported by the data; models that allow for mediation and colliders find that education does indeed contribute to growth and are much better fits to the covariances found in the data.

So to conclude, I hope to have shown at least two things. Treating the question of unrealistic models as a set of more specific questions about particular models using particular methods allows us to make progress on the issues in a way perhaps not possible if the issue is approached in perfectly general way. Similar conclusions hold about the role of mechanisms: their importance and place varies according to the kind of causal claims being made and the methods used to support them.

Bibliography

Binmore, K. (2007). *Does game theory work?* Cambridge: MIT Press.
Cartwright, N. (1989). *Nature's capacities and their measurement*. Oxford: Oxford University Press.
Clarke, K., & Primo, D. (2012). *A model discipline: Political science and the logic of representation*. Oxford: Oxford University Press.
Friedman, M. (1953). The methodology of positive economics. In *Essays in positive economics* (pp. 3–34). Chicago: University of Chicago Press.
Kincaid, H. (2011). Causal modeling, mechanism, and probability in epidemiology. In I. P., R. F., & J. Williamson (Eds.), *Causation in the sciences* (pp. 70–91). Oxford: Oxford University Press.
Kincaid, H. (2014). Mechanisms, causal modeling, and the limitations of traditional multiple regression. In I. P., R. F., & J. Williamson (Eds.), *Oxford handbook of the philosophy of the social sciences* (pp. 46–64). Oxford: Oxford University Press.
Kirman, A. (2010). The economic crisis is a crisis for economic theory. *CESifo Economic Studies, 56*, 498–535.
Kuorikoski, J., Lehtinen, A., & Marchionni, C. (2010). Economic modeling as robustness analysis. *British Journal for the Philosophy of Science, 61*(3), 541–567.
Mäki, U. (2009). MISSing the world: Models as isolations and credible surrogate systems. *Erkenntnis, 70*, 29–43.
Morgan, M. (2012). *The world in the model*. Cambridge: Cambridge University Press.
Orzack, S., & Sober, E. (2001). *Adaptationism and optimality*. Cambridge: Cambridge University Press.
Reiss, J. (2005). Causal instrumental variables and interventions. *Philosophy of Science, 72*(5), 964–76.
Rodrik, D. (2015). *Economics rules: The rights and wrongs of the dismal science*. New York: W. H. Norton.
Smith, V. (2008). *Rationality in economics*. Cambridge: Cambridge University Press.
Sober, E. (1983). Equilibrium explanations. *Philosophical Studies, 43*(2), 201–210.
van Fraassen, B. (2008). *Scientific representation*. Oxford: Oxford University Press.
Wilson, M. (2006). *Wandering significance*. Oxford: Oxford University Press.
Ylikoski, P., & Aydinonat, E. (2014). Understanding with theoretical models. *Journal of Economic Methodology, 21*(1), 19–36.

Author biography. Harold Kincaid is Professor of Economics at the University of Cape Town and Visiting Professor at the Finnish Center of Excellence for Philosophy of Science at the University of Helsinki. Early books were *Philosophical Foundations of the Social Sciences* (Cambridge 1996) and *Individualism and the Unity of Science* (Rowman and Littlefield 1997). He is the editor of the *Oxford Handbook of the Philosophy of Social Science* (2013) and coeditor of *Scientific Metaphysics* (Oxford 2013),

What is Addiction? (MIT 2010), *Distributed Cognition and the Will* (MIT 2007), *Toward a Sociological Imagination* (University Press, 2002), *The Oxford Handbook of the Philosophy of Economics* (Oxford 2009), *Classifying Psychopathology* (MIT 2014), *What is Addiction?* (MIT 2010), *Establishing Medical Reality*, (Springer, 2008), *Value Free Science* (Oxford 2007), the *Routledge Companion to the Philosophy of Medicine* (forthcoming), and numerous journal articles and book chapters. Kincaid has also been a left wing trade unionist, shop steward, and union executive board member of the Communications Workers of America.

20 Modelling failure

USKALI MÄKI [*]

Abstract. Philosophy of science is largely inclined to portray science as a success story. While not in the least denying the great successes of science, it would seem there is an important and interesting other side of the story, also worth the systematic attention of philosophy of science: developing accounts of the nature, conditions and dynamics of both failure and success should be on the philosophy of science's agenda. In this article I focus on one prominent style of scientific inquiry, that of modelling, and on one part of philosophical study of science, that of offering philosophical accounts of models and modelling in science. A sound philosophical account of modelling should contain resources for identifying and diagnosing modelling failure. The ability to articulate (at least rudiments of) a systematic account of modelling failure can be used as a test of one's account of model and modelling. Here I expose my own account to such a test using an example from economics (its alleged failure in anticipating and conceiving the 2008 financial crisis), showing how my account provides an encompassing framework for identifying and analyzing failures in modelling.

Keywords: models, models in economics, modelling failure, economics, philosophy of economics, financial crisis.

1 Science fails too

Philosophy of science is largely inclined to portray science as a success story. It takes as its task to explain how science manages to acquire truthful information about the world or to generate successful predictions and explanations of phenomena, and it aspires to do this by articulating the procedures and principles of science that make this success possible. While not in the least denying the great successes of science, it would seem there is an important and interesting other side of the story, also worth the systematic attention of philosophy of science – not just as an unfortunate accidental

---

[*]University of Helsinki

residuum of failures soon to be corrected, but rather as a massive enduring part of scientific practice with its own regular features. Developing accounts of the nature, conditions and dynamics of both failure and success should be on the philosophy of science's agenda. Ability to produce such accounts should be one of the criteria of success of philosophy of science itself.

As my illustrations here will derive from the discipline of economics, it is interesting to note that there is a striking analogy between conventional economics and conventional philosophy of science. Both are fascinated by the successes of their target domains, the market economy and science, respectively. Consider the following statement by the economist Nouriel Roubini. In promoting an unorthodox approach to the study of the economy ("crisis economics") he stresses the difference between studying failures and studying successes. "Crisis economics is the study of how and why markets fail. Much of mainstream economics, by contrast, is obsessed with showing how and why markets work – and work well." (Roubini & Mihm, 2011, p. 39) My paper is an exercise in the philosophy of science analogue of crisis economics – while much of conventional philosophy of science is analogical to the urge of mainstream economics to show how wonderfully markets function.

Naturally, the ambition to understand failure in science must be divided into manageable portions. Here I focus on one prominent style of scientific inquiry, that of modelling, and on one part of philosophical study of science, that of offering philosophical accounts of models and modelling in science. There are many such accounts available in the literature, and the challenge is to compare them for their credentials. One obvious way to proceed is to check them against empirical evidence concerning actual models and actual modelling practices. And provided we take these practices to include failures, then the capacity of the philosophical accounts in dealing with such failures may be taken as a major criterion of the success of those accounts. A sound philosophical account of modelling should contain resources for identifying and diagnosing modelling failure. The ability to articulate (at least rudiments of) a systematic account of modelling failure can be used as a test of one's account of model and modelling. Here I expose my own account to such a test, showing how it provides an encompassing framework for identifying and analyzing failures in modelling.

2 Failure of model, failure of target: Economics and the financial crisis

'Modelling failure' is ambiguous between *modelling <failure in the target system>* and *failure in <modelling>*. In the first category, the failure lies in the functioning of the target system rather than in an attempt to model the target system. The possibility of such a failure applies to target systems to which we are entitled to ascribe some idea of proper functioning. Scientists can then model such failures of their target systems to function properly, such as heart failure in the human body, business failure, market failure, failure of materials, failure of machines in manufacturing systems, failure of engineered arrangements such as energy systems and jointed rock slopes to resist landslides, and so on.

Scientists may be successful in modelling failure. Or they may fail in modelling failures in their target systems. Or they may succeed or fail in modelling the proper (non-failing) functioning of their target systems. It is an instance of the second combination that I will use as an illustration in what follows. I will be discussing *tools and acts and strategies of modelling that allegedly fail to model the failures of some target systems* – such as this example from materials physics: "Traditional hyperelastic models of materials ignore the fact that no material can sustain large enough deformations without failure." (Volokh, 2010, p. 684) Here models fail to model failure due to employing the excessively idealizing assumption of hyperelasticity. The illustration I will use later in the paper is of this sort of double failure: failure in economics to model the failures of the financial system. The possible sources of failure, however, will be shown to be far more varied than just a single idealizing assumption.

Just a few years before the outbreak of the financial crisis of 2008 and the subsequent persistent recession, there was a good deal of optimism in the academic air, celebrating the "great moderation" of business cycles and praising macroeconomics for having solved the "problem of depression prevention" by developing theories and methods and policy recommendations that were disputed by few in the economics profession. Here is Nobel Laureate Robert Lucas from Chicago in 2003, in his presidential address to the American Economic Association:

> "The term ['macroeconomics'] then referred to the body of knowledge and expertise that we hoped would prevent the recurrence of that economic disaster. My thesis [...] is that macroeconomics [...] has succeeded: Its central problem of depression prevention has been solved, for all practical purposes, and has in fact been solved for many decades." (Lucas, 2003, p. 1)

Soon thereafter, things went badly wrong (in fact they were getting wrong at the time of Lucas's statement). In a few years, triggered by the subprime mortgage crisis that burst the bubble in the US housing markets, the global financial system would collapse – without anticipations, warnings, or recommendations as to how to prevent it offered by the profession that had just a while earlier congratulated itself for having solved the problem.

The generation of the crisis was no simple process. It is therefore no surprise that the aftermath of the financial crisis of 2008 has exhibited an almost proverbial blame game. Whom, or what, to blame for the crisis? The candidates have ranged from the government (for regulating too little, too much, or wrongly) through the design of the global financial market system (for its inherent instability and susceptibility to systemic risk) to the credit rating agencies (for massive mistakes in their assessments) and, of course, the human nature in general (greed and all that).

The discipline of economics has received its share of the blame. Even those who charge central bankers like Alan Greenspan and Ben Bernanke for having made fatal mistakes consider that those mistakes were shared by the establishment of the economics profession and so individuals like them cannot bear the main responsibility (e.g. Posner, 2009, p. 286). It is a matter of collective and institutional failure, and this is where the discipline of economics enters the picture, with its collectively held and institutionally ingrained conventions and convictions, principles and practices. Indeed, many people, including many leading economists, started questioning the performance of economics as a scientific discipline, including the explanatory and predictive capacities of the highly abstract mathematical models that have become so popular in the discipline. In July 2009, the cover of *The Economist* asked: "What went wrong with economics?" A bit later, Nobel Laureate Paul Krugman (2009) famously wrote in *The New York Times Magazine*:

> "[...] the economics profession went astray because economists, as a group, mistook beauty, clad in impressive-looking mathematics, for truth. [...] When it comes to the all-too-human problem of recessions and depressions, economists need to abandon the neat but wrong solution of assuming that everyone is rational and markets work perfectly. The vision that emerges as the profession rethinks its foundations may not be all that clear; it certainly won't be neat; but we can hope that it will have the virtue of being at least partly right."

The above passage provides rudiments of a methodological diagnosis of the alleged failures of economics. It claims that economists have been preoccupied with the beauty and neatness of their models, expressed in impressive mathematics, while they have forgotten the task of looking for truths about the real world. As to the contents of their models, the claim is that economists have envisaged a fantasy world of perfectly rational agents in perfectly self-regulating markets, and this fantasy world is too far removed from the imperfections of the real world to be helpful for acquiring truthful information about the latter. The economics profession is criticized for holding an all too strong faith in the powers of the invisible hand, manifesting itself in the widely accepted and applied but allegedly failed DSGE (dynamic stochastic general equilibrium) models in macroeconomics and those of efficient markets in finance.

The remarks in the following sections will provide an elaboration of such rather popular Krugman-style allegations. This will be done by dealing with the alleged failures of economics vis-à-vis the financial crisis as modelling failures and by offering a series of possible partial diagnoses of different kinds and sources of modelling failure. Modelling is a multi-stage and multi-faceted cognitive process, so there are multiple sources of, and multiple opportunities for, possible failure – as well as multiple ideas of what constitutes failure. Some of these will be mapped. For this task, we need an account of model and modeling that is rich enough for exhibiting several such opportunities for failure.

3 An account of model and modelling

In the last couple of decades, philosophy of science has recognized models as among the key cognitive tools in science and modelling as one of the key activities in scientific practice. It has become a major industry within philosophy of science to produce accounts of model and modelling in science. These accounts have often been generated and applied (sometimes even tested) using historical and contemporary case studies in a variety of scientific disciplines, from physics and biology to the social sciences. I will now briefly summarise my own account and then put it into use for highlighting failure rather than success, with economics and the financial crisis serving as the illustrative case. Its capacity in this role provides a test of its general adequacy.

Philosophers of science have moved beyond the idea that models involve a simple two-place representational relationship between the model and its target: *M is a model of target R*. It is now commonplace to conceive of representational models as more complex, involving in addition an agent and a purpose: *Agent A uses (builds, employs) M as a model of target R for purpose P* (see e.g. Giere, 1999). Some philosophers wish to add further components in the modelling relationship, such as some idea of an interpretation (see e.g. Weisberg, 2013). I have proposed that representational modelling is still more complex (e.g. Mäki 2009b, 2011b, 2011a). My account portrays model representation as a multi-faceted activity that can be schematized as [ModRep][1]:

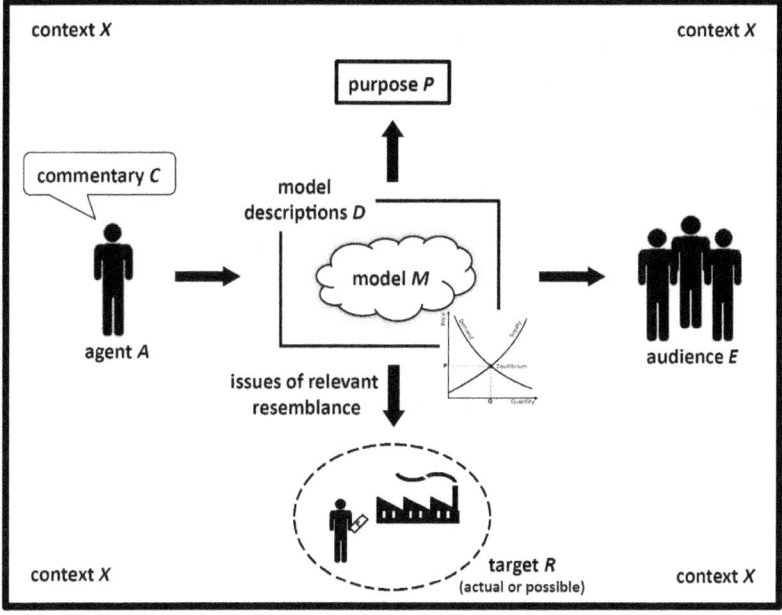

---

[1] Thanks for brilliant help in designing the figure go to Ilmari Hirvonen

[ModRep] can be put in words only:

> Agent A
> uses multi-component object M as
> a representative of (actual or possible) target R
> for **purpose P**,
> addressing **audience E**,
> at least potentially prompting genuine **issues of relevant resemblance** between M and R to arise;
> describing M and drawing inferences about M and R in terms of one or more **model descriptions D**;
> applies **commentary C** to identify and coordinate the other components;
> and all this takes place within a **context X**.

Among the novelties here we can list the involvement of an audience; the idea that merely potentially giving rise to genuine issues of relevant resemblance is needed for representation to be in place; a rich functionally loaded notion of commentary; and a generous placeholder for various contextual factors.

A simple version of the idea of *success* in modelling would say that success occurs when relevant resemblance between M and R becomes secured. This requires, for example, that the causal structure or causal forces captured in the model relevantly resemble those in the target; the model descriptions are of the sort that enable the needed inferences; and that the components of [ModRep] become coordinated and communicated so that purposes P and audiences E are sufficiently served. Finally, context C must be such that it is sufficiently supportive of those accomplishments.

We can then say that *failure* in modelling occurs when relevant resemblance between M and R fails to be secured, in one way or another, for one or another of the various possible reasons that [ModRep] helps identify. The components [ModRep] and their relations will next be investigated as potential loci and sources of modelling failure. It appears that many existing critiques of economic modelling can be construed as focusing on some specific component in the structure of [ModRep] and that some other possible critiques can also be envisaged within this framework. In both cases, it seems obvious that the claims about possible sources of failure can be made sharper than without the framework of [ModRep] and that the framework will also enable more focused assessments of the credibility of such claims.[2] I will next rush through the framework component by component.

---

[2] I will use the structure of [ModRep] for outlining *possible* sources of failure without making any strong claims about actual failures. The usefulness of the framework for organising focused scrutiny is not dependent on the correctness of such claims. There is an analogy between the exercise here and much of economic modelling: the latter often describes possible ways in which a pattern of phenomena comes about, and so do the examinations outlined in the present remarks.

4 Multiple sources of modelling failure

I will exhibit the usefulness of the [ModRep] framework for organizing some of the popular misgivings about economic modelling – for identifying and localizing the various criticisms of economic models for their failures regarding the financial crisis (and surely beyond this case as similar criticisms are a chronic feature of debates around economics). It appears that all components listed in [ModRep] have been or can be charged for not having done well.

Agent A

Identifying the agent of modelling as a separate component, and attributing characteristics to it, is somewhat awkward since this is the melting pot where the influences of the other components get together. This is fortified by the fact that individual and collective agency interact and depend on one another. The individuals inventing and proposing models have been educated and socialized in the collective disciplinary culture and prevalent fashions of research practice. On the other hand, the individual modeller must persuade the collective to join her in using an object as a model. The identity and properties of A make a difference for whether A qualifies as a (credible) economist; what sorts of models are being built and examined, shaped by A's skills and background beliefs; what is taken seriously as a model worth some further attention; and so on.

Economists are generally recognised as intelligent people. Yet the critics argue that this is not sufficient for successful modelling and that the failures regarding the 2008 crisis are one indication of this. They say economists are too narrowly educated (mainly just in contemporary economics, math and statistics), too ignorant about history (of the economy and of their own discipline), about the other social sciences, about culture and human psychology. Some say their competences and epistemic preferences are ill suited for modelling the complexities of social reality. Their mathematically inclined style of inquiry encourages them to streamline the nuances of the real world in epistemically harmful ways: they are extremely skilful in mathematical puzzle solving when reasoning about the model worlds, but relatively speaking clumsy and uninformed in connecting their formulas to the detailed complexities of real world economies. These capacities and their limitations may also nurture overconfidence, hubris, and arrogance – characteristics often attributed to the economics profession and conducive to the sorts of failure witnessed in connection to the 2008 crisis (see e.g. Posner 2009; Fourcade et al. 2015).

Regarding the contents of their worldviews, there is empirical literature suggesting that economists are more self-seeking than other professions, either due to economics education or self-selection (see e.g. Carter & Irons, 1991). This may be suggested to result in systematic biases in favour of models that put too much stress on self-seeking behaviour amongst the populace at large.

... uses multi-component object M as

Even though models are systems with multiple components, they are necessarily partial: models isolate only a subset of potentially relevant causal factors in the economy while leaving the rest out of the model. Such isolations – inclusions and exclusions of factors – can be accomplished simply by omitting them without any mention, or by means of idealizing assumptions that assume, directly or indirectly, those factors to be absent, constant, or otherwise ineffective (Mäki, 1992).

An important potential source of modelling failure is, of course, the contents of the model system itself. Failure occurs if causally, explanatorily and/or predictively relevant factors are excluded from the model. Virtually all criticisms of the performance of economics regarding the 2008 crisis make this charge. Here is Nobel Laureate Joseph Stiglitz: "the conventional models inadequately modelled – and typically left out – many, if not most, of the key factors that played a central role in this crisis" (Stiglitz, 2011, p. 172).

Models of efficient financial markets rely on strong idealizing assumptions such as zero transaction costs and perfect and symmetrical information between the agents. These idealizations help produce an image of the financial system in which market prices fully incorporate all relevant information and in which there can be no bubbles in asset prices such as those of stocks or houses. This is a model system that has the features of being self-regulated and having the capacity of containing all relevant risks without external regulation (so Alan Greenspan and others believing in these models were able to relax). Critics argue that such an exclusion of real-world imperfections from the models is fatal, suggesting that the properties of real-world markets may in fact be the reverse of those of the model-world markets: "...where the Efficient Markets Hypothesis suggests that financial markets provide a way of managing economic risk, the evidence suggests that they are actually a major source of risk." (Quiggin, 2010, p. 51)

Macroeconomic models, in their turn, have employed the highly idealized notion of representative agents. These models have chronically missed the crucial causal factors such as animal spirits, herd behaviour, informational asymmetries, structure of financial markets, and corporate governance, therefore failing to recognize phenomena such as excess indebtedness, debt restructuring, bankruptcy, and agency problems. The critical verdict is that any model with these characteristics "leaves out much, if not most, of what is to be explained; if that model were correct, the phenomena – the major recessions, depressions and crises that we seek to understand – would not and could not have occurred" (Stiglitz 2011, 168; see Akerlof and Shiller 2009). So the failure regarding this component is a matter of leaving out from the models important (e.g. bubble-generating) factors and mechanisms that are responsible for the sort of crisis we have just experienced.

...a representative of (actual or possible) target R

The key point of modelling is that models stand for their targets as their representatives, and that models are directly examined for their properties and behaviour so as to indirectly acquire information about their targets. This involves a sort of objectification or reification of model systems as such direct subjects of inquiry. One simple failure lured by this is the failure to distinguish between the model and the target in one's reasoning, to proceed, without further systematic inquiry, as if the model system is the target system, and to believe that the properties discovered to be possessed by a model are also properties of its targets. This is a very old criticism of economic models, levelled already in the 19th century. It is still topical today, also in relation to the modelling failures regarding the 2008 crisis.

The targets of models can be either actual or merely possible objects or systems, and whether they are one or the other has consequences for what kind of information modelling can deliver, and what sorts of failure may occur. Even though models can be required to ultimately stand for some actual target objects or systems (such as really existing macro economies or the current global financial system), economists often treat their models as representatives of some possible targets, using them as tools for how-possibly explanations. Often an economist simply takes on the task of examining whether it is possible to derive a given stylized fact from the assumptions of individual optimizing behaviour, that is, whether it is conceivable that the stylized fact is an outcome of the functioning of a mechanism involving such behaviour. Or an economist may examine the conditions under which a simple imagined market system is stable. It may happen that no one (or no contemporary economist, or no sufficient number of economists credible enough to have their voices heard) takes the task of establishing whether these connections also obtain in some actual target system. This means that no information is generated about actual systems. Such a situation may miss important facts about the relevant actual target systems, such as the global financial system.

One can conceive of the very idea of a model's target with different degrees of specificity, in terms of a varying magnitude of attributes. With a sufficiently rich notion of target, one can then argue that the failure of economic modelling was to be preoccupied with relatively stable, bubble-free economies as the main or sole targets, and this meant missing the actually emerging dramatic instabilities in the bubble-generating crisis-prone economies. This is to say those models deal with "special cases where market inefficiencies do not arise" (Stiglitz, 2011, p. 166) or that they do not apply to economies that are capable of generating bubbles. In other words, the targets of the criticized models were too narrowly conceived, especially as they crowded out actual targets that mattered greatly at the time. Ben Bernanke (2010) acknowledges the premise of this reasoning, but uses it for defending macroeconomic models, arguing that they work fine under ordinary conditions: "Economic models are useful only in the context for which they are designed. Most of the time, including during recessions, serious financial instability is not an issue. The standard models were designed for these non-crisis periods, and they have proven quite useful in that context." (2010, p. 17) In other words, the 2008 crisis was not a feature of the proper targets of the then-dominant economic models. Thus the problem was not with those models but with the modelling practices that had failed to address targets with bubble-generating capacities.

...for purpose P,

There is a vast multiplicity of possible purposes that models – or any given model – can serve, or intended or presumed to serve. These include explaining this or that phenomenon or aspect of the phenomenon; predicting this or that phenomenon with this or that accuracy of timing, magnitude etc.; exploring the properties of possible scenarios; solving this or that puzzle (so as to get yet another paper published); elaborating this or that new or old technical tool; designing an institution or policy with these or those desired features; educating or persuading students (or lay people, media, politicians); and so on.

It makes no sense to talk about successes or failures without specifying the relevant purposes. Success and failure are functions of purpose, among other things. This provides part of the pragmatic context of models. Modelling can serve multiple purposes, such that any particular model is designed for – or is expected or found to serve - one or more particular purposes and can only be evaluated in terms of its success in serving those purposes. It is worth noting that the issue of target specification or the proper domain of a model's applicability (that we just briefly discussed above) can often be transformed into an issue of the purposes of modelling. This is what Stiglitz suggests: "Is the purpose of an economic model to help us predict a little bit better how the economy is performing in 'normal' times – when things do not matter much? Or, is the purpose of an economic model to predict, prevent and manage big fluctuations and crises?" (Stiglitz, 2011, p. 168)

Stiglitz and many others charged economics for *predictive failure*, but this is very ambiguous. Of the 2008 financial crisis, one may have failed to predict (a) *that* it is coming; (b) *by what pathways* it is coming; (c) *when* it is coming; and (d) *with what (e.g. quantitative) characteristics* it is coming. These tasks are concerned with the anticipation of *actual* events, and they are not equally easy. While (a) is surely the easiest task, (c) and (d) are far more difficult – while their degree of difficulty depends on the desired accuracy of the predictions. What is important is that many economists defended their models and modelling practices by arguing that prediction is not among their proper and reasonable goals after all.

An even heavier charge is for a *failure of conception*: economists not only failed to predict the crisis, they failed to conceive it, which is even more serious. As much as economists are fond of modelling possibilities, it is notable that the sort of systemic crisis exhibited by the financial system in 2008 was not among the possible scenarios entertained by economists. They failed to conceive what is *possible* within the system given the sorts of causal mechanism that the system contains. And they failed to conceive what the system has a *propensity* to do, given those mechanisms.

Many critics have pointed out that the dominant purposes of large and most respectable parts of economic modelling may have been removed too far from the timely needs of effective economic policy – such as the anticipation of certain dramatic kinds of real world development or even the recognition of their possibility based on accounts of the complex system of underlying mechanisms. The suspicion is that economists are inclined to address relatively small and easy specialized intra-academic puzzle-solving challenges so as to maximize their academic performance measured in terms of publication output, rather than to tackle big and difficult issues that take more time and effort and are more risky in their capacity to yield an impressive publication record. Perhaps the dominant purposes of modelling have become rather fragmented and inward looking.

These observations inspire two further remarks. The first is that given the large variety of possible purposes, there is also a large variety of ways of failing (so it is easy to fail) as well as of ways of escaping charges of failure (so it is difficult to fail). One can defend a model or a model format by varying the associated purposes. In response to (alleged) failure in relation to (intended or presumed) purpose P_1 the modeller makes an excuse or escape by proposing an allegedly more appropriate purpose P_2 and by claiming success in relation to P_2 (and in response to the possible further failure in relation to P_2 the modeller escapes by proposing another purpose P_3, and so on). No doubt this structure has been in place in the debates around economics. The second remark is that there is a limit to the room that can be permitted for escape. The purposes of modelling themselves are and should be subject to critical scrutiny and debate. Not all possible purposes are equally appropriate. Some purposes just cannot be ignored without damaging the legitimacy of a modelling discipline.

... *addressing audience E,*

Modelling is always addressed to some audience or set of audiences. Models are used to convey information to audiences, to educate them, to impress them, to persuade them, to use them as test partners. The possible audiences include likeminded experts in the modeller's specialised field of inquiry and those in disagreement, other members of one's own discipline and those in other disciplines, students and journal editors, media and policy makers, and so on. The expectations and beliefs of particular target audiences partly determine what is offered to them by the modellers. This provides another pragmatic aspect of modelling. (One can of course try to embed the roles of audience in the category of purpose, but this would make the latter intractably extensive and variegated and internally structured, so I prefer keeping audience as a separate component.)

I mentioned using an audience as a test partner. Any model, in the course of its lifetime, travels through a private-public dimension, from private conception to public acceptance or rejection. At various stages, audiences as test partners can be internal (modeller herself) or external (other scientists or non-scientists). The internal audience as a test partner examines the conceived model for its public recognisability and acceptability, at least for further study and discussion, before submitting it to public scrutiny. Anticipations of the forthcoming public reception of the model are shaped by the dominant modelling conventions in the discipline or research field, so the model had better be constructed and proposed without violating these social conventions to have a chance. The public examination of models will therefore virtually always restrict itself to a limited and relatively homogeneous set of candidate models (see Mäki, 1993, p. 97). This is one mechanism resulting in the sort of intellectual herd behaviour that has been mentioned among the sources of modelling failure regarding the 2008 crisis.

Among the external audiences, there is the primary audience consisting of other academic economists attending specialised seminars and reading top journals in the same field. One of the important functions of this audience is to serve as a pool of referees used for assessing paper submissions for publication in journals. Given how academic performance is nowadays measured, this is crucial for academic survival and promotion, so modellers had better be successful in persuading this audience. This takes place within a narrow specialism, which means that critical or corrective feedback tends to be limited, focused mostly on technical details. Broader issues and those of societal relevance play side roles at most. The big picture easily gets lost.

Then there are secondary audiences, such as economists and non-economists in other fields, and policy makers and the general public as well as journalists mediating between academic research and these non-academic audiences. The demand for the big picture may be pressing, but the demand is not very likely to be met due to the overriding pressures of specialised academic performance. In general, there tends to be little effective feedback on the relevant models from these secondary audiences, and this may be due to a lack of competence (e.g. insufficient mathematical literacy), or perhaps in some cases ideological coherence or political convenience.

So it is possible that much of the most respectable academic work in economic modelling has been directed predominantly to other academic economists within some limited field of specialization rather than to audiences busy with policy work concerned with concrete complexities of real world issues. One consequence of this may have been a failure to develop sound comprehensive accounts of the functioning (or malfunctioning) of the system as a whole. Or, in case economic research has been addressed to people in policy practice, the latter may have been happy to receive them without complaints because the economy was believed to be running its course sufficiently smoothly and perhaps because the message from the academia was in line with their political prejudices. Indeed, one of the suspicions presented by the critics is that the belief in the self-stabilizing capacities of unregulated markets has been strong both among leading economists and among leading politicians, and that these beliefs have fortified one another (see e.g. Stiglitz 2010 and Quiggin 2010).

... at least potentially prompting genuine issues of relevant resemblance between M and R to arise;

While many different things can stand for, or can be used as a representative of, a given target thing, only those that resemble the target in relevant ways are useful for generating true information about the target. To find out whether this is so, the issue of relevant resemblance must arise. The notion of relevant resemblance involves both objectively ontological and subjectively pragmatic aspects. Resemblance is a matter of objective matter of fact, while relevance is a matter of relativity to some purposes, goals, interests, and audiences. One and the same resemblance relation between a model and its target may be sufficiently relevant for some purpose and audience but not so for another. The critic may then argue that the dominant available models do not resemble their targets relevantly for conceiving, anticipating, and avoiding the sorts of crisis that was witnessed in 2008. It is important to see that there are two ways in which models may fail to relevantly resemble their targets (Mäki, 2009a, 2011a).

One kind of failure occurs when relevant resemblance is pursued by the modeller, but the attempt fails. In this case *the issue of relevant resemblance arises*, but it will be resolved in support of the conclusion that a given model does not relevantly resemble some target. Or, it may be that this would be the right conclusion, but it is not actually drawn by economists for one reason or another. In either case, *surrogate modelling* is exercised: models are conceived as *bridges* to their targets, so that finding about the properties of models are intended and hoped to be informative about the respective properties of the targets. But the bridge may fail, and the intended resemblance is not established. The model may be based on isolations that serve to exclude, rather than include, some important crisis-generating mechanisms in operation in the real economic system. This would be one type of failure of resemblance, based on *trying and failing*.

Another kind of failure occurs when there is *no trying at all*, unlike in the case of surrogate modelling. It is not a failed attempt but rather a failure to attempt. In what I call *substitute modelling*, models are conceived as *islands* (rather than bridges), with *no issue of relevant resemblance being prompted at all*. Imaginary model worlds are examined for their own sake, as it were, with economists learning a great deal about the properties of their models but nothing about any real world targets – not even about what these targets are not like. The study of models substitutes for the investigation into anything beyond them. Criteria other than those pointing towards resemblance between models and their targets dominate the exercise. This is probably what Paul Krugman had in mind when writing, "economists, as a group, mistook beauty, clad in impressive-looking mathematics, for truth" (see also Hodgson, 2009). Krugman is not alone with this suspicion.

The distinction between surrogate modelling and substitute modelling is not crystal clear though, and this creates some room for justifying modelling practices that do not immediately address issues of resemblance. As I said before, it is part of the nature of modelling that attention and research effort are directly focused on examining the properties of models. There are two dimensions along which to consider the further issue of whether such attention and effort are a matter of surrogate or substitute modelling. One dimension is that of *individual-collective*: the task of addressing the issue of resemblance is delegated to other researchers within the scientific community in line with some principle of division of scientific labour. Even if one modeller may appear to practice substitute modelling only, this is supplemented by others who turn the exercise into one of surrogate modelling. Another dimension is the historical one of *now-later*: there is an appropriate temporal order of research tasks, and the task of addressing issues of resemblance will be taken up and accomplished by some later (generations of) researchers. Both of these dimensions have been utilised by economists in defending their models. (See Mäki 2009a, 2011a.)

The critical diagnosis, however, is to argue that major parts of economic modelling may have degenerated into the investigation of imaginary model worlds only, with no issues of resemblance being raised under the prevalent division of scientific labour or along a natural temporal sequence of intellectual effort. This situation results in treating models as substitute systems that are investigated in their own right with no concern whatsoever with how they might connect with real-world systems. Such a tendency may have been supported by a variety of factors, such as the increasing specialization of economic inquiry, its over-mathematization, and perhaps in some cases by the fact that economists and some of their audiences have been charmed by the smooth functioning of model economies within the model worlds, apparently justifying non-interventionist attitudes in policy making.

... describing M and drawing inferences about M and R in terms of one or more model descriptions D;

A given model can be described variously, such as in terms of verbal means, algebra, geometry, diagrams, etc. It is in terms of whatever medium is used for describing the model world that inferences are drawn about the properties and behaviour of that model world. Each type of medium has consequences for many other things in the modelling exercise, such as the range of inferences and claims that can be made (and are likely to be made) about the models, the range of ingredients to be included in the models, and the sorts of audience that can be reached.

In economics education, addressing students from introductory to advanced levels, one and the same model can be described variously at different stages of the education process, beginning with using easily accessible mathematics and gradually employing ever more demanding techniques. In policy memos, one may explain the key message of a model verbally, perhaps using diagrams, and then adding an appendix with a full-blown mathematical description of the model (with goals such as exposing the model to expert scrutiny or impressing the policy makers with a façade of scientificity). The structure of journal articles in terms of model description techniques also resonates with issues of audience, such as the readability of published articles and the sales and citations attracted by the journal.

It has been argued that the popularity of sophisticated mathematical means of description may have fortified the allegedly great distance between some economic models and the real world. This charge can be made more precise by invoking the idea of formal tractability as a dominant guiding principle of modelling (see e.g. Hindriks, 2006). The claim then would be that economists may have been excessively and uncritically constrained and inspired by considerations of formal tractability rather than empirical adequacy or relevant resemblance, and that this has imposed serious limitations on how they view the world. The discipline suffers from "the obsession with technique over substance" (Hodgson, 2009, p. 1210).

Recall Krugman's claim that the charm of "impressive-looking mathematics" has

Modelling failure 395

taken an upper hand and suppressed an interest in truth about the complexities of real world economies. In his vigorous response to Krugman, Chicago economist John Cochrane Cochrane (2011, p. 39) acknowledges that real world economies are more complex than the models economists build and use, and that it would indeed be desirable to be able to model the frictions and other imperfections in the world – and that, unfortunately, the mathematical tools now available are limited for the purpose: "Frictions are just hard with the mathematical tools we have now." So it seems Cochrane admits that economic models have missed important features of real world targets and that part of the reason lies in mathematical descriptions of the models constraining the contents of those models. But he thinks the problem is not with mathematics per se but rather with the limitations of the sorts of mathematics now used: "The problem is that we do not have enough mathematics."

... and applies commentary C to identify and coordinate the other components

By themselves, model objects do nothing to serve a purpose, to reach an audience, or to link with a target object. Model objects themselves and their relations to such other things are complex and typically not fully transparent, so they require clarification and coordination. This is provided by the model commentary. Yet there is no guarantee that modellers always have the sort of complete understanding of the complexity of the modelling exercise that would enable them to provide a commentary that successfully spells out what is not transparent. Economists busy with mere substitute modelling might even not have a strong interest in providing a clarifying commentary accurate and comprehensive enough to deal with intricate connections with the real world.

A major challenge is to be clear about the roles of and interrelations between what easily appear as outrageously false idealizing assumptions used in describing theoretical models. Commentary has the important task of accomplishing a functional decomposition amongst model components (e.g. Mäki, 2011c). Idealizations are not to be interpreted literally at face value, they typically need first to be translated into claims whose roles are well understood in the structure of the modelling process, so that they can be more directly assessed for how well they serve their proper functions. These latter claims may turn out to be true statements about things such as negligibility (of some falsehood for some purpose) or applicability (of the model to one rather than another domain), or they may become justified as serving a useful purpose as early steps in a sequence of models leading to ever better ones. Or they may fail in playing their proper roles. To be able to check whether they fail or succeed, an informed commentary must have done its job of functional decomposition. Economists often fail to provide such a commentary.

Model commentary also resonates with audiences. Given the reputation of arrogance and self-confidence enjoyed by the economics profession, it is remarkable that when addressing *fellow economists*, they exhibit relatively more modesty and humility when presenting their models and making claims about them. They pay more attention to the complexity of the issues, and are more explicit about the various provisos, background assumptions, and uncertainties that are involved. When addressing *non-economists* on the other hand, there is less modesty and humility, and less attention is paid to the complexities hidden in the background. "In private discussions among ourselves we recognize this complexity, but we don't add the appropriate warning labels to our models when they are discussed in public. There, we pretend we understand more than we do." (Colander, 2011, p. 20) Another version of the story, often told by economists, is that their otherwise sound ideas are misinterpreted and ill applied by policy makers due to incompetence or the involvement of messy conflicts of political interest. This charge may be intended to redirect the blame for failure, but it can also be taken to emphasise the high importance of supplementing (sound or unsound) models with a sound commentary, but it is not obvious that economists are doing their best in this regard.

It should be part of the academic competences of trained economists to be able to be clear about what their models are for; what the models are about; what the models are capable of doing, and what not; how reliable the models are; what sorts of criticisms have been levelled against the models and how the criticisms have been responded; what alternative models there are; etc. The challenge is not easy, and it is clear that it has not been met with sufficient exuberance and success. The capacity of writing "warning labels" would be part of the needed professional competence. Colander suggests requiring "publications of models that seem to have policy relevance to include an explicit warning directed at the non-scientific users of the model" (Colander, 2010, p. 424). Such warning labels would alert the relevant audiences to the capabilities and limitations of the models.

It is clear that the commentary attached to popular economic models has failed in spelling out their potentials and limitations. This has led to existing models being used for inappropriate purposes and non-existing but important models not having been built at all. Existing models depict bubble-free economies, but they have been erroneously applied to the current financialized economies that are not bubble-free; and models that would be adequate for representing our bubble-prone economies have failed to be developed by a sufficient number of sufficiently credible economists. Avoidance of these errors would be helped by reliable commentaries informing what any given model can and cannot do. It should be added here that philosophers, methodologists and historians of economics might be expected to be well equipped to do their share in contributing to the construction of an enlightening commentary of economic models.

... and all this takes place within a context X

The final component of the context of modelling includes lots of various further ingredients that make a difference for models and modelling practices. It includes items such as intra-disciplinary conventions and practices, standards and incentives, arrangements of education, research and publishing, and so on. The context also includes various external (including non-academic) enabling and constraining conditions, expectations, pressures, resources, such as the ongoing transformation of the university institution and the societal status of economics. Such contextual factors make a direct or indirect impact on the other components of the system of modelling.

For example, the deficiencies in the competence of the economics profession to provide adequate commentary of the capabilities and limitations of the models it produces is likely to be an outcome of the rather narrow education nowadays offered to economics students – consisting mainly of recent economic theory and mathematical and statistical techniques. It is generally recognized that fatal consequences may follow from not systematically learning about the other social sciences, about the society at large, about past economic history, about the history and philosophy of the economics discipline itself.

Thinking of the allegedly flawed contents of the models behind the 2008 crisis more directly, the structure of the disciplinary division of intellectual labour may have played important roles. That the isolations of macroeconomic models have excluded causally relevant details of financial markets is partly a product of the structure of disciplinary practices within economics, namely the increased specialization that has deepened the divide between the two fields: "The fact that finance and macroeconomics have become separate fields with some difficulties of intercommunication may have been the inevitable result of the relentless pressure for ever-greater specialization in academic disciplines." (Posner, 2009, p. 328)

As to the epistemic values and conventions underlying disciplinary practices, the ideal of formal tractability may have favoured certain problematic idealizations and suppressed the pursuit of factual adequacy. A tendency towards substitute modelling may have been supported by failures in the division of labour in economic inquiry (economists supposedly bridging the gap between theoretical models and their targets not showing up); by the compulsory incentive to publish elegant modelling exercises in prestigious journal with little need to address issues of real-world connections; and perhaps in some cases by economists and some of their non-academic audiences being overly charmed by the smooth functioning of model economies, therefore reluctant to consider the importance of any real-world imperfections.

Exceptional amongst the social sciences is the role of the economics discipline in contemporary society, the intellectual and political authority economics enjoys regardless of its failures. Above, I cited Colander's confession, "we pretend we understand more than we do" and we could add that economists do so in order to – or with the consequence of – protecting and promoting their socially acknowledged authority. In the worst case, there is a nightmarish scenario on which the more economists are consulted for policy advice, the more they need to pretend to know, and so the higher the likelihood of policies going astray. Avoiding the nightmare would require some smart restructuring of the institutions of the economics discipline.

***

Models and modelling as such are powerful means of acquiring information about dynamically complex systems such as economies. While powerful, they are also prone to error and epistemic risk – and not just epistemic risk, but institutional risk as well in that the (academic and other) institutions of economic inquiry may fail to provide appropriate incentives and other preconditions for adequate modelling. Investigating the possibility and actuality of modelling failure is a matter of exercising disciplinary risk management (see Mäki, 2011a).

Bibliography

Akerlof, G., & Shiller, R. J. (2009). *Animal spirits. How human psychology drives the economy, and why it matters for global capitalism*. Princeton: Princeton University Press.

Bernanke, B. (2010). Implications of the financial crisis for economics. Remarks at a conference at Princeton University, 24.10.2010.

Carter, J. R., & Irons, M. D. (1991). Are economists different, and if so, why? *Journal of Economic Perspectives, 5*, 171–177.

Cochrane, J. H. (2011). How did Paul Krugman get it so wrong? *IEA Economic Affairs*, 36–40.

Colander, D. (2010). The economics profession, the financial crisis, and method. *Journal of Economic Methodology, 17*, 419–427.

Colander, D. (2011). How economists got it wrong: A nuanced account. *Critical Review, 23*, 1–27.

Fourcade, M., Etienne, O., & Yann, A. (2015). The superiority of economists. *Journal of Economic Perspectives, 29*, 89–114.

Giere, R. (1999). *Science without laws*. Chicago: University of Chicago Press.

Hindriks, F. (2006). Tractability assumptions and the Musgrave-Mäki typology. *Journal of Economic Methodology, 13*, 401–423.

Hodgson, G. M. (2009). The great crash of 2008 and the reform of economics. *Cambridge Journal of Economics, 33*, 1205–1221.

Krugman, P. (2009). How did economists get it so wrong? *The New Your Times Magazine*, 2 September.

Lucas, R. E. JR. (2003). Macroeconomic priorities. *American Economic Review, 93*(1), 1–14.

Mäki, U. (1992). On the method of isolation in economics. *Poznan Studies in the Philosophy of the Sciences and the Humanities, 26*, 319–354.

Mäki, U. (1993). Social theories of science and the fate of institutionalism in economics. In B. Mäki U. Gustafsson, & C. Knudsen (Eds.), *Rationality, institutions and economic methodology* (pp. 76–109). London: Routledge.

Mäki, U. (2009a). MISSing the world: Models as isolations and credible surrogate systems. *Erkenntnis, 70*, 29–43.

Mäki, U. (2009b). Realistic realism about unrealistic models. In H. Kincaid, & D. Ross (Eds.), *Oxford handbook of the philosophy of economics* (pp. 68–98). Oxford: Oxford University Press.

Mäki, U. (2011a). Contested modelling: The case of economics. In *Models, simulations, and the reduction of complexity*. Hamburg Academy of Sciences.

Mäki, U. (2011b). Models and the locus of their truth. *Synthese, 180*, 47–63.

Mäki, U. (2011c). The truth of false idealizations in modelling. In P. Humphreys, & C. Imbert (Eds.), *Models, simulations, and representation* (pp. 216–233). London: Routledge.

Posner, R. A. (2009). *A failure of capitalism. The crisis of '08 and the descent into depression*. Cambridge, MA: Harvard University Press.

Quiggin, J. (2010). *Zombie economics. How dead ideas still walk among us.* Princeton: Princeton University Press.
Roubini, N., & Mihm, S. (2011). *Crisis economics. A crash course in the future of finance.* London etc: Penguin Books.
Stiglitz, J. E. (2010). *Freefall. Free markets and the global economy.* London etc: Penguin Books.
Stiglitz, J. E. (2011). Rethinking macroeconomics: What went wrong and how to fix it. *Global Policy, 2,* 165–175.
Volokh, K. Y. (2010). On modeling failure of rubber-like materials. *Mechanics Research Communications, 37*(8), 684–689.
Weisberg, M. (2013). *Simulation and similarity. Using models to understand the world.* Oxford: Oxford University Press.

Author biography. Uskali Mäki: Academy Professor, Academy of Finland; Professor of Practical Philosophy, University of Helsinki; Director of the Academy of Finland Centre of Excellence in the Philosophy of the Social Sciences; Former Professor and Director at Erasmus Institute for Philosophy and Economics [EIPE], Erasmus University of Rotterdam (1995-2006); Former Editor of Journal of Economic Methodology [JEM] (1996-2005); Former Chair of the International Network for Economic Method [INEM] (2007-2008).

Part E

International Council for Science (ICSU) special session: Future Earth

21 Transformative research for a sustainable future Earth

GORDON MCBEAN [*] AND HEIDE HACKMANN [†]

Abstract. The scientific consensus is that we have entered the Anthropocene, a new geologic epoch defined by our own massive impact on the planet. This paper addresses the broad scope of interconnected issues, such as biogeochemical flows and biodiversity integrity, land-system and climate change, which are all interconnected. There is need to identify and quantify the planetary boundaries that provide a "safe" operating space for humanity. In the present international context, the issues are being dealt with as disaster risk reduction (Sendai Framework), climate change (Paris Agreement) and Sustainable Development Goals. How can science best provide the inputs to these policy processes and more importantly to help governments and people address the issues? These questions require outputs leading to outcomes that address complex socio-economic, natural, health, engineering, philosophical and cultural issues and most challenging their intersections.

The Program Future Earth: Research for Global Sustainability has, as its goal: "To provide the knowledge required for societies in the world to face risks posed by global environmental change and to seize opportunities in a transition to global sustainability". The program has adopted a unique approach of both a Science Committee and an Engagement Committee to co-design and co-produce the scientific research program. The Science Committee: represents the full spectrum of scientific fields, as well as scientists from other sectors and the Engagement Committee includes representatives from business, civil society and government. The research theme of transformations to

---

[*]President, International Council for Science, Co-Chair, Governing Council, Future Earth, Professor Emeritus, Western University, London, Canada

[†]Executive Director, International Council for Science

sustainability will be a special challenge in dealing with issues such as transformation processes and global and regional governance, including incentives and international law. The challenge for Future Earth will be to bring together interdisciplinary, transdisciplinary teams of scientists to undertake transformative research providing outputs leading to outcomes that make a difference for global sustainability.

Keywords: sustainable development, climate change, disaster risk reduction, transformative research, international research collaboration, science for policy, science and society.

1 Introduction

The scientific consensus is that our planet has entered the Anthropocene, the age of humans, a new geologic epoch defined by our own massive impact on the planet (Steffen, Crutzen, & McNeill, 2007). This paper addresses the broad scope of interconnected issues, such as biogeochemical flows and biodiversity integrity, land-systems, ozone layer depletion and climate change, which are all interconnected. There is need to identify and quantify the planetary boundaries that provide a "safe" operating space for humanity (Syvitski, 2012). How far have we gone? Where are the global "tipping" points that may exceed the planet capacity of adaptation? What actions can turn us back? These are critically important policy issues for societies and governments that are challenging global realities. A global perspective of the relative risks of climate change, extreme weather and natural disasters was given in The Global Risks Report 2016 (World Economic Forum, 2016). The failure of actions on climate change mitigation and adaptation is now ranked as the global risk with the highest impact and one of the most likely ones. The Top 5 Global Risks in Terms of Likelihood are ranked to be: 1. Large-scale involuntary migration; 2. Extreme weather events; 3. Failure of climate change mitigation and adaptation; 4. Interstate conflict with regional consequences; and, 5. Major natural catastrophes. We are living in a world of converging crises and facing the imperative of profound social transformation. As we go forward, there will be more and more interdependencies, of which we understand less and less. The bringing together the typical ingredients for an upcoming crash means that it is essential to have coordinated efforts of public authorities, civil society, industry, and academia to avoid possible collapse of society as we know it (Lechner, Jacometti, McBean, & Mitchison, 2016).

An example of the issues and their interconnectedness is what we may call the "Disaster Risk-Poverty Nexus" (International Strategy for Disaster Reduction, 2009). There are global drivers of risk, such as uneven economic and urban development, climate change and weak governance and limited endogenous capacities. When these are imposed on the underlying risk drivers, such poor urban and local governance, vulnerable rural livelihoods, ecosystem decline and lack of access to risk transfer and social protection and the mix of everyday, extensive or intensive risk resulting in disaster impacts, the results are poverty outcomes which in turn accentuate the poverty and related factors, making the community's risks even higher. The result is a "feedback" loop in that communities that are already poor are more impacted by hazards, leading to further enhancement of their poverty and, hence risk, with even greater negative impacts when the next hazard strikes. Hence, weak governance, limited endogenous capacities and superimposed climate change, lead to uneven economic and urban development, further enhanced when a hazard strikes, demonstrating the linkages of the issues of poverty, health, environment, development and governance.

These interconnected issues are challenging, and changing, science, and all its aspects. Doing integrated science that makes a difference and pursuing excellence through engagement in open knowledge-action arenas is important and necessary to address these issues. There is need for science-based information as inputs to the issues of global security, intersections of cultures and societies and these are challenges for science policy and practice. Now is the time to create the 'conditions of possibility', to support science for a sustainable and just world. The characterization of today's global realities includes (United Nations Educational, Scientific and Cultural Organisation & International Social Science Council, 2009):

- the inseparability of social, political, cultural and environmental problems;
- the centrality of people;
- the inadequate social responses to date; and
- the urgent need for social transformation.

There is need for transformations recognizing the complex processes of profound social change, the altering of our social and economic systems and the values and lifestyles in ways that could put society on a fundamentally different development path. The global scientific community needs to, and is, responding to these changes.

2 The International Council for Science

The International Council for Science (ICSU)[1] is a leading non-governmental science organization that was created in 1931. The Council now has 122 National Members and 31 Unions or Associations of scientists by discipline. The scope of these disciplines include math, physics, chemistry, geology, biology, anthropology, sociology and the history and philosophy of science. The Mission of the International Council for Science (ICSU) is *"to strengthen international science for the benefit of society"*, for all societies. The vision of the Council is for a world where excellence in science (all sciences) is effectively translated into policy making and socio-economic development, with universal and equitable access to scientific data and information, where

---

[1] http://www.icsu.org

all countries have scientific capacity, enabling the generation of new knowledge and nations can establish their own development pathways in a sustainable manner. The Council's key priorities and associated activities are: Science for Policy (and policy for science); Universality of Science with the freedom to do science while recognizing the responsibilities of science and scientists; and International Research Collaboration.

The International Council for Science works closely with the International Social Science Council (ISSC)[2]. ICSU and ISSC are co-sponsors of the Integrated Research on Disaster Risk Program and the Future Earth Program. They also collaborate in other ways, such as through Science International[3]. In 2015 the International Council for Science (ICSU) initiated 'Science International' as a new series of action-oriented meetings bringing together major international science bodies: International Council for Science (ICSU); the International Social Science Council (ISSC); The World Academy of Sciences for the advancement of science in developing countries (TWAS)[4]; and the InterAcademy Partnership (IAP)[5]. The 2015 edition of Science International has developed an international accord on the values of open data in the emerging scientific culture of big data. The Accord recognises the need for an international framework of principles on *"Open Data in a Big Data World"* and proposes a comprehensive set of principles. The Accord has now close to 100 endorsements from academies and other international organisations. Negotiations are now underway towards a merger of the two councils to create a scientific organization for "all" sciences".[6]

Another example of the ICSU and ISSC working together was the proclamation of the 2016 International Year of Global Understanding[7] by the International Council for Science (ICSU, International Social Science Council (ISSC) and the Conseil International de la Philosophie et des Sciences Humaines (CIPSH)[8], based on an initiative of the International Geographical Union (IGU). The IYGU stems from the recognition that global understanding is an important and essential human condition. In the face of global change, it is important we think globally recognizing that local actions alter global processes and that global understanding clarifies the connections between the local and the global. The International Year of Global Understanding was established to address the ways in which we inhabit an increasingly globalized world and examine the questions: "How do we transform nature?" and "How do we build new social and

---

[2] http://www.worldsocialscience.org/

[3] http://www.icsu.org/science-international/science-international

[4] TWAS. http://twas.org/

[5] IAP. http://www.interacademies.org/ The InterAcademy Partnership is an umbrella organization that brings together IAP - the global network of science academies, the InterAcademy Medical Panel (IAMP), and the InterAcademy Council (IAC).

[6] http://www.icsu.org/general-assembly/extraordinary-general-assembly-oslo-2016/background

[7] http://www.global-understanding.info/

[8] http://www.cipsh.net/htm

political relationships for the emerging global reality?" It is recognized that societies and cultures determine the ways that we live with and shape nature. They influence how we perceive the global consequences of our everyday actions. Hence, we need to understand what our daily actions mean for the world as a whole in order to overcome global challenges. The IYGU is an important example of transformative approaches for a sustainable future Earth.

3 International environmental policy agenda and the international agreements of 2015

The year 2015 was a crucial year for the international policy agenda with the Third UN World Conference on Disaster Risk Reduction (WCDRR)[9] and the ensuing Sendai Framework on Disaster Risk Reduction 2015–2030[10], United Nations Sustainable Development Summit and ensuing Agenda 2030 and Sustainable Development Goals (SDGs)[11] (2015), the 21st Conference of the Parties (COP-21) of the UN Framework Convention on Climate Change and its Paris Climate Agreement[12] (2105) and the International Conference on Financing for Development (2015)[13]. For these international negotiations, the International Council for Science was the principal in the Science and Technology Major Group [14] [15] and, hence, had a major participatory role. To address these issues, there is need for integrated information on these topics and related issues. As we look ahead for the next decades, there is need for recognizing the responsibilities of global science to contribute to post-2015 frameworks, including the Sendai Framework, Agenda 2030, Paris Climate Agreement and the urban agenda at Habitat III Conference[16].

---

[9] www.wcdrr.org/

[10] UNISDR; www.unisdr.org/we/inform/publications/43291

[11] Sustainable Development Goals; https://sustainabledevelopment.un.org/?menu=1300. SDG/Agenda 2030 "Transforming our world: the 2030 Agenda for Sustainable Development" http://www.un.org/ga/search/view_doc.asp?symbol=A/RES/70/1&Lang=E

[12] Paris Agreement, UNFCCC; https://unfccc.int/resource/docs/2015/cop21/eng/l09r01.pdf

[13] Third International Conference on Financing for Development,
http://www.un.org/ga/search/view_doc.asp?symbol=A/CONF.227/20

[14] https://sustainabledevelopment.un.org/majorgroups/scitechcommunity

[15] www.icsu.org ›Science for Policy ›Sustainable Development Goals

[16] https://www.uclg.org/en/issues/habitat-iii

3.1 Sendai Framework for Disaster Risk Reduction 2015–2030

In 2005, nations agreed to the Hyogo Framework for Action on disaster risk reduction[17]. In 2015, at the meeting in Sendai, nearby to the location of the Fukushima disaster, negotiations were undertaken to provide a post-Hyogo Framework. As Representative of Science and Technology Community Major Group, I was invited to make presentations both on panels and as part of the negotiation process. The new Sendai Framework for Disaster Risk Reduction 2015–2030 takes into account the experience gained through the implementation of the Hyogo Framework for Action. The Framework states: *"In pursuance of the expected outcome and goal, there is a need for focused action within and across sectors by States at local, national, regional and global levels in the following four priority areas:*

1. *Understanding disaster risk;*
2. *Strengthening disaster risk governance to manage disaster risk;*
3. *Investing in disaster risk reduction for resilience; and*
4. *Enhancing disaster preparedness for effective response, and to "Build Back Better" in recovery, rehabilitation and reconstruction."*

These intergovernmental negotiations on the post-2015 development agenda, financing for development, climate change and disaster risk reduction provide the international community with a unique opportunity to enhance coherence across policies, institutions, goals, indicators and measurement systems or implementation, while respecting the respective mandates. It is important to ensure credible links, as appropriate, between these processes that will contribute to building resilience and achieving the global goal of eradicating poverty. The Council is collaborating with UN and other partners to make this a reality, including at the now completed International Conference, held in January. 2016 in Geneva, and specifically through the Integrated Research on Disaster Risk Program (IRDR)[18] and its projects on data and monitoring systems and methodologies for forensic investigations of disasters. The IRDR Program has a major role in addressing the Sendai Framework.

3.2 Sustainable development goals

Sustainable Development (World Commission on Environment and Development, 1987) (World Commission on Environment and Development, 1987) is defined as: *"Humanity has the ability to make development sustainable - to ensure that it meets the needs of the present without compromising the ability of future generations to meet their own needs"*. A principal key part of the sustainable development is the linking of

---

[17] https://www.unisdr.org/we/coordinate/hfa

[18] http://www.irdrinternational.org/

social, economic, technology, science and environmental issues and connecting the future with the present. It essentially leads to science-informed decision making. The concept of "seeing the future" (McBean, 2008) is an essential part of sustainable development, and disaster risk reduction. Another important aspect is understanding the interconnectivity between actions and responses across and around the planet with its major societal, logical and philosophical issues.

There are 17 Sustainable Development Goals[19] with 169 targets agreed to for the Post-2015 Development Agenda. They collectively address issues of: poverty; hunger and food security; healthy lives; inclusive and equitable quality education; gender equality; water and sanitation; sustainable and modern energy; sustainable economic growth; inequality; and sustainable consumption and production patterns. Specifically linked to the other pillars of the 2015 international agenda include: Goal 9. Build resilient infrastructure, promote inclusive and sustainable industrialization and foster innovation; and Goal 11: Make cities and human settlements inclusive, safe, resilient and sustainable; - linked to Sendai and Goal 13: Take urgent action to combat climate change and its impacts – obviously links to the Paris Agreement. Goal 14 is: Conserve and sustainably use the oceans, seas and marine resources for sustainable development; and Goal 15 is: Protect, restore and promote sustainable use of terrestrial ecosystems, sustainably manage forests, combat desertification, and halt and reverse land degradation and halt biodiversity loss. Goal 16 is about peaceful, inclusive societies and justice. It also includes, importantly, accountable and inclusive institutions at all level. Goal 17 is: Strengthen the means of implementation and revitalize the global partnership for sustainable development. The latter specifically addresses the challenges for transformative research for a sustainable future Earth.

Associated with the 17 Goals are 169 Targets and the International Council for Science and International Social Science Council convened a group of experts to review the targets from the philosophical perspective and issues of measurement, measurability, logic and methodology (International Council for Science & International Social Science Council, 2015). The Group concluded that the SDG framework was a major improvement on the Millennium Development Goals (MDGs) but also concluded that the SDG framework would benefit from an overall narrative articulating how the goals will lead to broader outcomes for people and the planet and that it does not identify the wide range of social groups that will need to be mobilized. Of the 169 targets, 49 (29%) were considered well developed, 91 targets (54%) could be strengthened by being more specific, and 29 (17%) require significant work.

Following from this, a draft framework for understanding SDG interactions (International Council for Science, 2016) was developed as part of a project led by the Council to explore an integrated and strategic approach to implementation of the SDGs. The

---

[19]Sustainable Development Goals; https://sustainabledevelopment.un.org/?menu=1300. *SDG/Agenda 2030" Transforming our world: the 2030 Agenda for Sustainable Development* "http://www.un.org/ga/search/view_doc.asp?symbol=A/RES/70/1&Lang=E

framework is a starting point for building an evidence base to characterize the goal interactions in specific local, national or regional contexts. The Council is currently convening research teams to develop thematic case studies, starting with the SDGs for health, energy, and food and agriculture. The case studies will be compiled into a report, expected to be published at the end of 2016. Through these initiatives, international science is playing a major role in addressing the international development issues.

3.3 Climate Convention

As part of the preparatory process for the 21st Conference of the Parties (COP-21) of the UN Framework Convention on Climate Change and its Paris Climate Agreement, ICSU convened on 6 July 2015 in Paris an event Science and the Road to Transformation: Opportunities in the post-2015 Global Climate Regime[20] which brought together leading scientists and journalists to examine the scientific issues. UNESCO, Future Earth, and ICSU convened in Paris the *Our Common Future Under Climate Change* 7-10 July 2015. The Outcome Statement[21] noted that: *"Science is a foundation for smart decisions at COP21 and beyond. Solving the challenge of climate change requires ambition, dedication, and leadership from governments, the private sector, and civil society, in addition to the scientific community"* and expressed the commitment of the scientific community *"to understanding all dimensions of the challenge, aligning the research agenda with options for solutions, informing the public, and supporting the policy process."*

A separate event, *Climate Summit of the Americas*[22] was held 7–9 July, 2015 in Toronto, Canada to bring together representatives from the western hemisphere. Included in the Summit was a climate science statement calling for governments, industry and community leaders to make risk- and science-based decisions to limit global warming. The result was the first-ever Pan-American action statement on climate change signed by 23 states and regions in the Americas. These scientific events contributed to the successful Climate Convention CoP21 Paris, 2015[23].

4 International research collaboration

The International Council for Science (ICSU) is very involved in initiating, organizing and leading international research collaboration, often partnering with other gov-

---

[20] http://www.icsu.org/events/ICSU%20Events/science-and-the-road-to-transformation-opportunities-in-the-post-2015-global-climate-regime

[21] http://www.commonfuture-paris2015.org/The-Conference/Outcome-Statement.htm

[22] https://www.ontario.ca/page/climate-summit-americas-retrospective

[23] https://unfccc.int/resource/docs/2015/cop21/eng/l09r01.pdf

ernmental and non-governmental organizations. The International Geophysical Year (IGY)[24] of 1957 is an important early example. Recognizing the increasing societal concerns about the climate system, the International Council for Science (ICSU) and the World Meteorological Organization (WMO) joined in 1980 to create the World Climate Research Programme (WCRP)[25] with the scientific objectives: to determine: *the predictability of climate; and the effect of human activities on climate.* The Intergovernmental Oceanographic Commission of the UN Educational, Scientific and Cultural Organization (UNESCO) became a co-sponsor in 1992, to most effectively connect the global oceanographic community to the WCRP. By the mid-1980's, the level of international concern regarding climate change and broader issues of global environmental change, plus the discussions on sustainable development, led the International Council for Science (ICSU) to initiate the global change program, International Geosphere-Biosphere Programme (IGBP)[26] to: *study earth system science and to help guide society onto a sustainable pathway during rapid global change.* In 1992, the International Council for Science (ICSU), with one of its unions, the International Union of Biological Sciences, and the ICSU Scientific Committee on Problems of the Environment, with UNESCO, recognizing the concerns about the state of biodiversity on the planet, created the program DIVERSITAS[27], as an integrative biodiversity science, that links biological, ecological and social disciplines to address the complex scientific questions posed by the loss in biodiversity and ecosystem services and to offer science-based solutions to this crisis. The International Human Dimensions Programme on Global Environmental Change (IHDP)[28] was established in 1996 by its two scientific sponsors, the International Council for Science (ICSU) and the International Social Science Council (ISSC). IHDP was an international, non-governmental, interdisciplinary research programme addressing the coupled human-natural system in the context of global environmental change (GEC). It fostered high quality research aimed at describing, analysing and understanding the human dimensions of GEC. Human dimensions are the ways in which individuals and societies contribute to global environmental change, are influenced by global environmental change and mitigate and adapt to global environmental change.

The impacts of natural hazards continue to increase around the world with hundreds of thousands of people killed and millions injured, affected, or displaced each year because of disasters, and the amount of property damage has been doubling about every seven years over the past 40 years. To address the shortfalls in current research on how science is used to shape social and political decision-making in the context of hazards and disasters, the International Council for Science (ICSU) initiated the Integrated

---

[24] http://www.icsu.org/publications/about-icsu/the-international-council-for-science-and-climate-change-2015/the-international-council-for-science-and-climate-change-2015-1

[25] http://www.wcrp-climate.org/

[26] http://www.igbp.net/

[27] http://www.diversitas-international.org/

[28] http://www.icsu.org/what-we-do/past-interdisciplinary-bodies/hdp

Research on Disaster Risk (IRDR) Programme[29]. The IRDR mission is to develop trans-disciplinary, multi-sectorial alliances for in-depth, practical disaster risk reduction research studies, and the implementation of effective evidence-based disaster risk policies and practices. The IRDR Programme objectives are: 1) Characterization of hazards, vulnerability and risk; 2) Understanding decision-making in complex and changing risk contexts; and 3) Reducing risk and curbing losses through knowledge-based actions. Attainment of these objectives through successful projects will lead to a better understanding of hazards, vulnerability and risk; an enhanced capacity to model and project risk into the future; better understanding of decision-making choices that lead to risk plus how they may be influenced; and how this knowledge can better guide disaster risk reduction. The IRDR Programme is now co-sponsored by the International Council for Science (ICSU), the International Social Science Council (ISSC) and the UN Office for Disaster Risk Reduction (UNISDR, International Strategy on Disaster Reduction)[30].

With the increasing growth of populations in cities and the accompanying health issues, the International Council for Science (ICSU), in partnership with the UNU International Institute for Global Health[31] and the Inter-Academy Medical Panel (IAMP)[32] created the Health and Wellbeing in the Changing Urban Environment: a Systems Analysis Approach Programme[33] to promote systems approaches to understanding health and wellbeing in urban settings by understanding the functioning of the urban system as a whole. The systems approaches involves one or more of the following elements: 1) development of new conceptual models that incorporate dynamic relations; 2) use of systems tools and formal simulation models; and 3) integration of various sources and types of data including spatial, visual, quantitative and qualitative data. The overarching vision for the Urban Health and Wellbeing Programme is the development of aspired levels of wellbeing for people living in healthy cities.

5 Science system realities

The realities of the global scientific system include the persistent funding pressures and the unfortunately continued, and in some place growing, public mistrust of science. These lead to the new sense of urgency, and unrelenting pressure, for science to make a difference to real-world problem-solving. The grand challenge is: to urgently contribute transformative solutions to a converging set of global crises and to work simultaneously to protect planetary resources, safeguard social equity and hu-

---

[29] http://www.irdrinternational.org/
[30] https://www.unisdr.org/
[31] iigh.unu.edu
[32] www.iamp-online.org/
[33] http://urbanhealth.cn/

man wellbeing. These issues bring forward the needs for integrated science, which can be seen as:

- Works across disciplines and fields – (inter-disciplinarity)
    - Supporting the joint, reciprocal framing, design, execution and application of research
- Works globally – (international collaboration)
    - Including the agendas, perspectives, approaches, methods and models of scientists from all parts of the world
- Works with society – (trans-disciplinarity) (Mittelstrass, 2011)
    - Engaging decision makers, policy shapers, practitioners, as well as actors from civil society and the private sector as partners in the co-design and co-production of solutions-oriented knowledge, policy and practice

6 Future Earth: Research for Global Sustainability

In recognition of these issues, the International Council for Science joined with others for the Science and Technology Alliance for Global Sustainability[34]. Strategic planning sessions were held to develop the concepts (Reid et al., 2010). A transition team was formed in 2011 and the result was Future Earth: Research for Global Sustainability[35], launched in 2015, as major international research platform, with the Goal: *"To provide the knowledge required for societies in the world to face risks posed by global environmental change and to seize opportunities in a transition to global sustainability"*. Future Earth will advance Global Sustainability Science, build capacity in this rapidly expanding area of research and provide an international research agenda to guide natural and social scientists working around the world. It is also a platform for international engagement to ensure that knowledge is generated in partnership with society and users of science, connecting closely with the Sustainable Development Goals and climate and biodiversity agreements (United Nations Framework Convention on Climate Change and the Convention on Biological Diversity[36] and the Intergovernmental Platform on Biodiversity and Ecosystem Services[37]).

Future Earth brings together and, in partnership with existing programmes on global environmental change, coordinated new, interdisciplinary approaches to research on

---

[34] http://www.icsu.org/future-earth/who

[35] http://www.futureearth.org/

[36] https://www.cbd.int/

[37] http://www.ipbes.net/about-us

three themes: Dynamic Planet; Global Sustainable Development; and Transformations towards Sustainability. DIVERSITAS, the International Geosphere-Biosphere Programme (IGBP) and the International Human Dimensions Programme (IHDP) have been merged into Future Earth and the World Climate Research Programme (WCRP) is a partner. Partnerships with START[38] and other programs are also being undertaken.

Future Earth is a platform for international engagement to ensure that knowledge is generated in partnership with society and users of science and will bring together scientists of all disciplines, natural and social, as well as engineering, the humanities and law. The governance structure of Future Earth embraces the concepts of co-design and co-production of science with relevant stakeholders across a wide range of sectors (Mauser et al., 2013).

Future Earth is led by a Governing Council and supported by two advisory bodies: a Science Committee and an Engagement Committee. The Governing Council of Future Earth is composed of the International Council for Science (ICSU), the International Social Science Council (ISSC), the Belmont Forum of funding agencies, the United Nations Educational, Scientific, and Cultural Organization, the United Nations Environment Programme, the United Nations University, World Meteorological Organization, Sustainable Development Solutions Network (SDSN)[39] and the STS Forum[40]. The Future Earth Engagement Committee is a strategic advisory group, comprising thought-leaders from stakeholder groups including business, policy and civil society. Working together with the Future Earth Science Committee and the Secretariat, its primary purpose is to foster in-depth and innovative interactions between science and society. The Engagement Committee provides leadership and creative thinking on how to bridge the gap between knowledge and solutions for sustainable development. Through their joint actions the research program of Future Earth is developed to co-design the themes, priorities and approaches with the stakeholder community so that the co-produced knowledge, technologies and approaches with better address societal needs.

Future Earth, and all ICSU related programs, have a role in outreach, communication, regional activities. The Future Earth program is building Open Knowledge Action arenas in specific socio-ecological settings that focus on concrete challenges and address specific transformation needs or opportunities (Cornell et al., 2013). The arenas are to: traverse boundaries between different disciplines, perspectives, approaches, and types of knowledge; bring knowledge partners—academic and non-academic—together in networks of collaborative learning and problem solving; and to contribute to a global knowledge trust that can support transformations to a sustainable and just world.

---

[38] http://start.org/programs

[39] http://unsdsn.org/

[40] www.stsforum.org/

The Future Earth science co-design, co-produce and co-deliver model changes the science-policy-practice interface from the linear model, with its impacts and uptakes, in which "science proposes, society disposes"(Guston, 2001)(Guston & Sarewitz, 2002) and dualistic mechanisms of production and use (policy briefs, assessments, some advisory systems) to iterative interaction, with feedback loops and sometimes messy processes on all sides.

7 Science for Policy and policy for science

An important area of action for the International Council for Science (ICSU) is Science for Policy, which includes the international research programs described above, and policy for science. In the global policy arena, there are the intersecting issues of climate change, disaster risk reduction and sustainable development and their applications for cities, energy, resilience, health, populations and security. As shown schematically in Figure 1, there is the need to bring the integrated science together for policy so that the issues of technology and society can be addressed for the benefits of future societies. There is also the need to address fully global science capacity so that science benefits of all societies.

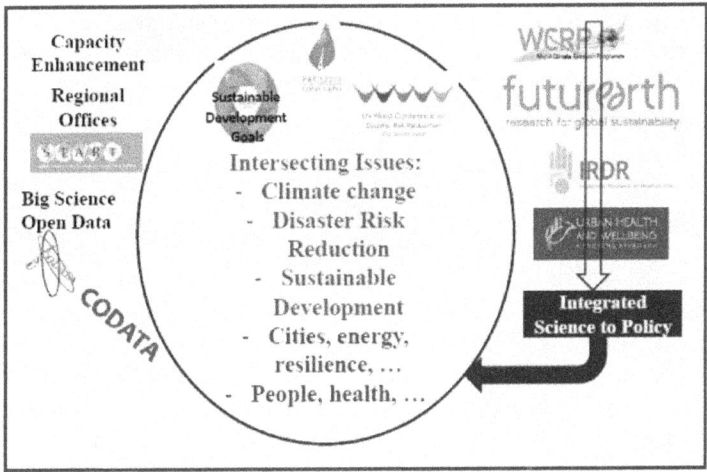

Figure 1. Intersecting issues and the need for integrated science to policy

8 Relevance, autonomy and science-society relationships

It is important that the relevance of science be established and highly considered in its planning and implementation. There is a need to change from "taken-for-granted promise of science" to approaches that:

- do strategic basic or use-inspired fundamental research
- drive innovation, economic growth and competitiveness
- address complex global challenges that require science to step into the transformative 'solutions spaces' of, for example, open Knowledge-Action Networks

The principle of "academic autonomy" has always been important but it is important to move from being unfettered by external constraints: autonomy as a right to autonomy as "collaborative assurance": working with society in securing the public good of science. This will result in changing conceptions of who society is, i.e., who has a say in steering science, shifting from only governments to governments and industry and then to the full range of 'users' and stakeholders, including funders, practitioners, citizens and social movements. The knowledge that counts will shift from only ivory tower expertise to multiple knowledge actors and a diversity of valid knowledge claims as we shift to science with society relationships.

There are challenges which will continue. Disciplines will still dominate in academic training, funding priorities and mechanisms, evaluation, rewards and career advancement. The integration of the natural, social (and human and other) sciences remains a challenge, including incentive mechanisms. The added value of international collaboration is still questioned, and when it is supported, historically institutionalized hegemonic systems and practices persist. The co-design and co-production of knowledge is not well understood, let alone supported in real dollar terms. There is also a need for researchers to have the process and communication skills needed to facilitate and manage the processes involved.

9 Conclusions

Although we are aware of the challenges and complexity of transformative research, the important "BUT" is, it is very important and essential to address issues of climate change, reduce risk, sustainable development for people and the planet. It also is essential towards addressing and solving the issues of intergenerational and international equity and ethics. Through the Future Earth Program and its co-development in integration with the Integrated Research on Disaster Risk, Urban Health and initiatives such as another science-policy issue of "big" science and open data Accord. The Accord is an example of international science, the global voice of science, addressing issues of policy for science.

This paper has discussed the building on transformative research to address global issues and bringing all the sciences together, to enable through collective actions to have the "future we want". It has focussed on the role of the International Council for Science (ICSU) as a key, leading international non-governmental organization that works with many partners to achieve these common goals. We look forward to working together for the benefit of all societies.

Acknowledgements. Thanks to our colleagues in the International Council for Science (ICSU) and around the world for the initiatives and leadership for international science.

Bibliography

Cornell, S., Berkhout, F., Tuinstra, W., Tàbara, J. D., Jäger, J., Chabay, I., ... van Kerkhoff, L. (2013). Opening up knowledge systems for better responses to global environmental change. *Environmental Science & Policy, 28,* 60–70.

Guston, D. H. (2001). Integrity, responsibility, and democracy in science. *SciPolicy: The Journal of Science and Healt Policy, 1*(2), 167–340.

Guston, D. H., & Sarewitz, D. (2002). Real-time technology assessment. *Technology in Society, 24*(1), 93–109.

International Council for Science, & International Social Science Council. (2015). *Review of the sustainable development goals: The science perspective.* Paris, France: International Council for Science (ICSU).

International Strategy for Disaster Reduction. (2009). *Global assessment report on disaster risk reduction.* Geneva, Switzerland: United Nations. Retrieved from http://www.preventionweb.net/english/hyogo/gar/report/index.php?id=9413

Lechner, S., Jacometti, J., McBean, G., & Mitchison, N. (2016). Resilience in a complex world–avoiding cross-sector collapse. *International Journal of Disaster Risk Reduction, 19,* 84–91. Retrieved from http://dx.doi.org/10.1016/j.ijdrr.2016.08.006

Mauser, W., Klepper, G., Rice, M., Schmalzbauer, B. S., Hackmann, H., Leemans, R., & Moore, H. (2013). Transdisciplinary global change research: The co-creation of knowledge for sustainability. *Current Opinion in Environmental Sustainability, 5*(3), 420–431.

McBean, G. A. (2008). Role of prediction in sustainable development and disaster management. In H. G. Brauch, O. Ú. Spring, C. Mesjasz, J. Grin, P. Dunay, N. C. Behera, ...P. H. Liotta (Eds.), *Globalization and environmental challenges* (Vol. 3, pp. 929–938). Hexagon Series on Human and Environmental Security and Peace. Berlin, Germany: Springer.

Mittelstrass, J. (2011). On transdisciplinarity. *Trames, 15*(4), 329–338. Retrieved from http://dx.doi.org/10.3176/tr.2011.4.01

Reid, W. V., Chen, D., Goldfarb, L., Hackmann, H., Lee, Y. T., Mokhele, K., ... Schellnhuber, H. J., et al. (2010). Earth system science for global sustainability: Grand challenges. *Science, 330*(6006), 916–917.

Steffen, W., Crutzen, P. J., & McNeill, J. R. (2007). The anthropocene: Are humans now overwhelming the great forces of nature. *AMBIO: A Journal of the Human Environment, 36*(8), 614–621.

Syvitski, J. P. M. (2012). Anthropocene: An epoch of our making. *Global Change, 78,* 12–15. Retrieved from http://www.igbp.net/news/features/features/anthropoceneanepochofourmaking.5.1081640c135c7c04eb480001082.html

United Nations Educational, Scientific and Cultural Organisation, & International Social Science Council. (2009). *World social science report 2013, Changing global environments.* Paris, France: UNESCO and OECD. Retrieved from http://www.worldsocialscience.org/activities/world-social-science-report/the-2013-report/changing-global-environments/

World Commission on Environment and Development. (1987). *Our Common Future*. Retrieved from http://www.un-documents.net/our-common-future.pdf

World Economic Forum. (2016). *Global Risks Report* (11th ed.). Geneva, Switzerland: World Economic Forum.

Author biographies. Professor Gordon McBean is President, International Council for Science, Co-Chair of the Governing Council for Future Earth: Research for Global Sustainability and Professor Emeritus of Geography at Western University, London, Canada. He was Professor of Geography at Western (2000-2015) with joint appointments in Political Science and Physics. From 1994 to 2000, he was Assistant Deputy Minister, Environment Canada. Previously, he was Professor, Atmospheric-Oceanic Sciences at University of British Columbia and a research scientist, Environment Canada. His recent research projects include: Coastal Cities at Risk: Building Adaptive Capacity for Managing Climate Change in Coastal Megacities; and Integrated Strategies for Risk Reduction Research Project. He is a Member of the Order of Canada (2008), Order of Ontario (2010) and is a Fellow of the: Royal Society of Canada; American Geophysical Union; Canadian Meteorological and Oceanographic Society; International Union of Geodesy and Geophysics; and American Meteorological Society.

Dr. Heide Hackmann is Executive Director, International Council for Science, since March 2015, following eight years as Executive Director of the International Social Science Council. Heide read for a M.Phil in contemporary social theory at the University of Cambridge, United Kingdom, and holds a PhD in science and technology studies from the University of Twente in the Netherlands. She has worked as a science policy maker, researcher and consultant in the Netherlands, Germany, the United Kingdom and South Africa. Before moving into the world of the international councils, Heide worked as Head of the Department of International Relations and Quality Assessment of the Royal Netherlands Academy of Arts and Sciences. Her career in science policy dates back to the early 1990s when she worked at the Human Sciences Research Council in South Africa. Heide holds membership of several international advisory committees, including the Scientific Advisory Board of the Potsdam Institute for Climate Impact Research in Germany, and the Swedish Research Council's Committee for Development Research, Sweden. She has been a Board member of START, as well as Cape Farewell in the UK.

22 Biodiversity and triage

MARK COLYVAN *

Abstract. We are currently in the midst of one of the largest mass extinction events in Earth's history and there is no end in sight. Many species have little or no chance of survival and for many others their fate depends on the development of promising, implementable conservation management strategies. These strategies include, for example, captive breeding programs, habitat protection, habitat recreation, and reversing various forms of land degradation. An important step in the development of such conservation strategies is deciding where to best focus conservation efforts. The situation is not unlike an under-resourced war-zone hospital, facing regular massive influxes of casualties. Sadly, it's not possible to even attempt to save all the species currently classified as threatened. Like the war-zone hospital, triage measures need to be implemented to determine where we should spend our time and resources. Such measures are controversial for a number of reasons, not least of which is that they sometimes recommend allowing a particular threatened species to go extinct. I will give a qualified defence of triage, outlining its theoretical underpinnings and discussing its limitations.

Keywords: decision theory, triage, optimisation, conservation management, extinction, climate change.

---

*University of Sydney, A14 Main Quadrangle, Sydney, NSW, 2006, Australia.
Email: mark.colyvan@sydney.edu.au.

1 The sixth great extinction

There is no doubt that we are in the midst of one of the largest mass extinction events in the Earth's history. This is the sixth such major extinction event and it is clear that humans are largely responsible for this one. Biodiversity is being lost at an alarming rate.[1] According to leading ecologist and conservation biologist Stuart Pimm:

> Current rates of extinction are about 1,000 times the background rate of extinction. These are higher than previously estimated and likely still underestimated. (Pimm et al., 2014, p. 987.)

To put this in perspective, the International Union for Conservation of Nature (IUCN) claims that:

> 17,291 species out of the 47,677 assessed species are threatened with extinction. The results reveal 21 percent of all known mammals, 30 percent of all known amphibians, 12 percent of all known birds, and 28 percent of reptiles, 37 percent of freshwater fishes, 70 percent of plants, 35 percent of invertebrates assessed so far are under threat. (International Union for Conservation of Nature, 2009.)

In light of all this, many species have little or no chance of survival and for many others their fate depends on the development of promising, implementable conservation management strategies. But such conservation efforts need to be strategic.[2]

This raises a crucial question: what is the appropriate conservation strategy or strategies in the face of such an ecological crisis? Several possibilities spring to mind. We could focus conservation efforts on biodiversity hotspots[3] such as Madagascar and the Indian Ocean Islands. Alternatively we could focus conservation efforts on those species facing the greatest threat, wherever they are. Another option is to focus on rare ecosystems that provide unique services. Or we might instead opt for something less systematic and more ad hoc, such as attempting to save threatened species when they come to our attention. Current conservation prioritisation is something of a mix of all of the above and has no doubt met with some success. But there is an argument that we can do better by adopting a triage strategy. To see how this works, we must first rehearse some basic decision theory.

---

[1] Here I'll focus on species-level biodiversity (i.e. diversity of species). Alternatives would be to focus on biodiversity at lower levels such as sub-species or even the genetic level, or at higher levels such as that of ecosystems. See Maclaurin and Sterelny (2008) for a good discussion of how to understand biodiversity.

[2] Although the existence of the sixth mass extinction is not in doubt, there is disagreement about its magnitude. A great deal of the disagreement hangs on the difficult task of estimating the background extinction rate, with which the current rate is compared (Regan, Lupia, Drinnan, & Burgman, 2001; De Voz, Joppa, Gittleman, Stephens, & Pimm, 2015).

[3] These are areas where there is a great deal of biodiversity and which face current or future threats.

2 Conservation prioritisation

2.1 Decision theory

Decision theory is the formal theory of rational choice. It assumes that an agent has a number of actions at her disposal, A_1–A_n, and that the world might be in any number of different states, S_1–S_m (where S_1–S_m form a partition of the ways the world might be, or turn out to be). Outcomes are the result of performing a particular act when the world is in some state or other. That is, an outcome is simply an act-state pair: O_{ij} is the result of the agent choosing action A_i while the world is in state S_j. Probabilities, p_{ij} and utilities u_{ij} are then assigned to each outcome O_{ij}. The expected utility of act A_i is $EU(A_i) = \sum_{k=1}^{m} u_{ik} p_{ik}$, where $\sum_{k=1}^{m} p_{ik} = 1$ We calculate the expected utility of each act at our disposal and the decision rule is: *choose the action with the greatest expected utility, if there is such and action* (Jeffrey, 1990).

A special class of decision problems are known as *optimisation problems*. An optimisation is a decision under constraint where one is attempting to maximise one quantity, subject to a constraint of some kind. For example, consider the problem of enclosing the maximum rectangular area of land, subject to the constraint of a fixed length of fencing. Or, closer to our current purposes, consider the problem of trying to maximise some quantity with a fixed budget. In such situations some outcomes are inaccessible because the costs of the actions involved are too high. Although it is common for the constraint to be financial, it need not be; the constraint can be in terms of any resource. In its most general form, an optimisation problem is one of choosing the action with the greatest expected utility, subject to the constraint in question. Now we apply such methods to our conservation strategies.

2.2 Triage

In war-time hospitals the sick and wounded can arrive in such numbers that not all of them can be given the most thorough treatment. Moreover, some patients are in greater danger of dying than others. Medical staff must make decisions about who gets treated first and, in some cases, who gets treated at all. In the most difficult cases, patients are so seriously injured that little can be done to increase their chances of survival. In other cases, patients could be treated but their chance of survival would not be significantly increased and the treatment would seriously drain resources—both medical supplies and the time of the medical team. Since such resources are limited, decisions need to be made about where best to spend resources. We thus have an optimisation problem of maximising expected lives saved, subject to resource constraints.

The relevance of this to the current environmental situation is obvious enough. We are faced with a situation of limited resources (financial, temporal, etc.) so we cannot do all we would like to preserve all the biodiversity we currently find on our planet.

Suppose we have a ranking of the various threatened species from most to least threatened. We must answer the question: where should we spend our valuable resources? A tempting answer is to focus resources on the most threatened species—those higher up the list. But this is akin to treating the most seriously injured first in our war-time hospital, without considering the chances of success and without looking at the relevant opportunity costs (the lives that otherwise might have been saved). This is a bad strategy in the war-time hospital and it is a bad strategy for saving species.

The answer to our environmental prioritisation problem is to set up and solve the relevant optimisation problem. That is, we need to engage in *environmental triage* (Wilson, McBride, Bode, & Possingham, 2006). Here we try to maximise some appropriate quantity (e.g. reducing the expected number of threatened species), subject to existing resource and budgetary constraints.

It is worth pausing to note that one of the benefits of the decision-theoretic way of looking at such issues is that it forces decision makers to be explicit about what the quantity to be maximised is and how it is to be measured. For example, is the aim to reduce the number of extinctions in the wild, critically endangered species, endangered species or threatened species?[4] The quantity to be maximised (or in this case, minimised) needs to be decided and explicitly incorporated into the set up.[5] We are thus not able to get by with vague and imprecise aims such as "improving the natural environment" or "making the world a better place to live". We may still achieve such lofty goals but we do so by maximising a well-defined, measurable and explicitly-stated quantity such as numbers of species saved from extinction. This approach also forces us to think about the nature of the resource constraints we face. More on this later.

As in the war-time hospital sometimes the result will be that utility is maximised by directing resources to those other than the most serious cases. Often, although there are things that could be done to save a particular species, the optimisation strategy will advise doing nothing for that species. The strategy will advise, in effect, to give up on some species in order to focus resources elsewhere. As a notable defender of triage, ecologist and conservation biologist Hugh Possingham, puts it:

> [w]e should pick winners rather than struggling away with the ones on their last legs. (Possingham, from an interview in the *Sydney Morning Herald*, 6/9/14.)

Just as it must be hard in war-time hospitals to leave some critically ill patients to die, giving up on some of our most endangered species is no easy matter. Leaving species to go extinct is certainly not something we do lightly; it is forced upon us

---

[4] As is standard, I'm here appealing to the IUCN Red-list classification (International Union for Conservation of Nature 2008).

[5] Technically, it is aways expected utility we are maximising but in practice we often allow more easily measurable surrogates for utility such as money, number of species saved and so forth.

Biodiversity and triage 425

by the overwhelming number of threatened species and the relatively poor prognosis for many of them. Desperate times call for desperate measures. Although the triage model is firmly supported by standard decision theory, it should come as no surprise that triage has attracted some criticism. I will now look at some of this criticism and offer a (qualified) defence of environmental triage against the criticism.

2.3 Limitations of the triage model

There are a number of poorly motivated arguments against triage (e.g. the value of the environment cannot be represented in economic terms or that the environment is sacred and conservation biologists should not play God) but to my way of thinking the best arguments against triage attack the theoretical framework used: optimisation under constraint. In any case, these are the arguments I'll focus on here.

First, we might question the assumption of the triage/optimisation model that the resources are an absolute constraint. It is not always the case that resources are fixed. In at least some circumstances, if the budget or resources are negotiable, it might be better to spend some effort petitioning for a larger conservation budget rather than trying to get by with an unreasonable budget or inadequate resources. There is no room for such negotiations in the standard optimisation models under consideration here.

Another assumption of standard triage models is that there are no costs associated with the reallocation of resources. That is, it is assumed that were we to decide not to pursue conservation efforts in relation to some species, the Orange-bellied parrot (*Neophema chrysogaster*, currently classified as critically endangered), say, we could redirect those resources to conservation efforts for some other species, the Tasmanian Devil (*Sarcophilus harrisii*, currently classified as endangered), say. There are a number of difficulties with this assumption. First, sometimes the funds made available have strings attached. Resources from a bird conservation source may not be allowed to be reallocated to conservation efforts directed at a non-bird. In any case, the conservation efforts themselves are also going to be very different, so a group working on saving a bird may not be qualified to engage in conservation of, say, Tasmanian Devils. At the very least there will be transaction costs. There are also issues of geopolitical strings attached to resources: sometimes funds cannot cross national or state boundaries.

Finally, triage requires a standard probabilistic representation of uncertainty. This is not always available and sometimes probability may not even be the appropriate tool for representing the uncertainty in question (Colyvan, 2008a). For example, standard probability theory, is based in classical logic, which is appropriate only if the world is not gappy. That is, for any proposition P, either P or not-P is the case, even though we may be uncertain about which. But now consider cases where vagueness is involved. Take a question such as 'is a given level of risk of extinction of a particular species acceptable?'. Here the vagueness of the notion of 'acceptable risk' (which permeates much of risk analysis), means that there may not be a fact of the matter.[6] If this is the case, it can be argued that probability theory is not the appropriate tool for the representation of the resulting uncertainties (Regan, Colyvan, & Burgman, 2002).

I take all of these objections to the triage model seriously. I do not, however, take them to be decisive. Rather, they highlight limitations on the applicability of the model. As with any model, it has its domain of useful applications and one should exercise caution when using a model beyond its domain of intended applications. All the above objections draw attention to various idealisations in the triage model that are unrealistic in some circumstances. We can thus respond by acknowledging the idealisations in question and this, in turn, forces us to think about the appropriateness of each of the idealisations in question for the particular application of the model.

For instance, if the budget in question is not fixed, then we simply do not have an optimisation problem. If the budget is genuinely open for negotiation then we should move to another more-appropriate model (e.g. a Nash bargaining model) for the budget negotiation. But once the budget has been negotiated, we can then return to the optimisation model with the newly-bargained budget. Similarly, reallocation expenses can be included in the model—we do not need to assume zero-transaction costs. Dropping this idealisation makes for a slightly more complex model but there is no problem in principle incorporating such complications. Similarly, non-probabilistic representations of uncertainty can be accommodated in decision theory (e.g. vague probabilities Walley, 1991), but again it complicates things. This is not to give up on the basic triage model though. If there is no need to introduce such complications (as will sometimes be the case), then we should stick with the basic model.

So, while these criticisms of the assumptions of the triage model are correct, they do not undermine the basic approach. Rather, focussing attention on the assumptions is useful in determining the appropriate applications of the model and can even help in suggesting modifications of the model and different models for applications beyond the scope of the basic triage model. Now we turn to some more radical attacks on triage.

3 Infinite value

Some have suggested that the natural environment is infinitely valuable. If this were right, saving the natural environment would take priority over mere economic concerns. What is not often noted is that it would also mean any environmental trade-

---

[6]This, of course, depends on your preferred theory of vagueness. Epistemic accounts of vagueness (Williamson, 1994; Sorenson, 2001) hold that there is an unknowable fact of the matter, but others allow there to be truth-value gaps associated with vagueness (e.g. Fine, 1975). Such vague categories can be sharpened up or, rather, replaced with sharp categories but this brings its own problems (Regan, Colyvan, & Burgman, 2000).

offs would be impossible. Triage would not work but neither would very much else. Thankfully there is no reason to entertain infinite environmental values. The natural environment is valuable but not infinitely so. But lets's look at these issues in more detail.

Douglas McCauley is one who appeals to infinite values in order to prioritise the environment over other concerns.

> I suggest that the aggregate value of a chunk of nature—its aesthetic beauty, cultural importance and evolutionary significance—is infinite, and thus defies incorporation into any ecosystem service programme that aims to save nature by approximating its monetary value. (McCauley, 2006, p. 28.)

The idea is clear enough: if a chunk of nature is infinitely valuable, its value trumps any finite, economic concerns. But what reason do we have for believing that nature is infinitely valuable? McCauley seems to suggest (without further argument) that the infinite value of nature follows from accepting that nature has intrinsic value (McCauley, 2006, p. 28). There are several problems with this line of argument. First, it is anything but clear that nature is intrinsically valuable. And even if it were, it would not follow that its value would be infinite (Justus, Colyvan, Regan, & Maguire, 2009). So what reason is there to entertain infinite value for nature? Of course we can simply stipulate that this is so but this solves nothing. Apart from from being unmotivated, it leaves the door open for others to stipulate alternative uses of land to be infinite: coal-seam gas mines, car parks, and so on. We end up with value inflation by stipulation. If nature is to be seen as infinitely valuable, this needs to be motivated. Without such motivation, this line of argument is without merit. But even setting aside the motivational problem, there are other problems for this view.

The first big problem is that infinite value insufficiently discriminates the salient outcomes (Colyvan, Justus, & Regan, 2010). For example, if we hold that some piece of the natural environment is infinitely-valuable (e.g. mangrove forests), assigning meaningful values to larger regions of that habitat is problematic because there is no extra value to be had in larger areas of it. The problem is that all infinite values (at least in standard decision theory) are equal (and it is not clear that appeals to trans-finite values help here). This approach thus prohibits prioritisation of conservation goals—they are all equal!

The second big problem is that infinite value swamps probability. More formally: we violate the Archimedian condition.[7] Without some clever use of non-standard analysis, non-Archimedian decision theory, or the like, we end up with the expected

---

[7]The Archimedian condition states that whenever a rational agent prefers A to B and B to C, there exists a real-valued $0 < w < 1$, such that the agent is indifferent between B and the following lottery: a w chance at C and a $(1-w)$ chance at A. Informally: there exists a weighted average of the best and worst outcome which is equal in expected value to the middle outcome. When A is infinite, there is no such w.

utility of any action with a non-zero chance of the desired outcome being infinite. For example, if persistence of an endangered species is considered infinitely valuable, any action with a non-zero chance of the species' survival will have infinite expected value. We would need to be indifferent between hunting the species in question, initiating captive breeding programs, designing reserves for them, poisoning them, and so on, for all of these management strategies have *some* non-zero chance of outcomes that result in the preservation the species in question.[8] But this is clearly the wrong answer; we want to invest in conservation strategies with *higher* chances of success. For example, we want to be able to recommend designing appropriate reserves for rhinos rather than hunting them. Infinite values stand in the way of such common-sense judgements.[9]

It is worth noting explicitly that this does not mean that nature has only finite value. Rather, the lack of discriminatory power we have just seen gives us a pragmatic reason for rejecting appeals to infinite values. But there are also good reasons to reject that the environment has infinite value. Apart from anything else, one needs evidence that something has infinite value; it is not enough to simply assert this as McCauley does. What evidence is there that the natural environment has infinite values? In fact all evidence seems to suggest that its value is finite. If the values in question are instrumental, then we reveal these values via our behaviour. But our behaviour includes making precisely the kinds of trade offs that we have already seen would be impossible were the values in question infinite. The instrumental value of natures is surely finite. One might instead appeal to intrinsic values and insist that the *intrinsic* value of nature is infinite. On this account, value is disconnected from human behaviour so the fact that we behave as though nature is only finitely valuable is neither here nor there. But therein lies the problem with the intrinsic value account: if human behaviour does not reveal the values in question, what does? Intrinsic-value accounts of the value of nature face serious epistemic problems: the values in question are inaccessible and thus there is no evidence that they are infinite.[10] It seems that we cannot coherently let go of our anthropocentric perspective (Grey, 1993).

It might, instead, be claimed that the values in question are incommensurable or that we are fundamentally ignorant of the value of nature. But as Elliott Sober correctly points out, ignorance does not help motivating conservation efforts: "[i]f you are completely ignorant of values, then you are incapable of making a rational decision, either for or against preserving some species" (Sober, 1986, p. 175). If you don't know what something's value is, a fortiori you have no reason to think its value is greater than something else. Similar problems exist for incommensurable value. If something's

---

[8] This line of objection is due to (Hájek, 2003) where it is raised in relation to the infinite values found in Pascal's Wager.

[9] Some of these problems can be avoided by moving to a non-standard decision theory that allows for comparisons between various infinite outcomes (e.g. Bartha and DesRoches, 2017; Colyvan, 2008b).

[10] Such epistemic problems exist for intrinsic values more generally, either infinite or finite (Justus et al., 2009).

value is incommensurable with, say, economic values, then no trade offs are possible and there can be no rational motivation for allocating budgets to conservation efforts. (Justus et al., 2009.) In short, infinite values, incommensurable values, and unknown values are no basis for conservation.

4 Special pleading

Another line of attack on the triage model is to argue that sometimes, at least, special pleading is in order. Defenders of this line of thought then point to successful cases in support. One such case is sometimes thought to be the California Condor (*Gymnogyps californianus*).

In 1987 the California Condor was on the brink of extinction. At that stage there were only 6 birds remaining in the wild. The bird's cause was relentlessly championed by two conservationists, Noel and Helen Snyder, when others had given up on saving the California Condor (Snyder & Snyder, 2000). The remaining wild Condors were captured and placed in captive breeding programs at San Diego and Los Angeles zoos along with 16 other captive birds. The breeding program was very successful and by 1991 Californian Condors were released back into the wild. By October 2014 there were around 425 California Condors (about 219 in the wild, the rest still in captivity) (U.S. Fish & Wildlife Service, 2014).

Stuart Pimm, for one, presents the Condor case as flying in the face of triage.

> [N]ot all species are equal in their ability to inspire us—the condor is among the most spectacular birds—nor in their ability to extend our applied scientific skills to the limits necessary to save them. Were I endangered, I'd want the Snyders and their colleagues there to ensure I wasn't written off. (Pimm, 2000.)

Presumably, the thought is that because the resources used to save the condor were rather extensive (and hard to justify in cost–benefit terms), triage would have recommended that we should have given up on the California Condor. The success of the Snyders and the captive breeding program thus is supposed to tell against the triage approach.

But it is not clear that the condor case is the problem for triage that Pimm takes it to be. First, there is nothing in the triage approach that suggests that all species are of equal value. For whatever reason, some species might be assigned greater importance (e.g. iconic species), just as in wartime hospital, a General might be given greater weight than a Private. Given the cultural importance of the California Condor, it could well be argued that it should be assigned extra weight in any triage model. If so, the triage model may well recommend the same course of action as the Snyders' captive breeding program. But in any case, the triage calculations would need to be done. One shouldn't simply assume that triage always recommends against assigning resources to critically endangered species.

The second thing to say about this case is that it is misleading to focus on a single successful story. Decision theory does not—nor should it—guarantee the best outcome in a given case. Triage and decision theory in general is probabilistic at its core. This means that sometimes it recommends an action that does not result in the best outcome. But nothing can guarantee the best outcome in any given case. Just as casinos can suffer the odd big loss (but win in the long run), triage will sometimes recommend that the chance of a species recovering are so low as to not warrant investing resources and yet, against the odds, it recovers. The Californian Condor might be such a case.

Moreover, focussing on cases where decision theory does not result in the best outcome is to ignore the many cases where it does get it right. For instance, had the recovery program for the California Condor failed (and the game is not over yet—the Californian Condor is still critically endangered), what would Pimm say about the many species that went extinct in the meantime. But of course the recovery program for the Californian Condor does look to have been successful but the point remains that the resources spent on this single species was at the expense of others. There are always opportunity costs, even in successful cases.

5 Ethics versus decision theory

Triage adopts an explicitly consequentialist framework: the values in question are attached to outcomes, which in turn are products of actions of the agents and the states of the world. Indeed, this is how standard decision theory (including economic theory) proceeds. But it might be argued that this consequentialist framework is controversial. For example, it might be argued that we have duties to care for non-human species and this duty has nothing to do with consequences. At the very least, we ought to give due respect to more explicitly ethical motivations for conserving biodiversity, or so goes this line of thought.

The first thing to note is that although standard decision theory is consequentialist in the sense that it is outcomes that are the bearers of value, it is not consequentialist in any stronger sense. For example, the values attached to outcomes need not be anything to do with promoting the greatest good for the population at large. Indeed, the so-called utility function of decision theory is simply a mathematical function that satisfies the von Neumann-Morgenstern axioms (von Neumann & Morgenstern, 1944), and these axioms are, in many ways, rather liberal. They rule against intransitivity of the utility function[11] but they do not rule against valuing genocide. The theory fixes

---

[11] If the utility of outcome A is greater then the utility of B, and the utility of outcome B is greater than the utility of C, then the utility of A must be greater than the utility of C.

the structure of the utility function but the values this function takes are left open. In an important sense, decision theory is *amoral*. A natural thought is to appeal to ethics to place further restrictions on the values the utility function can realise (so, for example, we might want genocide to be highly disvalued). On this way of looking at things, at least, there is no conflict between ethics and the decision-theoretic framework that triage appeals to; ethics complements decision theory.

Moreover, any ethical theory cannot "go it alone" so to speak. Ethical theories, such as deontology simply do not give advice about what to do in an uncertain world (Smith & Jackson, 2006). Ethical theories need a decision theory. Now it might be thought that the consequentialist framework adopted by standard decision theory is not well suited to non-consequentialist ethical theories such as deontology or virtue ethics. But non-consequentialist ethical theories can be combined with standard decision theory (Colyvan, Cox, & Steele, 2010). The basic idea, for deontological decision theory, for example, is that further formal constraints are placed on the utility function so that the utility function assigns high values to outcomes arising from certain classes of actions (the "duties") and low values to outcomes arising from another class of actions (the "prohibitions").[12] The details need not concern us; the important point is that ethical theories on their own do not give advice about anything let alone about how best to preserve biodiversity. For this we need decision theory. On this way of looking at things, again there is no conflict between ethics and the decision-theoretic framework; decision theory complements ethics.

In short, ethical theories require decision theory in order to give any practical advice about conservation and non-consequentialist ethical theories can be accommodated in something like the standard decision theoretic account. A version of triage is thus compatible with non-consequentialist ethical theories. There is nothing essentially consequentialist (in the ethical sense) about triage. It is thus hard to see the sense in which triage might be thought to be in tension with ethical sensibilities.

6 Conclusion

I've provided a limited defence of triage as the core decision tool in conservation management and conservation planning. There are devils in the details of its application though. As with models, wherever they are applied, we need to be aware of, and critically examine, the various idealisations and assumptions. Understanding these idealisations and assumptions can help in ensuring that the model is used only in appropriate settings. For the triage model under discussion this means: not accepting

---

[12] Still, the framework remains consequentialist in that it is outcomes that are the bearers of value. If this is objectionable to the deontologist or to the virtue theorist, they need to develop their own non-consequentialist decision theory. Until that is done, they both remain unable to give practical advice in an uncertain world such as ours.

resource constraints as absolute when they are not; including transaction and reallocation costs when appropriate; employing careful and appropriate representations of the various uncertainties; and employing careful and appropriate representation of utility (which includes appropriate values for iconic species). With these caveats in place, triage is simply the rational way to ensure the optimum allocation of resources for the best conservation outcomes. As I've already stressed, it does not guarantee the best outcome in any single case—nothing can do that—but it does guarantee the best results in the long run. And it is the long run we are interested in.

Finally, a word about complications arising from climate change. One of the most serious threats to the Earth's biodiversity in the medium term is anthropogenic climate change. Various climate mitigation measures, such as carbon sequestration schemes, if managed carefully, may be beneficial for both reducing climate change and for the preservation of biodiversity (Bekessy & Wintle, 2008; Venter et al., 2009). But care needs to be taken here. Too narrow a focus on carbon storage can lead to perverse environmental outcomes. For example, quick-growing plantation forests are a useful measure for reducing atmospheric carbon but they contribute little to biodiversity and may even reduce biodiversity, especially if native grasslands and other native vegetation are cleared for plantation forests. There are also concerns about the effects of plantation forests: the plantation tree may become invasive or otherwise significantly alter ecosystem processes (Putz & Redford, 2009; Lindenmayer et al., 2012). Any plausible strategy for dealing with the preservation of biodiversity must include strategies for curbing the effects of climate change but this must be done in a holistic manner. Environmental goals such as biodiversity and carbon sequestration need to be carefully distinguished and ways of jointly optimising these and other environmental values need to be identified. Triage will be an important part of such holistic strategies but careful attention to the relevant environmental values is crucial to success here.

Acknowledgements. I am indebted to Jack Justus, Hugh Possingham, Helen Regan, and Katie Steele for valuable discussions on the topic of this paper. I am grateful to the audience at the "Future Earth" session at the 15th Congress of Logic, Methodology and Philosophy of Science in Helsinki in August 2015 for some helpful comments and interesting discussion. Work on this paper was supported by an Australian Research Council Future Fellowship grant (grant number: FT110100909).

Bibliography

Bartha, P., & DesRoches, C. T. (2017). The relatively infinite value of the environment. *The Australasian Journal of Philosophy*, *95*(2), 328–353.

Bekessy, S. A., & Wintle, B. A. (2008). Using carbon investment to grow the biodiversity bank. *Conservation Biology*, *22*(3), 510–513.

Colyvan, M. (2008a). Is probability the only coherent approach to uncertainty? *Risk Analysis*, *28*(3), 645–652.

Colyvan, M. (2008b). Relative expectation theory. *The Journal of Philosophy*, *105*(1), 37–44.

Colyvan, M., Cox, D., & Steele, K. (2010). Modelling the moral dimension of decisions. *Noûs*, *44*(3), 503–529.

Colyvan, M., Justus, J., & Regan, H. M. (2010). The natural environment is valuable but not infinitely valuable. *Conservation Letters*, *3*(4), 224–228.

De Voz, J. M., Joppa, L. N., Gittleman, J. L., Stephens, P. R., & Pimm, S. L. (2015). Estimating the normal background rate of species extinction. *Conservation Biology*, *29*(2), 452–462.

Fine, K. (1975). Vagueness, truth and logic. *Synthese*, *30*(3–4), 265–300.

Grey, W. (1993). Anthropocentrism and deep ecology. *Australasian Journal of Philosophy*, *71*(4), 463–475.

Hájek, A. (2003). Waging war on Pascal's Wager. *Philosophical Review*, *112*(1), 27–56.

International Union for Conservation of Nature. (2008). Wildlife in a changing world: An analysis of the 2008 IUCN Red List of Threatened Species. *IUCN Website*. Retrieved June 6, 2016, from http://www.iucn.org/theme/species/publications/iucn-red-list-publications

International Union for Conservation of Nature. (2009). Distinction crisis continues apace. *IUCN Website*. Retrieved June 6, 2016, from http://www.iucn.org/content/extinction-crisis-continues-apace

Jeffrey, R. (1990). *The logic of decision* (2nd ed.). Chicago: University of Chicago Press.

Justus, J., Colyvan, M., Regan, H. M., & Maguire, L. A. (2009). Buying into conservation: Intrinsic versus instrumental value. *Trends in Ecology and Evolution*, *24*(4), 187–191.

Lindenmayer, D. B., Hulvey, K. B., Hobbs, R. J., Colyvan, M., Felton, A., Possingham, H. P., … Gibbons, P. (2012). Avoiding bio-perversity from carbon sequestration solutions. *Conservation Letters*, *5*(1), 28–36.

Maclaurin, J., & Sterelny, K. (2008). *What is biodiversity?* Chicago: University of Chicago Press.

McCauley, D. J. (2006). Selling out on nature. *Nature*, *443*(7107), 27–28.

Pimm, S. L. (2000). Against triage. *Science*, *289*(5488), 2289.

Pimm, S. L., Jenkins, C. N., Abell, R., T. M, B., Gittleman, J. L., Joppa, L. N., … Sexton, J. O. (2014). The biodiversity of species and their rates of extinction, distribution, and protection. *Science*, *344*(6287), 987–997.

Putz, F. E., & Redford, K. H. (2009). Dangers of carbon-based conservation. *Global Environmental Change, 19*(4), 400–401.
Regan, H. M., Colyvan, M., & Burgman, M. A. (2000). A proposal for fuzzy IUCN categories and criteria. *Biological Conservation, 92*(1), 101–108.
Regan, H. M., Colyvan, M., & Burgman, M. A. (2002). A taxonomy and treatment of uncertainty for ecology and conservation biology. *Ecological Applications*, (2), 618–628.
Regan, H. M., Lupia, R., Drinnan, A. N., & Burgman, M. A. (2001). The currency and tempo of extinction. *The American Naturalist, 157*(1), 1–10.
Smith, M., & Jackson, F. (2006). Absolutist moral theories and uncertainty. *Journal of Philosophy, 103*(6), 267–283.
Snyder, N., & Snyder, H. (2000). *The Californian condor: A saga of natural history and conservation.* Princeton: Princeton University Press.
Sober, E. (1986). Philosophical problems for environmentalism. In B. Norton (Ed.), *The Preservation of Species: The Value of Biological Diversity.* Princeton: Princeton University Press.
Sorenson, R. (2001). *Vagueness and contradiction.* New York: Oxford University Press.
U.S. Fish & Wildlife Service. (2014). California condor recovery program. *USFWS, 31 October 2014.* Retrieved June 6, 2012, from https://www.fws.gov/cno/es/calcondor/PDF_files/2014/Condor%20Program%20Monthly%20Status%20Report%202014-10-31.pdf
Venter, O., Laurance, W. F., Iwamura, T., Wilson, K. A., Fuller, R. A., & Possingham, H. P. (2009). Harnessing carbon payments to protect biodiversity. *Science, 326*(5958).
von Neumann, J., & Morgenstern, O. (1944). *Theory of games and economic behavior.* Princeton, NJ: Princeton University Press.
Walley, P. (1991). *Statistical reasoning with imprecise probabilities.* London: Chapman and Hall.
Williamson, T. (1994). *Vagueness.* London: Routledge.
Wilson, K. A., McBride, M., Bode, M., & Possingham, H. P. (2006). Prioritising global conservation efforts. *Nature, 440*(7083), 337–340.

Author biography. Mark Colyvan is a professor of philosophy at the University of Sydney and a visiting professor at the Munich Center for Mathematical Philosophy at the Ludwig-Maximilians University in Munich. He holds a BSc(Hons) in mathematics and a PhD in philosophy (the latter from the Australian National University). His main research interests are in the philosophy of mathematics, philosophy of logic, decision theory, and philosophy of ecology and conservation biology. He is the author of *The Indispensability of Mathematics* (Oxford University Press, 2001), *Ecological Orbits: How Planets Move and Populations Grow* (Oxford University Press, 2004, with co-author Lev Ginzburg), *An Introduction to the Philosophy of Mathematics* (Cambridge University Press, 2012), and over 80 refereed papers in philosophy journals, biology journals, and edited collections.

Index

a priori principles, 237, 238
Ackermann, W., 93, 94
action, 18, 158, 161, 164, 166, 170, 176, 182, 185, 186, 288–297, 423, 428–431
    selective, 335
affordances, 288, 292, 295
agency, 4, 12, 17, 21, 22, 28, 30–33, 36, 37
agent, 4, 11, 13, 15–18, 22, 25, 27–34, 36, 37, 106, 114–116, 119–121, 164, 167, 350–355, 358–361, 363, 364, 369, 370, 372, 384, 385, 387, 388, 430
    rational, 351, 354, 356, 423, 427
algebra, 107, 108, 359
alpha particles, 234
analysis
    chemical, 227, 228
    cytology and genetic, 250
    of hazardous trades, 246
    of karyotypes, 251
anti-realism, 159, 304, 305
anti-reductionism, 185
approximation, 120, 160
Archimedian condition, 427
Aristotle, 138, 146, 147, 149, 150, 176, 181, 183–186
arithmetic, 43, 51, 54, 56–58, 74, 138, 238, 308
    false, 51, 52
atom-smashing, 233
atomic energy, 244, 247, 248
atoms, 227, 228, 230–235
Avicenna, 141–145, 147, 150, 151
axiomatic theory (of probability), 162, 164

background theory, 43, 55, 56
Backward Induction, 18–21, 25, 26, 28, 29
backward induction, 374
Banach-Tarski Paradox, 95
bargaining, 350, 358, 361, 363, 364, 426
bargaining solutions, 350, 362, 363
belief, 18, 20, 22–24, 26, 29, 115, 159, 163, 168, 182, 183, 186, 352, 356
    revision, 106, 111, 113, 116, 121
    metaphysical, 317
    of mainstream philosophy of science, 304, 307
    revision, 351
    system of, 312
belief , 121
belief revision, 20–24, 36, 37, 211
Beltrami, E., 48–52
Bernays, P., 54–58, 76

biodiversity, 404, 409, 411, 413, 422, 423, 432
    conservation of, 430–432
biomedicine, 323–325
    reshaping of, 249
bisimulation, 9
bootstrapping, 235

calculus, 109, 115, 119
    of pure geometry, 42
    propositional, 57
Californian Condor, 429, 430
Carnap, R., 4, 70, 191–196, 198, 199, 305, 306
Carnot, S., 314
Cartwright, N., 226, 306, 307, 371
categories
    of the operation of scientists, 246
category, 52, 196
    logical, 191, 192, 197
causal model, 375
causality, 180, 184, 309
cause, 183, 184, 262–270, 272, 276, 277, 279–282, 284, 369, 374–377
Cavendish, H., 230
certainty, 165, 181, 184, 304, 311
    of science, 317
    predictive, 304
Chalmers, A., 236
chance, 422–424, 427, 428, 430
Chemical Revolution, 228, 229
chromosomes
    the study of, 251
    visualization of, 251
circulation of knowledge, 245, 246, 254
climate change, 404, 405, 409–411, 432
clinical trials, 218
    proliferation of, 249
Cockcroft, 234
coding
    of DNA, 329
coherence, 49, 51, 52, 54, 100, 117, 159, 163, 408
    theory of truth, 312
common knowledge, 14, 15, 25, 354

communication, 14, 15, 32, 33, 414, 416
complexity, 20, 22, 24, 35, 37, 73, 416
    epistemological, 325
    of the phenomenon 'cancer', 338
    of tumour evolution, 332
    of tumour heterogeneity, 339
    ontological, 328
composition, 6, 226, 227, 229–233, 236, 238, 239
compositionism, 228, 229, 234, 236–239
computer-assisted learning, 177, 181
concept
    of line, 48, 51
concepts, 87, 88, 237, 238, 305, 309, 315, 317
    construction of, 305
    in patchwork narratives, 329, 335, 339
conditionalization, 351, 359
confirmation, 183
conjunction elimination, 126, 128, 129
consequentialist, 430, 431
conservation
    laws, 238
    of energy, 234
    of mass, 234
    of nature, 422–425, 427–429, 431, 432
conservation management, 431
consistency, 42–44, 53, 56, 95, 101, 116
    classical, 115
    pragmatic, 115, 116
    probabilistic, 115
    as satisfiability, 53
    deductive, 52
    demonstrations of, 42, 44
    of axioms, 42
    proof, 43
constitutive principles, 238, 310, 311, 316
construction, 4, 7, 48, 49, 106, 111, 303
    geometric, 50
    of knowledge, 245, 304–308, 310, 312, 315–317
    of shared languages, 254

of the 'Nabor Carillo' Nuclear Center, 248
  on the real numbers, 44, 55
constructive empiricism, 304
content, 115, 116, 289, 292–297, 311, 314, 315, 317
  empirical and theoretical, 311, 315
  mathematical, 47
context of discovery, 304, 306, 307, 316
context of justification, 306
contradiction, 43, 48, 50, 54
cooperative game theory, 358
correspondence, 306, 313
  rules, 191, 192, 194, 197, 199, 200
counterexamples, 126–128, 132, 133, 338, 356
Cournot, A., 161
credence, 350, 360, 363
cytogenetics, 251
cytology, 250
  techniques of, 251

Dalton, J., 228, 232, 233
data, 185, 304, 307–309, 312, 355, 356, 376, 377, 408, 412
  big, 406
  open, 406, 416
  scientific, 405
  experimental, 315
  models of, 307
  theories of, 307
data structure, 307
decision theory, 180, 351, 422, 425–428, 430, 431
decomposition, 95, 226–233
deduction, 14, 113, 166, 185, 306, 307, 310, 315
  rigorous, 45
  sentential, 53
deficiency
  test of, 253
definition, 57, 159, 164, 168, 193, 237, 238, 246, 315
  implicit, 47, 49, 56, 193
  of chemical elements, 228
  operational, 315
degree of belief, 116, 159, 163, 168
Descartes, R., 176
description, 19, 20, 32, 308, 309
  of physical phenomena, 316
  of space, 52
determinism, 157, 159, 178, 183
dialetheism, 127
disaster risk reduction, 408, 409, 412, 415
discreteness
  metaphysical principle of, 237
  ontological principle of, 311
disjunctive syllogism, 128
diversity, 20, 28, 33, 36, 185, 416
  genotypic and phenotypic, 328
  subclonal within a tumour, 332
division, 141, 147
  cellular, 333
  Saccheri's, 46
dominance, 26
Drosophila, 250, 252–254
  laboratories, 250
  mutants, 252
dual visual systems hypothesis, 291
Dupré, J., 226, 227
dynamic logic, 17, 23, 24, 35, 37
dynamic-epistemic logic, 27, 35, 351

economic, 404, 405, 409, 416
economic theory, 397, 430
economics, 368–372, 375, 376, 382–385, 387, 388, 390, 391, 394–397
electrolysis, 229–231
electrons, 231, 233–235
elementary particles, 227, 233–235
elements, 94, 100, 228, 229, 234
empirical facts, 304
energy, 234, 235, 317, 409, 410, 415
engineering sciences, 304
epistemic
  authority, 255
  universalism, 244
epistemic logic, 15, 113, 351, 352, 354

epistemic tool, 304, 305, 308, 313, 315, 316
epistemic use, 304–306, 308, 310, 312–315, 317
epistemic values, 209, 219, 397
epistemology, 30, 31, 120, 121, 157, 163, 170, 181, 184, 209, 305, 307–309, 316, 328, 329, 337, 339
    formal, 315, 350, 351, 353–356, 364
    of scientific practices, 303, 304
Euclid, 44–48, 51, 52, 54
Euclidean, 10
    axioms, 44
    geometry, 43, 49
    lines, 50
    metric, 48
    object, 47
    plane, 48
    properties of lines, 49
    space, 50
    triangles, 49
evaluation game, 5, 6, 8, 12
evidence, 117, 119, 121, 160, 167, 208–210, 212–219, 234, 374, 375, 428
evidence-based, 410, 412
evolution, 288–290, 292, 294, 317
    clonal, 330
    of tumour, 330–332, 336–338
exchangeability, 160, 161
experiment, 14, 180, 181, 227, 230, 304, 307, 312
    in physics, 234, 235
    theories of, 307
explanation, 304, 306, 307, 311, 313, 314, 317
    of experience, 311
    of genotypic and phenotypic diversity, 328
    of scientific events and processes, 328, 329
    of the construction of knowledge, 316
    of tumour heterogeneity, 325, 330, 333, 339

extinction, 422, 424, 425, 429

failure, 382–393, 396, 398
Field, H., 65, 66, 68
financial crisis, 369, 383–385, 387, 390
First Degree Entailment, 128
fixed-point logic, 6, 8, 9, 12, 19, 20, 27
France, 251
Frank, P., 236
Frege, G., 14, 33, 44, 45, 66, 130, 181, 201
frequency, 156, 158, 162, 164, 166–169

game equivalence, 10, 11, 28, 29
games, 27–29, 31–33, 35–37
generalized quantifiers, 72, 76
genetics, 332
    development of, 250
    in Mexico, 253–255
genetics and radiobiology, 249, 250
    in Mexico, 244, 249
Glashoff, K., 138, 139, 150
Global Reflection Principle (GRP), 77, 95, 97
glue, 236, 237
group theory, 165
Gupta, A., 193

Hempel, C. G., 192, 198, 208, 306
Hilbert, D., 42–45, 52–57, 190, 191, 193, 194, 201
history of logic, 128
Hitler, A., 303
homogeneity, 159, 164, 166
human genetics, 250, 251
Hume, D., 185, 216
Humean epistemology, 304, 309–313, 316
hydrogen, 229–232

idealization, 64, 227, 308, 317, 352–354, 388, 395, 397, 426, 431
identification, 64
identity, 97, 126, 129, 132, 133, 139, 184, 387
implication, 106–108, 113
incommensurable value, 428

independence, 42–48, 52–54, 56–58
    demonstrations of, 42, 44, 53, 54
    of axioms, 42
    of axioms for propositional logic, 54
    of irrelevant alternatives, 359
    of the parallels postulate, 44, 49, 52, 54, 58
    of the real axioms, 45
    preservation property, 359
    proof, 42, 43, 54, 55
indescribable cardinal, 76, 93
indigenous groups, 250, 253
induction, 158, 170, 353
    principle of, 311
inductive synthesis, 162
inference, 13–15, 31, 32, 37, 142, 185, 352, 386, 394
    rules of, 106, 111, 112, 117–119
    logical, 43
    principles of, 54, 55
    rules, 54
infinite value, 427–429
information, 4, 8, 13–17, 21–24, 28, 114, 180, 182, 185, 186, 263–266, 268–270, 276–282, 290, 291, 297, 329, 351, 405, 407
    perfect, 353
information, science-based, 405
instrumental variables, 369, 374, 375
instrumentalism, 194, 370
intensions, 139–141, 143, 144, 146–148
intention, 186, 252
interaction, 245, 246
International Atomic Energy Agency, 247
International Council for Science, 405–407, 409–416
international research collaboration, 248, 252, 254, 406, 410
International Union for Conservation of Nature (IUCN), 422, 424
intervention, 262–265, 267–269, 272, 274, 276, 277, 279–281, 297, 311, 314
    therapeutic, 333

intrinsic value, 427, 428
intuitionism, 127
intuitionistic logic, 107
invariance, 263–265, 267–269, 272–274, 280
ions, 231, 232, 236
isolation, 388, 393, 397
    geographical, 253
isomorphism, 87–89, 91, 97, 100, 307, 372

Jeffrey, R., 210, 211, 218
justification, 117, 158, 303–306, 308, 314
    a priori, 121
    circular, 117, 118
    epistemic, 117
    of a model, 314
    of theories, 307

Kant, I., 50, 51, 176, 183, 184, 304, 308–313, 315
Kantian epistemology, 304, 308–313, 315–317
Kilwardby, R., 146–150
Kuhn, T., 317

Lambert, J. H., 46, 47, 51, 53
Lavoisier, A.-L., 228, 229, 231, 234
laws of logic, 13, 16, 126–128, 133
Lego, 226, 234–237
Leibniz, G.W., 137–139, 150, 151
Leibniz-Glashoff analysis, 150
Levi, I., 208
Lewis, C. I., 118, 121, 237, 238
Lewis, D., 100, 116, 190, 194–198, 201–203, 281
lithium, 234
Lobachevsky, N., 48–52
logic as games, 11, 12
logic of games, 11, 12, 18
Logical Empiricism, 303, 306, 309, 311, 315
logical empiricism, 190, 197, 198, 203
Logical Positivism, 309, 313
logical validity, 66, 67, 72

Macbride, F., 202
Mahlo, 92
Mahlo cardinal, 92, 93
Mahlo, P., 92
Malink, M., 146
Manifesto, 308
Martin, D.A., 87–91
mass, 229, 234
mass-energy equivalence, 234
mathematical objects, 87, 100
mathematical structures, 47, 56, 307, 313, 315, 317
McAllister, J. W., 309
McCauley, D., 427, 428
meaning, 128, 158, 168, 170, 181, 185, 197
    logical, 315
    of new scientific theories, 254
    of non-logical terms, 45
    of the axiom of parallels, 50
measurable cardinal, 86, 90, 99
measurement, 162, 165, 184, 211, 304, 307, 312, 408, 409
    theory of, 180, 181
mechanisms, 247, 288–297, 350, 357, 364, 368–370, 372, 373, 375–377, 388, 390, 393, 416
    epigenetic, 325
    genetic, 252
    of idealization, 353
    of probability aggregation, 360
medicine, 249
    before Eisenhower's initiative, 247
mereology, 100, 239
metaphysics, 3, 184, 226, 235, 237, 305, 308–310, 316
    probabilistic, 183, 184
microreductionism, 226, 227, 230
minimal logic, 108
mirror neurons, 290–292
model construction, 46, 57, 314, 315, 329, 350, 355
    mechanisms of, 355
model-theoretic
    methods, 57

model-theoretic argument, 190, 199–202
modelling
    audience, 386, 391–397
    commentary, 386, 395–397
    context, 351, 352, 364, 386, 389, 390, 396
    purpose, 350, 385, 386, 390–392, 395, 396
    target, 53, 307, 314, 353–358, 364, 382, 383, 385, 386, 388–393, 395, 397
modelling failure, 382–384, 386, 388, 392, 398
models, 6, 8, 12, 22, 35, 46, 50, 51, 53, 57, 106–108, 111, 112, 114–119, 121, 126, 127, 130, 132, 307, 308, 314, 317, 329, 330, 333, 338, 339, 350–355, 357, 358, 364, 367–377, 382–385, 387–397, 412, 413, 426, 431
    background, 106, 111, 120
    fuzzy, 109, 110
    lattice logic, 108
    Beltrami-style, 52
    decision theoretic, 425, 426
    developmental, 325
    ecological, 325, 339
    empirical, 351, 352, 354–356
    epistemic functioning of, 313
    evolutionary, 325, 337–339
    Hilbert-style, 43, 53
    in formal epistemology, 351, 353, 354, 356, 364
    in meta-mathematics, 42
    mathematical, 306
    minimal, 338
    molecular, 236
    of bargaining, 350, 351, 358, 364
    of data, 307
    of non-Euclidean geometry, 46
    of the flow of knowledge, 245
    of the theory, 306, 307
    of tumour heterogeneity, 329, 335–337
    phenomenological, 329

pseudospherical, 50
role of, 48, 54
scientific, 305, 308, 313, 315, 352
triage, 425, 426, 429, 431
models in economics, 368, 370–372, 389, 396
molecular genetics, 250
moral
authority, 255

Nagel, E., 42, 170, 176, 178
national, 244, 246, 247, 250, 251, 254
natural kind, 28
naturalism, 235
necessity, 95, 294, 295
quasi-Kantial conceptual, 237
nihilism
logical, 125–129, 133
nihilist model theory, 130
nomological machine, 306
non-Euclidean geometry, 46, 48, 51
non-representationalism, 315, 316
normative models, 350–358, 364
normativity, 350–355, 357
Northcott, R., 236
nuclear energy, 248
Commission of, 244
uses of, 247, 252
nuclear physics, 234, 248, 249, 329
nuclear technology, 247

objective probability, 162–164, 167, 168
objectivity, 160, 168, 309
of scientific knowledge, 302–304
observation, 14, 15, 37, 191–202, 227, 304, 312, 315
passive, 304, 309, 313, 316
ontological principles, 237, 238, 311
ontology
compositional, 227
compositionist, 238
pre- and post-Kantian, 308
Oppenheim, P., 227, 230, 234
optimisation, 423–426
oxygen, 229–232

pair-annihilation, 235
pair-creation, 235
partial interpretation of theories, 190–195, 197–203
patchwork narrative, 325, 329, 339, 340
Pauling, L., 229, 231
Pauling, P., 229, 231
Peirce, C. S., 158, 170, 315
perception, 50, 87, 88, 288–297, 309, 313
of space, 50
permutation, 195, 196
phenomena, 11, 24, 27, 156, 161, 184, 226, 309, 314, 315, 329, 350, 353–355, 357, 363
hematopoietic neoplastic, 333
observable, 307, 309, 311
phenomenal character, 288, 293–295
philosophy of economics, 397
philosophy of science in practice, 302–304, 316, 317
phlogiston, 228, 229
photons, 235
Pimm, S., 422, 429, 430
plausibility, 22–26
pluralism
logical, 125
Poincaré, H., 50–53
Poincaré, H., 161, 170
Popper, K. R., 157, 338
population genetics, 250, 252, 253, 329
possible world, 107
Possingham, H., 424
power, 9, 10, 12, 22, 29, 32
predictive, 328
power set, 90, 95
practical decision, 209, 212–214
pragmatism, 159, 170
prediction, 158–161, 170, 355, 356, 370
preferences, 4, 20, 28, 29, 159, 372, 387
probability, 106, 108, 109, 113–115, 118, 120, 156–170, 180, 184, 309, 350, 358–363, 425–427, 430
conditional, 114
function, 113–115

standard, 425
probability aggregation, 350, 358, 360, 363, 364
probability distribution, 267, 270, 281, 360
program
    of international cooperation, 244
    of research, 244, 248–251
projective determinacy, 95
propositional logic, 9, 10, 54
    axioms of, 54–56
protons, 234, 235
Psillos, S., 202
Putnam, H., 190, 192, 199–203, 227, 230, 234, 310, 313
Pythagoras' Theorem, 88

quantum mechanics, 157
Quine, W. V. O., 70, 312

radiation
    cosmic, 248
    effects of, 244, 248–252
radiobiology, 250
    in Mexico, 244, 250
Ramsey, F. P., 159, 161, 167, 170, 192, 194
randomness, 156–158, 183, 184
rationality, 14, 19, 20, 25, 118, 183, 185, 210, 215, 304, 353, 354, 356, 423
real numbers, 43, 44, 55, 67, 68
realism, 200, 226, 304, 305
reductionism, 185, 226, 227, 234
    epistemic, 226, 227
    metaphysical/ontological, 226, 230, 239
reductive levels, 234
reference class, 158, 162–164
reflection principles, 78
regulative principles, 310–312, 316
Reichenbach, H., 156, 158, 162, 164, 166, 170, 306, 313
reliability
    of models, 314

representation, 4, 16, 37, 47, 51, 55, 113, 193, 209, 289, 295–297, 307, 308, 312–317, 328, 355, 368, 372, 385
    of tumour evolution, 331
    of uncertainty, 425, 426, 432
    of utility, 432
representation theorem, 160, 180
resemblance, 386, 392–394
Ritter, J. W., 230, 231
robotics, 288, 291
Russell, B., 54–56

sabotage game, 34, 35
science and society, 303, 414
science for policy, 406, 407, 415
scientific concepts, 305, 308, 309, 311, 314, 315, 317
scientific corpus, 210–212, 214, 218
second-order logic, 75
self-defeat, 126, 127
semantic view, 306, 307, 314, 316
sentential context, 130
set theory, 27, 66, 67, 72, 76, 78, 86, 88, 89, 91, 93–95, 97
Shelah, S., 71
similarity
    between ideas and the reality, 313
simplicity, 32, 121, 209, 228, 230, 314
simulation, 412
Sober, E., 176, 373, 428
social choice
    theory of, 351, 354
socio-technology, 303
soft information, 22–25, 37
specificity, 263–265, 267–270, 272, 273, 275, 278–281, 283, 389
spectroscopy, 232
stability, 263, 264, 272–275, 277–282
statistics, 157, 164, 250
structuralist, 88
    approach to mathematical theories, 56
structuralist thesis, 192, 195, 197
sub-compact cardinal, 99

subjective probability, 159–163, 167–169
substitution, 127–129
sustainable development, 407, 408, 411, 414–416
    goals, 409, 413
synthesis
    chemical, 228
synthetic a priori, 51, 53, 190, 237, 308–310, 312, 316
system
    of science and technology, 245

tautology, 352
technology, 235, 302
    for genetic sequencing, 325
theories, 24, 31, 254, 306, 307, 311, 314, 329, 330, 338, 352
    axiomatic, 49, 53, 307
    ethical, 431
    in biology and medicine, 328
    in physics, 235, 236
    mathematical, 42, 47, 53, 56
theory, 190–202
    atomic and field, 309
    mathematical, 191, 201
    meaning of, 328
    of nuclear physics, 329
    of numbers, 45, 53
    of population genetics, 329
    of probability, 425, 426
    of rationality, 354
    of real numbers, 43
    of reference, 313
    of relativity, 234, 309
    of tumour heterogeneity, 328
    S-matrix, 235
Theory of Play, 11, 20, 28, 30, 33, 36
theory-mediated measurement, 190, 203
threatened species, 422, 424, 425
transformative research, 409, 416
transnational, 244, 246, 250, 254
    perspective in the history of science, 244
triage, 422, 424–427, 429–432

truth, 5, 6, 22, 24, 26, 86–88, 96, 107, 108, 111, 117, 125–133, 157–159, 165, 183, 184, 192, 200, 201, 209, 210, 304
    a priori, 316
    approximate, 234
    as correspondence, 313
    coherence theory, 312
    empirical, 312
    geometrical, 50
    justification of, 313
    logical, 193
    of logic, 55
tumour heterogeneity, 324, 325, 328–330, 335–340
    causal relation to tumour microenvironment, 335
    inter and intra, 326, 330, 331

uncertainty, 13, 16, 31, 159, 166, 351, 363, 425, 426
universal, 100, 112, 118, 138, 142, 143, 148, 185, 186, 194, 405
    affirmative, 137–139, 147, 149, 150
    negative, 138, 147, 149
    dynamics of knowledge, 246
update, 22, 24, 26, 35, 37, 351
    function, 106
utilities, 209, 211, 350, 360, 361, 363, 423, 424, 428, 430, 432
utility function, 350, 360, 361, 363, 430, 431

vagueness, 111, 116, 425, 426
validation, 357
value-neutrality, 208
van Benthem, J., 199
van Fraassen, B., 304, 307, 316
vehicle, 295–298
verification, 161, 369
Vienna Circle, 170, 311
von Helmholtz, H., 50–53, 57, 309
von Neumann, J., 430

Walton, E., 234

Wiener Kreis, 303, 305, 308, 310, 313, 316
Winnie, J., 193–196, 199, 200
Woodin cardinal, 86, 95

Zermelo's Theorem, 8, 34, 35
Zermelo, E., 8, 34, 89
zoom level, 10, 13, 19–21, 27

www.ingramcontent.com/pod-product-compliance
Lightning Source LLC
Chambersburg PA
CBHW071327190426
43193CB00041B/900